MW01071595

WILD BY DESIGN

BY DESIGN

THE RISE OF ECOLOGICAL RESTORATION

LAURA J. MARTIN

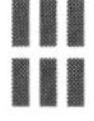

Harvard University Press

CAMBRIDGE, MASSACHUSETTS
LONDON, ENGLAND
2022

Printed in the United States of America
First printing

Library of Congress Cataloging-in-Publication Data
Names: Martin, Laura J., 1984– author.
Title: Wild by design : the rise of ecological restoration / Laura J. Martin.
Description: Cambridge, Massachusetts : Harvard University Press, 2022. |
Includes bibliographical references and index.
Identifiers: LCCN 2021041664 | ISBN 9780674979420 (hardcover)
Subjects: LCSH: Restoration ecology—United States—History—20th century. |
Restoration ecology—United States—History—21st century. | Nature—Effect of human beings on—United States—History—20th century. | Nature—Effect of human beings on—United States—History—21st century. | Environmental responsibility—United States—History—20th century. | Environmental responsibility—United States—History—21st century.
Classification: LCC QH104 .M38 2022 | DDC 508.73—dc23
LC record available at https://lccn.loc.gov/2021041664

CONTENTS

Introduction CULTIVATING WILDNESS 1

Part I: Reservations, 1900–1945

1 Uncle Sam's Reservations 19

2 Ecology in the Public Service 42

3 An Outdoor Laboratory 63

Part II: Recovery, 1945–1970

4 Atoms for Ecology 93

5 The Specter of Irreversible Change 113

Part III: Regulation, 1970–2010

6 Extinct Is Forever 139

7 The Mood of Wild America 168

8 An Ecological Tomorrowland 197

Epilogue DESIGNING THE FUTURE 223

Abbreviations 237

Notes 239

Acknowledgments 313

Index 315

Introduction

CULTIVATING WILDNESS

In 1982, George Archibald described to a rapt audience of the *Johnny Carson Show* the years he had spent mating with an endangered whooping crane. His relationship with the species had begun in graduate school, when he studied crane mating behavior, keeping detailed notes on the dramatic calls and intricate dances of these highly social birds. His relationship with this particular crane, a female named Tex, had begun in 1975, when biologists from the Patuxent Wildlife Research Center in Maryland sent Tex to Baraboo, Wisconsin, where Archibald had co-founded the International Crane Foundation. It was a last-ditch effort to get Tex to reproduce.[1]

When Tex arrived in Wisconsin, whooping cranes were in dire straits. Whooping crane territory had once extended from the Arctic to central Mexico and from Utah to New Jersey, but hunting and widespread draining of wetlands in the early twentieth century had caused whooper numbers to plummet. By 1937, only two perilously small populations remained: a stationary group of fifteen cranes in southwestern Louisiana and a flock of about twenty that migrated between coastal Texas and northern Canada. Decades of campaigns by environmental groups followed, and Congress eventually allocated money to the U.S. Fish and Wildlife Service in 1965 to establish a captive breeding program for whooping cranes at Patuxent. So began the Endangered Species Research Program, the first federal effort to propagate and restore an endangered species.[2]

The problem was that no one knew how to get whooping cranes to breed in captivity. Tex was only the second adult whooping crane to live at Patuxent, and for a number of years, Fish and Wildlife biologists tried and failed to get her to lay a fertile egg. It can take months for two cranes to bond with each other through elaborate ritual dancing, and the male crane at Patuxent, Canus, had a broken wing. In addition, crane chicks imprint on any moving object in their early environment. Tex, the biologists speculated, had imprinted on humans, the first organisms she had met as a chick at the San Antonio Zoo. Other cranes just did not excite her.[3] When the Patuxent biologists sent Tex to Wisconsin, they hoped Archibald would figure out a way to get her to lay a fertile egg through artificial insemination.

Archibald spent several years acting as a male crane with Tex, walking, calling, and dancing in the hopes of shifting her into reproductive condition. In the first year of Archibald's courtship, Tex laid a single unfertilized egg. This was progress, but Tex didn't produce a fertilized egg until Archibald made an "all-out effort" in the spring of 1982. He and Tex took turns bowing, jumping, and tossing small sticks in the air. They slept in the same space, a shed subdivided by chicken wire into an office for Archibald and a shelter for Tex. From March through May, Patuxent biologists sent fresh whooping crane semen from Patuxent to Baraboo twice weekly, and in late May, Tex finally laid a fertilized egg. Members of a crane pair alternate incubation, so Archibald took his turn every few hours, shuffling his sleeping bag over to the nest when Tex left to forage. "I was so excited!," Archibald recounted; "I felt like a father."[4]

Archibald's success with Tex was evidence that intensive human intervention might pull a wild species back from the brink of extinction. Encouraged, biologists at Patuxent brought more adults and eggs into captivity from the two tiny remaining wild flocks. They developed egg incubators and experimented with nutritional formulas. Soon they were tending more eggs and chicks than their adult cranes could care for, and so staff and volunteers stepped in to help raise the chicks. To prevent the chicks from imprinting on humans, the caregivers followed a strict protocol. They donned crane costumes, guided the birds with crane-shaped hand-puppets, and never spoke around them. Costumed technicians taught the chicks how to find food and how to explore their penned surroundings. By 1995, biologists were caring for a flock of seventy-two captive whooping cranes at Patuxent and a flock of 39 at the International Crane Foundation.[5] Although, sadly, Tex was killed by a pack of raccoons shortly after the hatching of her one and only chick, that chick went on to sire an astounding 178 whooping cranes by the time of his death in 2021.[6]

The goal of whooping crane restoration, however, was not just to create more individuals but to restore wild populations, and introducing captive-bred birds to the wild, it turned out, was just as difficult as breeding them. In May 1989, a female captive-reared whooping crane from Patuxent was released into Grays Lake National Wildlife Refuge in southeastern Idaho. She interacted with other cranes, but it seemed she was unable to fly long distances or to follow them as they migrated, and she disappeared before scientists could recapture her. Into the 1990s, biologists continued to attempt carefully controlled and monitored releases of captive-reared whoopers into sites in the Rocky Mountains and central Florida.[7] While the introduced Rocky Mountain population failed, the Florida population thrived—but it did not migrate. The trouble was that wild cranes learn their migration route from the previous generation: chicks follow their parents to wintering grounds. Figuring out how to teach captive-bred whooping cranes to migrate was going to take some ingenuity.

Beginning in 2001, the Whooping Crane Recovery Team, a collaboration between the U.S. Fish and Wildlife Service and the Canadian Wildlife Service, attempted to establish a new migratory population that would breed in central Wisconsin and winter in central Florida. Chicks reared at Patuxent were brought to Wisconsin's Necedah National Wildlife Refuge in early summer and were taught the migration path by costumed human pilots in ultralight aircraft. Seven birds made it to Florida on the first run, of which five returned to Wisconsin the following spring. The project, dubbed "Operation Migration," was enormously popular, "an uplifting story for the nation that was struggling with the aftermath of the 9–11 attack," as one reporter put it, a story that portrayed the aircraft as a tool of restoration rather than of terror.[8]

But while the public celebrated Operation Migration, biologists were deeply divided over the status of the flock. Skeptics felt that whooping cranes guided by ultralight aircraft could not be wild cranes. In their view, captive breeding had been a necessary intervention, albeit a risky and resource-intensive one, to save the species. But teaching released whooping crane fledglings to migrate by guiding them with an airplane was no longer wildlife management; it was too much like domestication or farming. These skeptics pointed to the fact that the new migratory population fledged only ten chicks in its first fifteen years. While the flock grew to nearly one hundred individuals, the vast majority were captive-bred, and it seemed likely that the flock would never become self-sustaining, because these cranes, having been raised by humans, were not learning to parent as wild cranes did. Humans would have to continue to intervene, and if that happened, then *wild* whooping cranes might well go extinct, even as whooping cranes

0.1 International Crane Foundation staff teach a young whooping crane to fly, donning costumes to prevent it from imprinting on them, at Horicon Marsh in Wisconsin in 2011. Tom Lynn

physically still existed and reproduced in captivity. In contrast, some proponents of Operation Migration argued that whooping crane restoration only counted as successful if cranes were migrating as they had evolved to do. The restoration of wild whoopers, they claimed, might really depend on Operation Migration's success.

In October 2015, the Fish and Wildlife Service came down on the side of the skeptics by announcing that it would no longer condone the use of ultralight aircraft in whooping crane restoration; going forward, the agency's goal would be "minimizing anthropogenic effects of captive rearing on whooping cranes."[9] Further, captive breeding programs would aim to reintroduce only chicks raised by whooper parents. Understandably, this decision pained the many supporters of Operation Migration, who petitioned the Fish and Wildlife Service to reverse its position on guided crane flights. Joe Duff, Operation Migration's CEO and migration pilot, lamented that while the agency deemed the ultralight method to be "the most artificial" restoration option because cranes were with humans the longest, "we say it is the most natural because it replicates the process. We protect them as their parents until they're ready to be released."[10] What is most significant and striking about this debate is that the two sides had not been divided over whether to care for a wild species, but over how to care for their wildness.

MANY OF US have cared for a pet or a houseplant—species that we choose to live with, that we cherish, that we might even consider family—species that we treat very differently from laboratory animals, or species we eat, or those we consider pests. We have some sense of what counts as caring for these "companion species," as feminist scholar Donna Haraway calls them.[11] We shield them from harm. We provide them with food, shelter, affection, entertainment, medical care, and, sometimes, monogramed sweaters. These are accepted ways of expressing care for our creaturely companions.

But how does one care for a "wild" species? We imagine wild species to be autonomous and self-reliant, living outside the boundaries of human society, unentangled with our built landscapes of work and rest.[12] In the eyes of many, human actions—even acts of care—diminished the wildness of whooping cranes. And yet in an age of habitat destruction, anthropogenic climate change, and the redistribution of species at a global scale, many wild species will not survive without ongoing human involvement.[13] How intensively, then, should we intervene to help wild species recover from harms that humans inflicted? This is the core question of ecological restoration, a science and practice that confounds the distinctions we expect to find between the autonomous and the managed, the wild and the designed. Ecological restoration is a mode of environmental intervention that seeks to respect the world-making, and even the decision-making, of other species. The international Society for Ecological Restoration, the main professional society for restoration ecologists today, defines restoration as "the process of assisting the recovery of an ecosystem that has been degraded, damaged, or destroyed."[14] In this definition, the verb "to assist" is key. Restorationists strive to cede some control of the restoration process to other organisms. Restoration, in other words, is an attempt to co-design nature with nonhuman collaborators.

The Society for Ecological Restoration's definition is purposefully capacious, as ecological restoration encompasses a wide range of interventions. Across Australia, land managers are killing introduced rabbits in order to restore populations of endemic plants. In Mozambique, a public-private partnership is replanting millions of trees in Gorongosa National Park. Members of the Coral Restoration Foundation tend to an underwater nursery of staghorn coral in the Florida Keys until the coral are large enough to be transplanted into protected areas.[15] Elsewhere, restoration involves breeding species in captivity, constructing wetlands, burning prairies to stimulate regrowth, and spraying herbicide from helicopters to make room for desired plants. The damages that ecological restoration seeks to undo are just as diverse: they include deforestation, overhunting, nonnative species invasions, wetland filling, and, increasingly, climate change. What unites these geographically, methodologically, and institutionally diverse projects is that they are motivated by a desire to undo human-caused ecological damage while striking a balance between human care and nonhuman autonomy.

Today ecological restoration is one of the most widespread and influential forms of environmental management in the world. Public and private organizations spend billions of dollars per year on restoration projects.[16] Some of these projects are megaprojects. The U.S. federal government, for example, plans to spend $8 billion dollars on wetland restoration in the Florida Everglades. Others are smaller scale: the annual budget of the Finger Lakes Native Plant Society in central New York, to which I once belonged, is approximately $3,600. Meanwhile, a suite of international agreements, including the Ramsar Convention on Wetlands and the UN Framework Convention on Climate Change, promotes and mandates restoration. Recognizing restoration's potential—and its centrality to international environmental negotiations—the UN General Assembly recently declared 2021–2030 to be the UN Decade on Ecosystem Restoration.[17]

To understand both the promises and the perils that ecological restoration holds for the future of biodiversity, we must understand its history. And yet little has been written on the topic.[18] Ecologists typically frame restoration as an endeavor that began with the establishment of the Society for Ecological Restoration in 1988, barely the stuff of history. Those histories that look earlier cite Aldo Leopold, former Forest Service biologist and author of the classic 1949 collection of essays, *A Sand County Almanac,* as the sole "father" of restoration ecology.[19] Environmental historians of the United States, meanwhile, describe the twentieth century as an era of contest between two other modes of environmental management, conservation and wilderness preservation. Whereas conservationists like President Theodore Roosevelt and Gifford Pinchot, first head of the U.S. Forest Service, believed that experts could guide the efficient management and use of natural resources for the greater good, preservationists adhered to a Romantic belief in the sanctity of undeveloped nature, opposing conservationist projects to harvest forests and dam rivers. Preservationists led by John Muir, for instance, were famously opposed to the construction of the

Hetch Hetchy dam in Yosemite National Park, which conservationists supported, and which was eventually built.[20]

Conservation, in this sense, concerns the sustainable use of natural resources like trees, water, and fish. The goal is to extract, and perhaps actively replenish, economically valuable species and materials in a way that minimizes environmental damage and maintains resource availability into the future. Conservation is a management vision for "working landscapes," as they are now called. In contrast, preservation is an effort to take wild species and wild places out of the human economy, to reserve them as "protected areas." Preservation began in 1872 with Yellowstone National Park, the world's first national park, which the U.S. military established by forcibly removing Nez Perce, Bannock, Shoshone, Crow, and Blackfeet communities from their ancestral lands.[21] In 1960, there were approximately 1,000 protected areas worldwide. Today, there are a staggering 267,170, encompassing 15.6 percent of terrestrial area and 7.6 percent of marine area.[22]

Whereas conservation embodies a management ethos, preservation is a "hands-off" approach to an entire place. A conservationist might plant trees in an arrangement best suitable for harvest, or kill predators that prey on the furbearing species she is trying to conserve. In a protected space, in contrast, nature is imagined to be not only unmanaged, but actually unaffected by humans—although climate change and introduced species challenge this ideal. At the Cabo Blanco Strict Nature Reserve in Costa Rica, for example, only forty visitors are allowed per day, and it is illegal to take even a single shell from the beach.

Dividing U.S. environmental management into these two movements, conservation and preservation, is instructive but ultimately reductive. Since the early twentieth century, restorationists have offered a third way, distinguishing themselves from both conservationists and preservationists. Restorationists, from the start, have grappled with the question of how to intervene in the lives of wild plants and animals while also retaining their "wildness." Indeed, ecological restoration challenges the idea that a place is either untamed or managed, wild or designed. This challenge has become only more salient with anthropogenic climate change and other global-scale environmental transformations.[23] As we will see, many of the debates among restorationists have concerned where restoration should fall between the poles of preservation and conservation, in terms of intensity and duration of human intervention. At first many restorationists believed that hand-rearing whooping cranes was too little intervention too late, for instance. But by the early 2000s, many held that hand-rearing was detrimental to the wild character of whoopers.

Built into the pursuit of all three modes of environmental management—preservation, restoration, and conservation—are assumptions about which is more malleable, human behavior or the lives of other species. Preservation, in its simplest terms, assumes that extractive capitalism and development will continue unabated, and that reserving places from these forces is nature's best shot at survival. Preservationists do not seek to control nonhuman species within the confines of protected areas. Nor do they try to control human behavior, instead excluding people from certain places entirely and allowing them free rein everywhere else. Conservation, in contrast, assumes that humans can develop enlightened ways of using nature more gently. Conservationists seek to control both human decisions and nonhuman lives. Restoration pursues a middle ground: it asserts that human care can help to undo some forms of human-caused environmental damage, while also respecting the autonomy of other species. Ecological restorationists strive to enable other species to thrive while, ideally, minimizing human intervention.

This book illuminates how ecological restoration emerged in the United States and how it transformed over the course of the twentieth century from a diffuse, uncoordinated practice into a scientific discipline and an international and increasingly privatized undertaking. By ecological restoration, I refer specifically to efforts to restore biotic relationships among species.[24] Such work today involves national and subnational government agencies, nongovernmental organizations, corporations, and private landowners. Regulatory mechanisms such as Section 404 of the Clean Water Act encourage restoration but do not outline how it should be pursued.[25] There are also many instances of regional initiatives, such as Louisiana's Coastal Protection and Restoration Program, that are enabled through a synthesis of legislation and partnerships at multiple levels of government.[26] And then there are the countless restoration initiatives pursued by nonprofits, corporations, and institutions as a means of reaching sustainability goals. The archives of ecological restoration are thus widely dispersed. This book assembles an archive of correspondence among ecologists; of unpublished and published scientific manuscripts; of field notebooks; and of the records of organizations such as the Ecological Society of America, The Nature Conservancy, the Fish and Wildlife Service, and the Society for Ecological Restoration.[27] Together these sources illuminate the networks of people, organizations, and technologies that gave rise to ecological restoration. This particular history moves across diverse geographies—including the Great Plains, Washington streams, Pacific atolls, and the Florida Everglades—to analyze key moments in the history of ecological restoration as an idea and as a scientific discipline grounded in applied practice.

In analyzing this archive, I build on scholarship at the intersection of environmental history and science and technology studies (STS). Many environmental historians study material change, relying on work done by environmental scientists to argue that humans are not the only agents of historical change.[28] But STS reveals that scientific knowledge is neither a stable nor an objective resource; rather, it is contested and constantly in flux.[29] Any piece of scientific knowledge is situated in a specific place and time, and both human and nonhuman actors play significant roles in the production of that knowledge. For example, the identification of anthropogenic climate change depended on the emergence of new social systems for knowledge production, such as intergovernmental scientific boards, as well as new technologies, such as Earth-observing satellites and particular computer models.[30] Understanding how restoration science has reshaped species and ecosystems worldwide, from rainforests to deserts, thus requires an STS approach. At the broadest scale, the study of ecological restoration is an occasion to examine how scientific and cultural ideas about nature come to shape material environments and, reciprocally, how material environments come to shape ideas about nature. By trying to reassemble communities of species, restoration ecologists came to new understandings of interspecies relations. Applied science shaped ecological theory at the same time that theory shaped practice.

I was a wetlands ecologist before I was a historian. My move from the field to the archive was motivated by a desire to tackle some of the questions that too easily fall between the cracks when the sciences are divided from the humanities. Fieldwork offered insights into how to reestablish and care for particular species, but it did not offer solutions to ongoing ecological degradation, which many days felt—and still often feels—insurmountable. Wading through marshes in central New York, I wondered: How did today's ecological restoration practices come to be? Could they have been different? Could the future of restoration be different? As I tell my students, one of the most important powers of the discipline of history is that it teaches us that the present was not inevitable—and, therefore, that the future is not predetermined. Ultimately, I turned to history with hope.

Teaching has motivated me to seek out hopeful environmental stories, in this time when almost every environmental news story is one of catastrophe. When facing the vastness of climate change, it can feel as though the actions of an individual, or even a nation, are inconsequential. Moreover, many of my students share the narrative that bad decisions have already locked us into a doomed future. The narrative that environmental harm is irreversible is, by definition, disempowering. The history of restoration teaches us, however, that, with work and will, some harms can be remedied. Of course, some types of ecological damage cannot be undone. The warbler that has died from colliding with a window is forever lost. Putting aside, for the moment, the technological possibility of de-extinction, an extinct species is forever gone. Other types of ecological damage, though not permanent, will last for untold generations. Nuclear pollution and "forever chemicals" like PFAS will persist in the environment for thousands, even tens of thousands, of years. But not every environmental harm is so final. When an ecosystem is damaged, it is impossible to create a perfect replica of what came before. (If nothing else, ongoing climate change prevents this.) Nevertheless, it is possible to design and create places where other species can thrive, to manipulate the physical and biological environment to reverse some of the harms—intentional and unintentional—people have caused other species. Restoration can be thought of as a mode of justice in which a wider community endeavors to repair some of the harm caused to other species by particular human offenders.

There is hope, then, to be found in ecological restoration. Restoration is, by definition, active: it is an attempt to intervene in the fate of a species or an entire ecosystem. If preservation is the desire to hold nature in time and conservation is the desire to manage nature for future human use, restoration asks us to do something more complicated: to make decisions about where and how to heal. To repair and to care. To make amends for the damage we have done, while learning from nature even as we intervene in it. The desire to protect whooping cranes from extinction led researchers to develop intricate and intimate relationships with another species. According to the International Crane Foundation, as of 2017 there were 757 whooping cranes in the world. While this is nowhere near the species' prior numbers, the work of dedicated individuals and organizations has greatly improved the whooping crane's odds of survival. It has undone, to some extent, a harm.[31]

THE VERB to restore and its relatives—to reinstate, to reestablish, to repair, to reconstruct—all suggest a return to a past state. Indeed, in 1990, the founders of the Society for Ecological Restoration initially defined their practice as "the process of intentionally altering a site to establish a defined, indigenous, historic ecosystem."[32] Today many land managers continue to base their restoration goals on historical baselines, inferring the ecological history of an area from documentary sources—written descriptions, historical photographs, maps, and even paintings—and from analyses of "biological archives" such as tree rings and fossil pollen.[33] In the United States, restoration ecologists have often endeavored to establish a precolonial ecological community, eradicating invasive "nonnative" species that arrived after European colonization.[34]

But in the past two decades, ecologists and historians have criticized the use of historical baselines in restoration from multiple angles. The first critique is empirical: archaeologists and ecologists have shown that present ecological communities are shaped by past human land uses extending to antiquity. For example, in northern France, the intensity of Roman-era agriculture still influences the number of species found in forests that have not been farmed for nearly two millennia. Similarly, traces of ancient Mayan gardens can be found in modern Belizean forests.[35] Thus contemporary ecosystems reflect a very long human history, which makes it difficult or impossible to identify a prehuman baseline. The second critique is political; Indigenous peoples managed American ecosystems long before European colonists did, and to mark 1492 as the beginning of anthropogenic landscape change is to erase that history.[36] To imagine precolonial lands as empty or pristine, as restorationists sometimes do, perpetuates a foundational and pernicious myth of settler colonialism, namely that Europeans came upon unsettled lands. A handful of historians have also critiqued ecologists for using historical records uncritically, that is, for failing to interrogate the conditions of archival production and reception, the ways in which scientific practice has changed over time, or the fragmentary nature of archives.[37]

Despite the importance of these points, it was not primarily critiques from the humanities that motivated restoration ecologists to shift away from historical baselines, but rather research on the trajectory of global environmental change. It is increasingly difficult to imagine reversing that change on a large scale. Scientists predict that climate change taking place because of increases in atmospheric carbon dioxide will be largely irreversible for at least a thousand years after emissions stop.[38] Already species are no longer found where they used to be; shifts in temperature and rainfall patterns have led many species' ranges to shift poleward or to higher elevations. Such geographical shifts could render place-based species conservation obsolete and result in the uncanny scenario, as one journalist recently put it, in which Joshua trees survive only outside of Joshua Tree National Park.[39] Meanwhile, the current rate of species extinction is estimated to be one hundred to one thousand times higher than the background extinction rate, an estimate of the standard rate of extinction before humans became a primary contributor.[40] A 2019 intergovernmental report posits that roughly one million species face extinction because of human actions.[41]

The loss of the narrative that environmental change is reversible has led to epistemological crises in both ecology and the environmental humanities.

In 2002, the Society for Ecological Restoration revised its definition of restoration to "the process of assisting the recovery of an ecosystem that has been degraded, damaged, or destroyed," removing any mention of "a defined, indigenous, historic ecosystem" and leaving open the question of what, exactly, restoration aims to restore. More recently, the International Principles and Standards for the Practice of Ecological Restoration defined ecological restoration as "any activity with the goal of achieving substantial ecosystem recovery relative to an appropriate reference model"; the reference model can include "native ecosystems" as well as "traditional cultural ecosystems."[42] Indeed, in the past two decades, ecologists have proposed more than twenty ways to reorient restoration. These proposals have looked both forward and back in time. Proponents of "Pleistocene rewilding," for example, have argued for the reintroduction of Pleistocene megafauna—large animals that disappeared some 13,000 years ago—to the western United States. Proponents of "novel ecosystems," meanwhile, contend that restoration should look to the future; they argue that we should accept and even design new ecosystems to provide specified functions and services. Whether restored ecological communities should be modeled on those that once inhabited an area, or on those that would best thrive under future conditions, remains a matter of heated debate.[43]

Like restoration ecologists, scholars in the humanities are grappling with the seeming irreversibility and magnitude of contemporary environmental change. In the past few years, they have taken up the idea of the Anthropocene, a term coined in 2002 by geologist Paul Crutzen to suggest that the Earth is transitioning out of the Holocene into a new geological epoch, one defined by human activity of such magnitude that it constitutes a global geological force.[44] Historians and geographers have understandably invested in defining a moment of historical rupture; proposed Anthropocene beginnings include the rise of agriculture, European colonialism, the Industrial Revolution, and the detonation of hundreds of atomic weapons.[45] Some scholars suggest that the Anthropocene represents not only a geological epoch, but also a rupture in Western thought, a challenge to traditional ways of understanding cause and effect.[46] Others have suggested that the epoch should be called the Capitalocene, to acknowledge that a specific economic system, rather than all of humanity, has been responsible for global change.[47] As the parallel investments of so many academic disciplines indicates, the question of how to live with other species, and how to do so in a changing world, transcends the division between the sciences and the humanities. Ecological restoration operates as both a form of care for nonhuman species and as a mode of reconciliation with the human past.

As ecological restoration becomes an ever more widespread practice, we must consider carefully what social and political conditions it presupposes, and what forms of care it perpetuates.[48] Caring for one species can involve neglecting or harming another. By investing in the restoration of whooping cranes, governments and NGOs chose not to allocate resources to other species. And caring for biodiversity can involve harming people, especially marginalized people. All too often, environmentalists promote patently racist means of protecting biodiversity under the cover of a blanket misanthropy. As a result, recent estimates of the worldwide number of "conservation refugees," people forcibly displaced from their homes in the name of biodiversity protection, range from five million to tens of millions.[49] Some restoration practices and ideas, too, have racist and colonial roots. Consider, as Chapter 1 reveals, that the establishment of the first wildlife restoration sites in the United States was part of a broader campaign to erode Native American sovereignty. It is my hope that, in addition to broadening our understanding of the history of ecological restoration, this book will help us discover how to coexist ethically with other species, and how to do so in a way that foregrounds human social justice.

Part I opens with the "game restoration" movement that preceded the emergence of ecology as a scientific discipline. Beginning in the early 1900s, a small group of conservationists distinguished themselves from those who were lobbying for further hunting restrictions. If game animals could be bred in captivity and then released onto federal lands, they argued, then it would be possible to increase population sizes while still allowing hunting for sport, and even for profit. The work of the American Bison Society typifies early restoration work in the United States, work carried out before the establishment of today's major federal land management agencies: the U.S. Forest Service (founded 1905), the National Park Service (1916), the Fish and Wildlife Service (1940), and the Bureau of Land Management (1946). The advocacy of game restorationists led to the establishment of the nation's first wildlife restoration sites on the sites of Indian reservations that the federal government was systematically dismantling in order to erode tribal sovereignty. These sites were the core of what became the National Wildlife Refuge System.

Part I next shows how the idea and the practice of ecological restoration, as opposed to game restoration, emerged from ecologists' efforts to secure permanent study sites in the 1930s. As the Dust Bowl escalated and Roosevelt's New Deal reformed federal land management, ecologists recast succession theory as a species management tool; they also developed "nature reservations," which they distinguished from game reservations and wilderness preservation areas, as lands for experimental restoration. The work of these scientists and practitioners would inform the design of "naturalistic gardens," including the Vassar College Ecological Laboratory and, later, Aldo Leopold's famous restoration project at the University of Wisconsin–Madison. Their aesthetic of letting a garden "direct itself," along with their methods of propagating native plant species, would be widely adopted by those advocating for restoration of native plants in degraded areas.

Part II analyzes how the Atomic Age simultaneously shaped ecological theory and restoration practice. After World War II, when the United States began detonating atomic weapons at the Marshall Islands, fisheries biologist Lauren Donaldson was put in charge of biological fieldwork there. At the Pacific Proving Grounds, he and his collaborators used radioisotopes from nuclear detonations to follow the circulation of elements among marine organisms. This was a new way of visualizing connections within ecological communities, and it led to the concept of the ecosystem. Donaldson's laboratory also applied atomic technologies to fish breeding, recasting radiation as a tool for restoration. By irradiating salmon, scientists hoped to breed fish that would survive in degraded environments. Notably, this was an attempt to restore a species' ability to thrive by altering the organisms themselves, rather than by restoring their environment; it was one of the first efforts to genetically modify an organism in the name of restoration, a precursor to today's efforts to use CRISPR-Cas9 and other genome editing systems for the "de-extinction" of species like the passenger pigeon and wooly mammoth.

As the Cold War escalated, the government charged ecologists with imagining ecological recovery after Doomsday. Ecologists anticipated a period of environmental and economic recovery after World War III and considered how the government could hasten that recovery—how they could pursue ecological restoration. Ecologists and military strategists revisited studies of past ecological disasters, including the Dust Bowl, in their attempt to plan for apocalypse. Their Doomsday imaginings drew on ecological succession theory, expanding the category of "environmental disturbance" beyond windstorms, fires, and floods to include nuclear bombs—and, ultimately, any human action. Meanwhile, in order to simulate the effects of World War III, ecologists began to destroy ecological communities intentionally. They irradiated forests and fumigated islands, trying to measure how intentionally stressed communities responded. Through these ecosystem destruction studies, ecologists came to see anthropogenic environmental damage as both systemic and irreversible. The now commonplace assumption that species diversity leads to ecosystem stability has its origin in these Doomsday experiments.

Part III reveals how the uneven propagation of the ecosystem concept through federal agencies and natural areas management in the 1970s and 1980s gave rise to the practice of killing nonnative species and the emergence of a new scientific subdiscipline—restoration ecology—with its own professional society, conferences, and journals. With atomic fieldwork, ecologists materialized ecosystems as objects of study and concern, facilitating 1960s grassroots environmentalism and heralding the "Age of Ecology." During this period, the goal of ecological restoration began to shift from restoring single species like bison or salmon to restoring habitats, assemblages of species, ecosystems, and, ultimately, "ecosystem functions" like water purification and nutrient cycling.

The influence of restoration thinking on the management practices of the Fish and Wildlife Service was particularly profound; during the 1970s, the agency haltingly transformed itself from a predator-killing service to a leader in restoring those same species. The Endangered Species Act of 1973 played a significant role in this transformation, as did a report on wildlife policy led by Aldo Leopold's son, Starker Leopold. The history of the Endangered Species Act of 1973 is often told as a legislative history, but the act was importantly shaped by on-the-ground species management. Section 7, in particular, led to policies and practices that helped professionalize restoration ecology. As Fish and Wildlife Service ecologists scrambled to respond to their new regulatory obligations, they leaned heavily on captive breeding programs and on requiring restoration as "compensatory mitigation" for ecological damage caused by other federal agencies. And as the act consolidated captive breeding within the Fish and Wildlife Service, state and private environmental organizations turned to restoring wild plant species on their lands.

Growth in the 1980s in the number and size of land trusts—nonprofit organizations that own and manage land—drove a rapid increase in the number of designated "natural areas" in the United States. As private and state natural areas programs expanded, managers exchanged stories about their experiences propagating native plants and eradicating nonnative ones. In 1988, a group of managers established the Society for Ecological Restoration and Management with the goal of "promoting the scientific investigation and execution of restoration." Today the Society for Ecological Restoration (as it is now called) has more than 3,000 members in seventy-six countries, and restoration ecology is an independent scientific discipline with its own academic journals and concerns. Early members of the society and environmental NGOs such as The Nature Conservancy distinguished themselves from conservationists through fieldwork and rhetorical work, framing nonnative or "invasive" species as disturbers of native ecosystems—

rather than opportunistic responders to human disturbance—at a time when fears of globalization were peaking. As we will see, this is the period during which restorationists embraced a precolonial baseline for American projects, naturalizing Native American land management and implicitly absolving contemporary politics and people of blame for ecological damage. It is also when The Nature Conservancy began to shift from "hands-off" preservation to interventionist restoration.

The 1990s saw the emergence of "off-site" mitigation—the practice of compensating for ecological harm in one location with restoration at another. Whereas the first restoration projects in the United States occurred on sites damaged by plowing, industrialization, or other human actions, off-site mitigation uncoupled sites of harm from sites of care. The Disney Wilderness Preserve is an early example of such a project; when looking to expand its parks in 1989, Walt Disney World was required to mitigate wetland destruction by Section 404 of the Clean Water Act. In a novel arrangement brokered by The Nature Conservancy, Disney funded a massive restoration project in central Florida, one of the first off-site mitigation projects in the country. The Disney Wilderness Preserve enabled Disney to build the Animal Kingdom Theme Park, among other attractions. There today you can visit one of the ultralight gliders used in Operation Migration. Crucially, off-site wetland mitigation paved the way procedurally and conceptually for today's carbon offsetting market and the international treaties and regulations that govern it. Off-site mitigation would reconfigure the geographic distribution of ecosystems, along with relations between the Global North and the Global South, fashioning the latter as a source of "wild" nature.

Over the past century, ecological restoration has become more interventionist, as restorationists shifted from the expectation that nature could essentially recover on its own, once left to its own devices, to the recognition that society and nature are so pervasively intertwined that ecological thriving will always depend on active management. In its ideal form, ecological restoration places simultaneous value on wildness and design, the nonhuman and the human. It recognizes the necessity of human intervention, but it aims to have a light touch. It has never fully embraced environmental justice, but it should. To do so requires reckoning with restoration's history, which in the United States begins in the early 1900s with efforts to breed and release bison for the sole benefit of white settlers.

PART I

Reservations, 1900–1945

1

Uncle Sam's Reservations

It was William Temple Hornaday's interest in dead animals that led him to work with live ones. When Hornaday was pursuing taxidermy at Iowa State Agricultural College in the 1870s, natural history museums were proliferating across the United States—government-sponsored institutions like the Smithsonian Institution as well as "dime museums" like P. T. Barnum's American Museum in New York.[1] Hornaday would found the Society of American Taxidermists in 1880 with the goal of elevating the craft "to a permanent and acknowledged position among the fine arts." American museums were, in his view, "storehouses of monstrosities"; at the time it was possible to purchase such taxidermized displays as a toad ice-skating, two squirrels playing euchre, and a kitten hauling a cart of corn to a model gristmill. Throughout his career, Hornaday would promote lifelike, naturalistic taxidermy displays.[2] When designing exhibits, he sought animal specimens with the largest bones, the healthiest coats. This prioritization of an idealized "wild" species form would guide his later bison restoration work.

Increasingly recognized as a skilled and serious taxidermist, Hornaday was appointed chief taxidermist of the newly established National Museum in Washington, DC, in 1882. Inventorying the museum's holdings, he found only two "sadly dilapidated" bison hides and a mixture of skulls and bones.[3] Convinced that market hunters would soon exterminate the bison, Hornaday wanted to acquire specimens for the museum and, through taxidermy, make them accessible to the public and thus render them "comparatively immortal."[4] His primary concern was not the continued existence of the species but, rather, maintaining scientists' access to quality specimens. In his 1891 handbook, *Taxidermy and Zoological Collecting*, Hornaday quipped, "If you really must kill all the large mammalia from off the face of the earth, do at least preserve the heads."[5]

Hornaday's doubts about the future of the species were well founded. In the decade after the Civil War, American settlers killed millions of bison. Some hunted for sport, others for profit, and some hunted because they believed that exterminating bison would force Native Americans into the burgeoning reservation system. Various technological changes hastened this violent colonial work, including the completion of the transcontinental railroad in 1869 and the production of large-bore rifles. With new methods for tanning bison hide, bison were processed at an industrial scale into blankets, furniture, military uniforms, and machine belts. Bison belts drove

1.1 Bison skulls piled at the glueworks in Rougeville, Michigan, c. 1892. Courtesy of the Burton Historical Collection, Detroit Public Library

water-powered mills, and bison bones were harvested to make glue, fertilizer, and refined sugar.[6]

The killing of bison to erode Indian sovereignty, while not an official national policy, was nevertheless an explicit plan. After the Civil War, the primary aim of the Department of the Interior was to transform lands expropriated from Indigenous nations and Mexico into land available to white settlers.[7] In this context, bison were considered an impediment to an America owned and governed by cattle-rearing settlers of northern European descent. When General William Tecumseh Sherman assumed command of the Military Division of the Missouri in 1866, he instructed settlers to kill bison indiscriminately. Secretary of the Interior Columbus Delano maintained in an 1872 report that destroying bison would aid "our efforts to confine the Indians to smaller areas, and compel them to abandon their nomadic customs."[8]

Only at this juncture did politicians and naturalists begin to debate the nation's practice of bison slaughter. Congressman Richard McCormick of Arizona argued to the House of Representatives in 1872 that bison extermination had failed to "quiet the Indians"; it had only made them, in his words, "more restless and dissatisfied."[9] Others maintained that bison should be protected because bison meat had become an important food source for white settlers. Henry Bergh, who had recently founded the American Society for the Prevention of Cruelty to Animals, circulated letters from advocates that described the bison as "a noble and harmless animal," "timid" and "defenseless," one in need of protection from "wicked and wonton waste."[10] A Santa Fe newspaper article read on the congressional floor in 1874 argued that "there is quite as much reason why the Government should protect the buffaloes as the Indians."[11] Such concerns signaled a change in settler attitudes toward bison and other large game species. They did not, however, signal a rejection of Indian subjugation. Rather, early attempts to restore bison were as much a part of the settler project as the bison's initial destruction. In a series of events that has been too often overlooked, Indian reservations were converted into the nation's first wildlife reservations.

The campaign to eradicate bison was so successful that, by 1880, the number of bison in the United States had been reduced from tens of millions to fewer than one thousand. When Hornaday and two assistants embarked for Montana on a bison specimen collecting expedition in May 1886, they did not find herds of bison, but rather thousands of skeletons, bleached by the sun, strewn across the landscape. Their party managed to capture only one calf, which they brought with them back to Washington, DC. Hornaday named it Sandy. Fortunately for Hornaday, a second expedition later that year yielded better results. On that trip his party killed or purchased a total of twenty-five bison skins, sixteen skeletons, fifty-one skulls, and two bison fetuses. "I am really ashamed to confess it," Hornaday later reflected, "but we have been guilty of killing buffalo in the year of our Lord 1886."[12]

Returning to Washington after the second expedition, Hornaday worked on mounting six of the best specimens—including Sandy, who survived on the national mall for only two months—in a mahogany and glass case. While preparing the National Museum's bison exhibit, he befriended a civil service commissioner who stopped by regularly to view its progress, Theodore Roosevelt, a connection that would prove key to later restoration efforts. The following year, Hornaday published *The Extermination of the American Bison,* in which he estimated that there were only 456 bison left in the United States: 256 in private herds and 200 within the boundaries of Yellowstone National Park. In keeping with many of his contemporaries, Hornaday did not question this outcome; he viewed the decline of bison as an "absolutely inevitable" step in the colonization of North America, writing, "From the Great Slave Lake to the Rio Grande the home of the buffalo was everywhere overrun by the man with a gun; and, as has ever been the case, the wild creatures were gradually swept away, the largest and most conspicuous forms being the first to go."[13]

It was around this time that Hornaday became interested in assembling what he called a "living zoological collection," a step toward his eventual involvement in bison restoration, but not yet a break with the idea that white settlement condemned certain species—and societies—to extinction. In the fall of 1887, he and George Goode, then secretary of the Smithsonian, organized what zoology student Harvey Brown called "a little tryout zoo" on the Smithsonian grounds.[14] The first zoological parks in the United States had opened in the decade prior. Based on a European model, these institutions were meant to educate the public, foster civic pride, and demonstrate America's global reach.[15] Perhaps because today we think of zoological parks as mostly "domesticated" places, historians have tended to study them separately from the history of in situ conservation and restoration, but the two histories are closely entwined. Hornaday became the first curator of the newly established Department of Living Animals, a position he held until 1890. Now, in addition to mounting and stuffing skins, Hornaday would be responsible for caring for living animals. By the spring of 1888, the collection had grown from fifteen to 172 animals, including four bison contributed by a rancher in Nebraska. Between two thousand and three thousand people viewed the collection each day.[16]

1.2 A group of schoolchildren viewing a bison behind the United States National Museum in Washington, DC, 1899. Smithsonian Institution Archives

As it turns out, Hornaday's tenure at the Department of Living Animals was short-lived. Hornaday helped plan for an expanded National Zoo at Rock Creek, but he resigned from the National Museum after disagreements with the incoming Smithsonian secretary Samuel Langley. Hornaday spent the following six years working in real estate and writing a novel about an American man who escaped urban life by taking refuge among headhunters in Borneo. Then, in January 1896, he received a letter from Henry Fairfield Osborn, chairman of the New York Zoological Society's executive committee, inquiring whether he would be interested in directing the New York Zoological Park (today the Bronx Zoo).[17] From this position Hornaday would become involved in founding the American Bison Society, the first game restoration organization in the United States.

Upon moving to New York City in 1890, Hornaday joined the Boone and Crockett Club, an elite men's hunting club founded by Theodore Roosevelt and George Bird Grinnell in 1887 to promote "manly sport with the rifle" and "exploration in the wild and unknown." The club's first meeting, a dinner for twelve wealthy and well-connected white men, was at Roosevelt's sister's home on Madison Avenue. The club was not purely social, however. In keeping with their interest in hunting, Boone and Crockett Club members worked with other game conservationists—as I will call them in order to distinguish them from restorationists—to reform state and federal hunting laws. Indeed the list of Boone and Crockett Club members will be familiar to students of the American conservation movement: besides Roosevelt and Hornaday, it includes Albert Bierstadt, Madison Grant, Clarence King, and, later, Gifford Pinchot and Aldo Leopold.[18] Wildlife conservationists sought to restrict the length of hunting seasons and the number of animals that hunters could kill; to outlaw certain hunting methods, especially those used by nonwhite hunters; to ban the sale or transport of game; and to establish state game commissions.

The Boone and Crockett Club was in many ways a typical early conservation organization; it was situated in a city center, gender segregated, and white. Responding to the growing concern that modern urban environments produced effeminate and degraded men, Boone and Crockett Club members put forth a vision of American masculinity that prized physical activity and revered frontier life. Meanwhile, the woman-led National Association of Audubon Societies and the Wild Flower Preservation Society (founded in 1901) endeavored to convince women to stop purchasing clothing decorated with feathers and wildflowers, respectively, by appealing to notions of proper femininity. By the 1890s approximately five million birds per year were killed in the United States for their plumes, and women's conservation groups decried declines in dogwoods, gentians, laurels, and other species used for centerpieces and Christmas decorations.[19] "Weddings, by the way," botanist Elizabeth Britton wrote in 1913, "are a new menace to our native plants."[20]

Despite their different formulations of how to care for wild species (and which species to care for), men's and women's conservation clubs shared the goal of restricting the environmental access of nonwhites. Between 1880 and the passage of the Immigration Act of 1924, approximately twenty-five million people immigrated to the United States, mostly from Southern and Eastern Europe. The so-called new immigrants quickly became the subject of anxiety for Protestant "Nordics" or "Anglo-Saxons." White conservationists described the hunting practices of new immigrants as wasteful and greedy, and into the twentieth century they increasingly portrayed immigrants and African Americans as the main threats to plants and animals. In *Our Vanishing Wild Life* (1913), Hornaday denounced "Italian and negro bird killers" for "eating everything that wears feathers." "The gathering of woodland treasures for the city market is largely the work of Italians," claimed Mary Perle Anderson in a 1904 essay for the Wild Flower Preservation Society. "With no thought beyond the present need, they are a dangerous foe to plants."[21]

Between the 1880s and 1910s, conservation organizations successfully established a suite of state and federal hunting regulations, including the Lacey Act of 1900, which made it a federal crime to transport illegally hunted animals across state lines. Such restrictions were resisted by local subsistence hunters, as well as by market hunters, hunting guides, and gun manufacturers. Efforts to constrain hunting practices led to a number of unprecedented and violent state interventions, and as historians Louis Warren and Karl Jacoby have detailed, Native Americans, African Americans, Latinos, and southern Europeans accounted for a disproportionate share of those charged with hunting violations.[22] Local subsistence hunters resented the fact that wealthy Manhattanites were restricting their access to game. Commercial hunters and hunting guides believed their professions were under attack. Politicians were well aware of these constituents' grievances, especially in western states. And when conservationists turned against the use of automatic guns, gun manufacturers pressured them to find another cause.

Game restoration emerged as an explicit alternative to conservation. Beginning in the early 1900s, a small group of conservationists distinguished themselves from those lobbying for further hunting restrictions. If game animals could be bred in captivity and then released, they argued, then it would be possible to increase game population sizes while still allowing hunting for sport, and perhaps even for profit. In other words, restoration would work through the manipulation of animal bodies rather than through the manipulation of laws and regulations. Rather than seek to manage human behavior, or economies, game restorationists would seek to manage game species themselves—and, in turn, to manage the wildness of those species.

Founding the American Bison Society

In 1902, more than a decade after Hornaday published *The Extermination of the American Bison,* Congress directed the secretary of the interior and the secretary of agriculture to determine how many bison lived in the United States and Canada, to what extent they were "running wild" or were "being domesticated," and whether or not they were "of pure or mixed blood."[23] After surveying naturalists across the country, including Hornaday, the secretaries' committee determined that 1,143 buffalo were left in the United States, all in captivity except fifty "running wild" in the state of Colorado and twenty-two enclosed in a fence in Yellowstone National Park. The captive bison lived on private ranches in twenty-six states, in herds that ranged in size from a few individuals to a few hundred, with the highest numbers in Montana, South Dakota, and Texas.[24] Some ranchers harvested the bison for meat, while others experimented with crossbreeding bison and cattle. Scotty Philip, a South Dakota bison rancher, advertised, "We supply Buffalo for Zoos, Parks, Circuses, and Barbecues."[25]

The 1902 report concluded that it would be prudent to place a herd under governmental supervision, although there was not then a federal agency whose mission was defined to encompass caring for such a herd. An article in the *Washington Times* announced that the "once mighty monarch of the plains" was destined for extinction unless the government could acquire and manage a sizable herd.[26] It is likely that this report inspired writer and naturalist Ernest Harold Baynes to propose a society dedicated to bison restoration. The idea occurred to him in the summer of 1904, while Baynes was working as the conservator for Corbin Park, a private game preserve in central New Hampshire (established by Austin Corbin, developer of Coney Island), home to one of the largest bison herds at the time. Private game parks were common: there were about sixty such parks in the Adirondacks alone. Establishments like Corbin Park and the Adirondack League Club stocked privately owned ponds and streams with fish and imported exotic game animals like caribou and English deer—and the occasional native game species. Members could then stay on the property to fish and hunt.[27]

The animals that Baynes cared for at Corbin Park included some 140 white-tailed deer, 135 elk, a few Himalayan goats, and 160 bison acquired from ranchers in Texas, Wyoming, Manitoba, and Winnipeg. Like many private herd owners, the Corbin family soon decided that the bison herd was too costly to maintain. Baynes began a campaign to convince the American public that the federal government should purchase bison from private owners like the Corbins because it was possible to domesticate them, and thereby profit from them. Baynes toured the Northeast with a pair of bison calves, tellingly named War Whoop and Tomahawk, that he had tamed. He took his team on a tour of county fairs, agricultural exhibitions, and natural history societies. In Waterville, Maine, one of the calves was pitted against a domestic steer in a mile-long race, and Baynes recounted that it won easily. At the Boston Sportsmen Show, Baynes demonstrated to the crowd that "even ladies" were able to drive the team.[28]

1.3 Ernest Harold Baynes and an unidentified woman with a team of bison. Plainfield Historical Society, New Hampshire

In his campaign to popularize bison, Baynes published dozens of magazine and newspaper articles and wrote letters to prominent conservationists, including Roosevelt and Hornaday. After attending Baynes's illustrated lecture, "The American Buffalo—A Plea for His Preservation," at the Campfire Club in spring 1905, Hornaday tentatively agreed to work with Baynes to create a new society dedicated exclusively to bison restoration. On December 8, 1905, fourteen men assembled in the lion house of the New York Zoological Park to mark the establishment of the American Bison Society.[29]

The initial goal of the American Bison Society (ABS) was to convince the federal government to acquire and maintain bison herds and to raise money for this cause. Members wrote editorials in local newspapers, and they sent letters to federal representatives and wealthy industrialists in which they proposed a partnership between the federal government and the ABS. The extinction of the bison, Baynes wrote in a widely circulated newspaper article, would be "a loss to the world which can never be repaired, since an animal once extinct has gone forever."[30] This argument about extinction's irreversibility was unusual for its time: conservationists typically grounded their appeals in the ideals of sportsmanly conduct and scientific use of natural resources. The society's goal of establishing new herds was also novel. The ABS was not advocating for mere conservation, either the regulation of hunting or the establishment of a preserve where hunting was banned. Rather, they planned to breed new bison. It would be the first society in the United States dedicated specifically to species restoration.

Baynes and his colleagues tried out multiple arguments about the worth of bison. In a 1906 article, Baynes invoked the receding white pioneer past by reminding readers of "the debt we owe the buffalo for the great part he played in the winning of the West."[31] At the same time, society members explicitly avoided language that would be seen as "sentimental"—a critique frequently levied at animal humane societies and at women more generally. Hence, they argued that the bison's extinction would be financially wasteful: "Shorn of all sentiment," Baynes wrote, "and as he stands on his hoofs, he is the most valuable native animal in the country." Bison meat, he continued, was as good as that of cattle. The skin could be used for clothing and winter carriage robes.[32] Hornaday offered a different economic justification for bison restoration. A decade prior, he had argued that it cost the federal government more "to feed and clothe those 54,758 Indians" than it would have cost to maintain a bison supply for them.[33]

At first game restorationists did not emphasize the "wildness" of bison. Indeed, according to early ABS members, bison could prove to be as valuable as domesticated cattle, if only the species could be "improved" through crossbreeding, and Baynes continued to speculate that, if properly broken, bison could be used as draft animals. In magazines such as *Rod and Gun*, ABS members argued that bison should be preserved because the cattle-bison hybrid—the "cattalo"—might prove profitable to white ranchers in the future. They imagined the hybrids would retain the desirable characteristics of each species: the hardiness, strength, and size of the bison, and the docility, softer hair, and "sweeter taste" of cattle. Bison were better adapted to northern prairies than cattle, cattalo boosters argued, and so cattalo would be able to tolerate harsh winters and even avoid poisonous weeds "by instinct."[34] Charles B. Davenport, director of the Carnegie Institution's newly established Department of Experimental Evolution, argued that owning cattalo would make settlers "more fit to face the blizzard."[35]

Hornaday pursued the idea of domesticating bison as the first chairman of the Committee on Breeding Wild Mammals of the American Breeders' Association. The American Breeders' Association had been founded in 1903, three years after the "rediscovery" of Gregor Mendel's 1865 paper on inheritance. Its founding members included commercial breeders, professors at agricultural colleges, and researchers at the U.S. Department of Agriculture—those interested in breeding for practical applications as well as for contributions to evolutionary theory. Within a decade of its founding, the association was a large and respected institution, one that played an important role in Progressive Era debates about the relative roles of environment and heredity in shaping human nature and, thus, debates concerning eugenics. Members moved fluidly between conversations about plant and animal breeding and conversations about human breeding. Early committees included the Committee on Breeding Roses and the Committee on Heredity of Criminality.[36]

Indeed, many early game conservationists and restorationists were eugenicists and influential proponents of scientific racism. Eugenicists held that intellectual, moral, and temperamental traits were heritable, that the fitness of human populations as well as individuals could be defined in these terms, and that the human species could be divided into biologically distinct races, with "Nordics" or Anglo-Saxons at the apex of a hierarchy. Among those to transform the question from who was individually unfit to who was racially unfit was Madison Grant, a founding member of the American Bison Society and a close friend of Hornaday. In 1916, Grant published *The Passing of the Great Race,* in which he argued that those of his own racial type were threatened through their failure to breed, by a war that was devastating Western Europe, and by eastern and southern European immigrants to the United States. The views of Grant and other devotees of scientific racism enjoyed wide support; in the 1920s and 1930s, congressmen routinely read passages of Grant's book aloud to argue for immigration restriction.[37]

Eugenicists applied their belief in genetic perfectibility to wild animals as well as to humans. (Grant was an eager collector of hunting trophies, but he disparaged the moose for having a "Jewish cast of a nose.")[38] The stated purpose of the American Breeders' Association Committee on Breeding Wild Mammals was "to investigate and report on the methods and technique of improving wild mammals" and "to devise and suggest methods and plans of introducing, procuring, and improving such wild animals." The committee's 1908 report identified two reasons to breed wild mammals. The first was to create new, domesticated types to join the ranks of approximately twenty-five species already "serving man as beasts of burden or furnishing food, clothing, or companionship." This could prove especially important, the report's authors noted, if supplies of fossil fuel were depleted. "Our extensive use of machinery and electricity must inevitably exhaust the coal, petroleum, and natural gas from the earth's crust," they speculated, and "we shall again be forced to rely largely upon the labor of animals."

The second reason to breed wild mammals, according to the committee, was to save species from extinction. The "extinction of species is a process of nature," they wrote, "and from an economic point of view is not necessarily a misfortune to the world"; nevertheless, for some species there was "good reason for the intervention of organizations of men."[39] Fearing extinction—of the Anglo Saxon/Nordic race as well as of species like the bison—eugenicists argued that the laissez-faire approach to social organization was obsolete and that future progress, whether in political economies or in the care for species, would require rational, bureaucratic management by experts and elites.[40] Indeed, some of the same white men who sought to intervene in the evolution of society through forced sterilization also sought to intervene in the evolution of bison. The future "wildness" of bison, like that of white men, they argued, was threatened by a lack of "open" space; the domesticated city, they feared, would produce weak men and weak bison.

From Indian Reservation to Game Reservation

Hornaday and his colleagues believed that confinement was "detrimental to the character of the offspring" of bison and other large animals. In *The Extermination of the American Bison,* Hornaday argued that, in captivity, the bison "fails to develop as finely as in his wild state"—he gets "fat and short-bodied," and "with the loss of his liberty he becomes a tame-looking animal."[41] In 1902, a columnist for the *Washington Post* argued that because of the "enervating influences of captivity and inactivity," the "poor bison weakens as a self-perpetuating race and at once commences to die out."[42] Moreover, early restorationists and zookeepers found it difficult to keep bison and other large animals alive in captivity. For a period, the New York Zoological Park's orangutans lived in the Hornadays' living room, where Hornaday's wife, Josephine Chamberlain, fed them "a teaspoon of castoria daily in order to get digestive apparatus in good working order."[43] Many of the zoo's animals died, including the twenty-five bison that Hornaday had acquired for the New York Zoological Park from Corbin Park, which soon succumbed to disease. By 1908, Hornaday considered it a "well-known fact that no large species of quadruped can be bred and perpetuated for centuries in the confinement of zoological gardens and parks."[44]

Because of the belief that captivity weakened large mammals that should have been wild, the ABS began to argue that bison restoration required large parcels of land. For these nascent restorationists, wildness was conferred by environment and not by genetic history.[45] They maintained that a lack of access to "open" space threatened the virility of native fauna in the same way it endangered white men.[46] Bison restorationists sought large expanses of land in which they believed bison could develop traits that only a wild bison could possess, like a large stature and a full coat. And where were such expanses of land to be found? The *Washington Post* columnist continued: "I do not believe that a small group of buffaloes will ever grow into a large herd unless that group is practically given its freedom, is allowed the run of thousands of acres. This the small park does not contain and the individual owner can seldom afford. But how easy it would be for Uncle Sam to turn one of his reservations into a buffalo breeding ground and if it be at a cost of even $1,000,000 the race is rehabilitated, does not the result far outweigh the cost?"[47]

It is no coincidence that the restoration, conservation, and preservation movements emerged around the same time in the United States. Signed into law in 1862 by Abraham Lincoln, the Homestead Act enabled heads of household who had never borne arms against the U.S. government to claim 160 acres of federal land. Claimants were required to "improve" the plot by building and cultivating. The act was a main engine of colonization; while the law was in effect, the federal government granted to private citizens more than 270 million acres—10 percent of U.S. land. Claims under the Homestead Act peaked in 1913; more than 60 percent of the claims under the act were made between 1901 and 1930.[48] During this period, conservationists and wilderness preservationists argued that the federal government should retain land and manage it for timber production or recreation.[49] Meanwhile, restorationists maintained that the best use of federal land would be as game reintroduction areas. These propositions were novel; one of the main functions of the federal government up until this point had been to transfer land from the public domain to settlers. Thus, game restorationists, Progressive Era conservationists, and wilderness preservationists alike argued that they had a compelling reason to reserve federal lands.

As homesteading peaked, the federal government dismantled the Indian reservation system that only a few decades earlier it had brutally assembled. Here, too, game restorationists would see opportunity. In 1830, the United States passed legislation that legalized the forced relocation of tribes east of the Mississippi to the western territories. Politicians justified the Indian Removal Act by arguing that Indians were too closely tied to the environment to assimilate to white ways, and that therefore they faced extinction without government intervention. Nearly fifty tribes had been forcibly moved west by the end of the Civil War, their lands seized through purchase, coercion, claims of right of discovery, and military force.[50] By 1890 the Bureau of Indian Affairs and the military had established some three hundred Indian reservations.

It was around this time that the government began to favor a policy of assimilation rather than segregation, compelling Indian communities into individual landownership, farming, and Christianity. With the purpose of eroding tribal sovereignty and erasing reservation boundaries, the Dawes Act of 1887 authorized the president to divide communally held tribal lands into allotments for individuals. As the act was implemented over the next fifty years, land area owned by Indians was reduced from 138 million acres to fewer than 48 million acres. Reservation land was sold to non-Indian settlers, to railroads, and to other large corporations. President Theodore Roosevelt referred to such allotment as a "mighty pulverizing engine to break up the tribal mass."[51] As the federal government opened large parcels of Indian reservations to white settlers, bison restorationists argued that this land would be best used as "game reservations." As one system of reservations was undermined, a new one came into being.[52]

The Boone and Crockett Club and other conservation organizations initially supported the creation of national parks and forest reserves, as they believed the federal government would restrict hunting access on these lands. But early court rulings determined that national forests, although owned by the federal government, would follow the game laws of the states, and a 1902 bill providing for designated game refuges failed in Congress; the House Committee called it "the fad of game preservation run stark raving mad."[53] The establishment of bird and game reserves, distinct from national parks and forests, therefore proceeded through executive action. President Theodore Roosevelt ordered the establishment of the first federal bird sanctuary in 1903. His executive order made Pelican Island a federal holding "set apart for the use of the Department of Agriculture as a preserve and breeding ground for native birds." Pelican Island was soon followed by other bird sanctuaries, including sites at Breton Islands, Louisiana (1904); Stump Lake, North Dakota (1905); and Malheur Lake, Oregon (1908). By the time he left office in 1909, Roosevelt had established fifty-one bird sanctuaries.[54]

The establishment of game reserves proceeded more slowly. When the Kiowa-Comanche Reservation in what is now Oklahoma was thrown open to non-Indian settlement in 1901, Congress placed 61,500 acres under the control of the Department of the Interior, establishing the Wichita Forest Reserve. This land of round hills, blackjack oaks, and mesquite grass was where General Sheridan had viciously campaigned against the Kiowa, Comanche, and Wichita in the 1850s and 1860s. It was where Geronimo had been held prisoner at Fort Sill. In January 1905, President Roosevelt designated areas in the Wichita National Forest to be set aside for the protection of game animals and birds. That June, Roosevelt re-designated the

INDIAN LAND FOR SALE

GET A HOME OF YOUR OWN

EASY PAYMENTS

PERFECT TITLE

POSSESSION WITHIN THIRTY DAYS

FINE LANDS IN THE WEST

IRRIGATED IRRIGABLE — GRAZING — AGRICULTURAL DRY FARMING

IN 1910 THE DEPARTMENT OF THE INTERIOR SOLD UNDER SEALED BIDS ALLOTTED INDIAN LAND AS FOLLOWS:

Location.	Acres.	Average Price per Acre.	Location.	Acres.	Average Price per Acre.
Colorado	5,211.21	$7.27	Oklahoma	34,664.00	$19.14
Idaho	17,013.00	24.85	Oregon	1,020.00	15.43
Kansas	1,684.50	33.45	South Dakota	120,445.00	16.53
Montana	11,034.00	9.86	Washington	4,879.00	41.37
Nebraska	5,641.00	36.65	Wisconsin	1,069.00	17.00
North Dakota	22,610.70	9.93	Wyoming	865.00	20.64

FOR THE YEAR 1911 IT IS ESTIMATED THAT **350,000** ACRES WILL BE OFFERED FOR SALE

For information as to the character of the land write for booklet, "INDIAN LANDS FOR SALE," to the Superintendent U. S. Indian School at any one of the following places:

CALIFORNIA: Hoopa.
COLORADO: Ignacio.
IDAHO: Lapwai.
KANSAS: Horton. Nadeau.
MINNESOTA: Onigum.
MONTANA: Crow Agency.
NEBRASKA: Macy. Santee. Winnebago.
NORTH DAKOTA: Fort Totten. Fort Yates.
OKLAHOMA: Anadarko. Cantonment. Colony. Darlington. Muskogee, Pawnee.
OKLAHOMA—Con. Sac and Fox Agency. Shawnee. Wyandotte.
OREGON: Klamath Agency. Pendleton. Roseburg. Siletz.
SOUTH DAKOTA: Cheyenne Agency. Crow Creek. Greenwood. Lower Brule. Pine Ridge. Rosebud. Sisseton.
WASHINGTON: Fort Simcoe. Fort Spokane. Tekoa. Tulalip.
WISCONSIN: Oneida.

WALTER L. FISHER, Secretary of the Interior.

ROBERT G. VALENTINE, Commissioner of Indian Affairs.

1.4 Poster circulated by the U.S. Department of the Interior, 1911. Library of Congress

forest reserve as a "game reservation" by proclamation, a space where hunting, trapping, or capturing game animals and birds would be unlawful unless permitted by the secretary of agriculture.[55]

Learning of this development, Hornaday offered to provide the Wichita Forest Reserve with a nucleus herd of bison from the New York Zoological Park, on the condition that the government take responsibility for the herd's maintenance. The American Bison Society described the arrangement as a type of "corporate sacrifice" and "partnership agreement" with the federal government. In November 1905, the U.S. Forest Service invited the New York Zoological Society to send a representative to survey the new Wichita Game Reserve. Anticipating the donation, the ABS purchased a small herd of bison from Charles Goodnight, a rancher and Texas Ranger, and placed it in the New York Zoological Park. On the federal side, an item was inserted in the annual agricultural appropriation bill, providing for $15,000 with which to purchase winter hay and to erect sheds and a substantial fence around the proposed bison reservation. The Forest Service appointed Frank Rush, "Oklahoma's cowboy naturalist," as caretaker of the new herd. They hired a contractor to construct corrals, sheds, telephone poles, and a specially designed 74-inch-high steel woven wire fence to encompass 8,000 acres. They filled open prospecting holes and constructed a firebreak. Consistent with their focus on breeding game animals, they rounded up and killed coyotes and wolves. Meanwhile, Hornaday studied the wooden crates used at the Zoological Park to ship exotic animals, and for the rail trip from New York City to Cache, Oklahoma, he designed crates just big enough to fit a bison, with feed doors and wooden water boxes.[56]

1.5 Bison at the Bronx Zoo being prepared for transport to the Wichita Forest and Game Preserve, October 1907. William Temple Hornaday appears on left.

On October 11, 1907, fifteen bison and three humans left Fordham Station bound for the Wichita Game Reserve. The American Express Company had offered to transport the two railcars needed for free between New York and St. Louis, and the Wells-Fargo Express Company did the same from St. Louis to Cache. "It was a bit awe-inspiring," bison handler Elwin Sanborn reported, "to realize that in the midst of this vast station with its multitudes of people, its coughing, booming trains, in the center of the greatest city of the new world, were fifteen helpless animals, whose ancestors had been all but exterminated by the very civilization which was now handing back to the prairies this helpless band, a tiny remnant born and raised 2,000 miles from their native land."[57]

They made the 1,858-mile trip in seven days. It was purportedly miserable. The bison were disgruntled. The men slept alongside the crates in the unheated cars. Curious and pushy crowds greeted them in Cleveland, Indianapolis, St. Louis, and Oklahoma City. "The word 'Zoological' was pronounced in more ways than I thought ever possible," Sanborn recounted. All fifteen bison survived the journey, and a caravan of wagons hauled the crates across the final twelve miles of prairie to the Wichita Game Reservation. "All persons who never have had an opportunity to become familiar with the difficulties involved in shipping a herd of large hoofed animals by rail should be advised that such an undertaking involves very serious difficulties," Hornaday later reflected. Upon arrival, each animal was sprayed with crude oil to kill ticks. Among the names bestowed on the bison were Comanche, Lottie, Geronimo, and Hornaday.[58]

After the successful establishment of the Wichita herd, the ABS rapidly resolved to have the Flathead and Crow Indian Reservations in western Montana examined "with a view to having suitable portions of them set apart as buffalo ranges." The federal government had opened the land, which belonged to the confederated tribes of the Flathead, Kootenai, and Upper Pend d'Oreille, to non-Indian settlement in 1906. At their second annual meeting, the ABS resolved to raise a special fund with which to purchase bison for a Montana game reservation, provided the government supplied land and fencing. Most of the donations came from residents in eastern states. A number of ranchers offered gifts of bison. "Unfortunately," the ABS wrote in newspapers across the country, "the Indians will have to be paid for any land that may be set aside for a bison range."[59]

Restorationist appropriation of Indian lands was affirmed in March 1908, when Senator Dixon of Montana sponsored a bill to create a 24-square-mile buffalo reserve from Flathead Indian Reservation lands. Roosevelt signed the bill into law in mid-May, and the American Bison Society began to survey portions of the reservation near the Northern Pacific Railway, ultimately

1.6 Train car holding bison at New York's Fordham Station, bound for Wind Cave Reserve in South Dakota, November 1913. © Wildlife Conservation Society. Reproduced by permission of the WCS Archives.

selecting land north of Ravalli, Montana. "The reasons for this are quite obvious," ABS member M. J. Elrod stated: "To ship animals in and out will be necessary from time to time. There may be need for transportation of forage. Fencing material must be procured, and long hauls by wagon are expensive. Lastly, the public will want to visit the animals and see them on the range, and will desire to reach them easily from the railway."[60] Thus one purpose of bison restoration was to provide an exhibition for a white audience. This was not so different from the project of the Department of Living Animals, which brought bison to the audience. Now, the audience would come to the bison. The railroad, so central to the destruction of the bison, had become a mechanism of its restoration.

The thirty-four bison donated by the American Bison Society reached Ravalli, Montana, on October 16, 1909, after being taken by railroad to Flathead Lake, where they were put on boats and shipped to Poison. From there they were hauled about 25 miles over land to the newly designated National Bison Range. The *Wellsboro Gazette* reported that naturalists had decided to "set apart a tract of land specially for the use of the defrauded monarch of the plains, just as reservations have been made for the Indians."[61] The parallel between the two reservation systems was explicit.

Hornaday resigned as president of the ABS after the establishment of the National Bison Range, but this did not occasion a change of mission. The first recommendation of his successor, Franklin Hooper, was that the society seek to establish a national bison herd in South Dakota. In February 1911, Hooper traveled to Washington to meet with members of Congress to discuss the plan. In consultation with the Departments of Agriculture and the Interior, the ABS decided on five possible sites: Standing Rock Indian Reservation, Rosebud Indian Reservation, Pine Ridge Indian Reservation, the Black Hills, and Wind Cave National Park. After surveying the sites, they decided on Wind Cave, South Dakotan lands that had been withdrawn from the Great Sioux Reservation in 1877.[62] The next national bison herd would also be situated on the former Great Sioux Reservation. On November 14, 1912, Roosevelt established the Fort Niobrara Game Preserve through executive order.[63] By 1915 five national game reservations—Yellowstone, Wichita, Flathead, Wind Cave, and Niobrara—had been established on appropriated Indian lands.

A National Wildlife Refuge System

Unlike the bird refuges established by Roosevelt's executive orders, these five game reservations were sites of active species reintroduction. The Department of Agriculture's Bureau of Biological Survey provided reservation managers, whose various responsibilities included repairing fences, putting out coyote poison, patrolling for fire, taking notes on weather and animal sightings, and corralling the bison so that they were visible to tourists. They also planted alfalfa and corn to feed the bison.[64] This earliest federal wildlife management was modeled on the burgeoning field of agricultural science; indeed, as we will see, wild animals would be managed as "crops" until the rise of ecosystem ecology in the 1970s.

Federal game managers soon began experimenting with introducing other large game species onto the reservations, including elk, pronghorn antelope, and mule deer, although these proved even more difficult to handle than bison. The Wichita Game Reservation failed to keep alive the eleven pronghorn antelope sent by the Boone and Crockett Club from Yellowstone National Park, or the fifteen elk sent by the U.S. Biological Survey from Jackson, Wyoming, but they did rear twenty-two wild turkeys from eggs by feeding the chicks cottage cheese.[65] The difficulties of transporting large animals long distances and keeping them alive once they arrived at their destination called for new types of expertise. The endeavor of restocking game reservations with animals was "a new one," noted Smith Riley, U.S.

district forester for Colorado, one that would require "some study and careful work."[66] A 1912 article in *Science* magazine outlined a new undertaking for zoological scientists: saving species from extinction. Though only politicians could establish hunting laws, Scottish zoologist P. Chalmers Mitchell argued, zoologists could learn how to breed endangered species and introduce them into managed reservations. There were many potential sites, Mitchell argued: "There exist in all the great continents large tracts almost empty of resident population" that "could afford space for the larger and better-known animals."[67] Of course, few places were actually "empty of resident population"—if they were, it was because of centuries of colonial violence. The first game reservations in the United States were established in places where people had lived and wanted to continue living.

While recognizing the difficulties of transporting and breeding large mammals, game restorationists imagined that once a few individuals were established on a large tract of land, they would repopulate it rapidly and without further intervention. Once "wild life restoration" had occurred on a particular reservation, the animals would "colonize areas presumably once populated but long since denuded of game animals," Robert Sterling Yard predicted in a 1928 book with the remarkable title *Our Federal Lands: A Romance of American Development.*[68] Stocking, it should be noted, was not the same as restoring populations to their original size; not until later in the century would restorationists concern themselves with what we might call historical fidelity. "It should not be imagined that any plan is contemplated for covering the plains of the West with millions of buffalo, as of yore," a 1911 newspaper columnist explained. "Those vast grazing areas are wanted to-day for cattle and sheep."[69] Nor did early restorationists strive to confine game species to their historical ranges. At the first meeting of the American Bison Society's board of managers, in January 1907, Franklin Hooper proposed that the society establish a bison preserve on state lands in the Adirondack Mountains of New York, outside the bison's historical range. His proposal was passed by the New York legislature, though was ultimately vetoed by the governor.[70] Further, game restorationists focused on single species of economic value. They were generally uninterested in plant species—as long as there were already grasses at the site for grazing—and they were actively opposed to predatory animals. A draft of Hornaday's "General Scheme for National Forest Sanctuaries" stated, "Predatory animals: To be killed."[71]

The work of the American Bison Society typified early restoration work in the United States, work carried out before the expansion of today's major federal land management agencies: the U.S. Fish and Wildlife Service, the U.S. Forest Service, the National Park Service, and the Bureau of Land Management. Between 1900 and 1930, federal wildlife restoration, and wildlife management more generally, relied heavily on donations and volunteers from societies like the ABS and the Audubon societies. Volunteers from the Audubon Society patrolled bird refuges for poachers. The American Bison Society donated bison bred at ranches, game parks, and zoological parks. Until the 1960s, most naturalists believed that it was impossible to restock wild birds; Hornaday wrote that while there was "no royal road to the restoration of an exterminated bird species"—all that could be done was to "protect Nature, and leave the rest to her." In the case of large mammals, however, Hornaday and his colleagues maintained it would be possible to "restock depleted areas."[72] Those seeking to "improve" wild animals through scientific breeding pursued the twin goals of creating animals better suited for labor, fur, or meat, and protecting species like bison, deer, and elk from extinction.

The game reservations backed by the ABS would become nodes in a growing network of animal breeding, transport, and exchange, one that divided responsibilities among zoological societies, nongovernmental organizations, multiple federal agencies, and the states. Between 1913 and 1925, twenty-four states established wildlife refuges. By 1937, the U.S. Biological Survey oversaw eleven federal game reservations in Oklahoma, Montana, Nebraska, South Dakota, North Dakota, Wyoming, Nevada, Oregon, and Alaska.[73] These reservations, along with the bird reserves established by President Roosevelt and successors, were ultimately incorporated into the U.S. Fish and Wildlife Service National Wildlife Refuge System in 1966, which today comprises a network of 567 sites and more than 150 million acres, an area larger than California.[74] At these sites, managers acquired animals from an informal network of federal, state, and private game reservations. In the year 1940 alone, the animals that state game commissions relocated from federal refuges included 44 antelopes, 15 mule deer, 998 white-tailed deer, 51 elk, 2 beavers, 20 minks, 1,313 muskrats, 386 raccoons, 11,215 ring-necked pheasants, 50 wild turkeys, and 30,000 fishes.[75] Thus wildlife refuges became both sources and destinations for biotic remixing at a national scale.

Game reservations were ultimately successful in preventing a number of imminent extinctions. More than 450,000 bison now exist in North America—a remarkable comeback from a low of a few hundred animals in the late 1800s. By 1934, when the American Bison Society took its last census, the federal government owned 2,392 bison—more than twice the number in existence when the society was founded.[76] Their success inspired environmental management elsewhere, including the founding of the International Society for the Preservation of European Bison in 1923.[77] Indeed, federal

1.7 National Park Service ranger J. Estes Suter on top of car watching bison herd in Wind Cave, South Dakota, c. 1936. WICA 4329, Wind Cave National Park, Courtesy of National Park Service

game reservations soon found that they had *too many* animals. In 1919, just fourteen years after the founding of the ABS, Congress authorized the secretary of agriculture to sell "surplus" bison to meat processers, state wildlife commissions, and public parks (a bison herd roamed Golden Gate Park in San Francisco, and still does).[78] American Bison Society member Martin Garretson announced in 1920 that, through the establishment of game reservations, "the future of the American bison, as a species safe from extinction" was "now secure." The American people had made "the best atonement that they possibly could make for the sins of the past."[79]

Although advocates of bison restoration touted its benefits to the nation, those benefits accrued specifically to white citizens, who viewed the bison as railway tourists or purchased bison from federal reserves to start private hunting or ranching enterprises. Early restorationists imagined game restoration to be the best possible use of lands not desired by white settlers, with no regard for Native American primacy and sovereignty. When ABS member Martin Garretson wrote that "such a merciless war of extermination was never before witnessed in a civilized land," he was speaking of bison, not people.[80]

Indeed, most game restorationists deemed the colonization of North America by people of European descent to be an inevitability, one that would necessarily result in the decline of "wild" species like bison and elk. To a certain Anglo-American imagination, Native Americans and bison alike persisted only through the benevolence of the federal government. On a drizzling February day in 1913, President William Howard Taft began the groundbreaking ceremony for the National Memorial to the American Indian on Staten Island with a short speech. He explained that the memorial would "perpetuate the memory of the succession from the red to the white race of the ownership and control of the Western Hemisphere."[81] The U.S. Mint made the groundbreaking the occasion for distributing the new nickel, which bore an anonymous, composite Native American on one side, a bison on the other. Two sides of the same coin.

It is possible to tell a triumphant history of bison restoration, one in which passionate naturalists realized that a species was on the brink of extinction, acted to save it, and succeeded. To focus on the *who,* however, is to elide the thornier questions of *where* and *how.* In its first decade of operation, the American Bison Society established five "bison reservations" on Indian reservations that the federal government was systematically dismantling in order to erode tribal sovereignty. Game restorationists attempted to mitigate a destructive ecological effect of settler colonialism—bison decline—by furthering settler colonialism, appropriating Indian lands. The establishment of the National Wildlife Refuge System depended on the ideology and the administrative apparatus of settler colonialism. The two reservation systems, wildlife and Indian, shared not only a geography, but a logic.

Game restoration set important precedents for ecological restoration, which aimed to reestablish relationships among multiple species. Borrowing techniques from zoological parks, game restorationists hoped to breed large mammals in order to stock public lands with them. This vision was not one of passive land preservation, but rather one of active restoration. Deeming the future of bison secured, the American Bison Society voted to disband in 1935. Bison numbers were increasing, and at the same time, the environmental sciences were professionalizing. It was not, therefore, organizations like the American Bison Society and the Audubon societies that vied to be the experts on biotic recovery in the coming decades. When drought devastated the nation in the 1930s, the contest was instead between professional foresters and members of an even newer discipline, ecology.

2

Ecology in the Public Service

"The field of ecology is chaos," Henry Chandler Cowles complained in the journal *Science* in 1904; its practitioners could not agree "even as to fundamental principles or motives."[1] Thirty-two years later, the president of the burgeoning Ecological Society of America, Walter P. Taylor, asserted confidently that ecology was "one of the most useful and essential of the sciences," a science poised to influence soil conservation, national land use planning, resettlement projects, and "practically all phases of conservation of natural resources."[2] When game restoration emerged, in the first decades of the twentieth century, ecology had been the obscure pursuit of a handful of botanists. By the time of Taylor's 1936 speech, ecology was an organized discipline with its own questions, journals, and scientific societies—and with its own approach to restoration. Nationwide environmental change had changed the nature of environmental expertise.

The 1930s opened with dust and depression. Crops failed. Grasses grayed. Relentless drought transformed prairie loam into a fine powder, causing "black blizzards" that made it all the way to New York City. Dust settled into beds and flour bins, into eyes and ears and throats. One evening in Chicago in 1934, dust fell like snow, four pounds of it for each person in the city. Tens of thousands of families abandoned their homes.[3] As Americans suffered drought after drought—in 1930, 1931, 1934, 1936, and again in 1939—they sought explanations. Some speculated that drought had and would continue to occur in cycles in the Great Plains. Others contended that prairie soils had been damaged by new agricultural technologies, including gasoline-powered tractors, one-way disc plows, and combines. Still others maintained that poor farmers and immigrant farmers were to blame for lack of prudence in farming decisions. At its heart, the question of whether the causes of the Dust Bowl were climatic, technical, or social was a question over how to pursue national recovery. If the Great Plains were naturally a desert, then perhaps vulnerable areas should be permanently left fallow. If the problem was technical, the argument went, then the federal or state governments should regulate where and with what tools farmers worked. And if the problem was social, then federal reeducation programs might restore economic and agricultural prosperity.[4]

It was through these competing proposals for national recovery that restoration became both scientific and ecological. The Dust Bowl was hardly the first large-scale environmental crisis—Plato wrote of soil erosion in the fourth century B.C., and the American soil conservation movement can be traced to antebellum agricultural reformists.[5] But the Dust Bowl, and Franklin D. Roosevelt's New Deal, meant that substantial federal resources, including land, would be available to whoever offered the most compelling path to national recovery. Across the political spectrum, members of an expanding professional scientific class vied to influence federal environmental policy; they included agricultural scientists, landscape architects, foresters, and members of the new discipline of ecology.

Historians have tended to tell the history of ecology through the lives of individual scientists or by tracking the history of specific ideas like community or stability.[6] To focus too narrowly on ideational change, however, would be to miss the essential roles that widespread drought and the New Deal played in ecology's transformation from an obscure pursuit to an internationally recognized science.[7] Indeed, it is impossible to understand the rise of ecology without understanding how disciplinary ecologists developed ecological restoration on the ground—vying to ensure that federal management was "ecological" and actually reshaping material environments. To take a broader lens: histories of applied sciences like environmental management have much to tell us about the construction of scientific expertise, the interplay between theory and experiment, and the role of science in civic life.

Inspired by recently established game reservations, American ecologists sought to establish "nature reservations" or "nature reserves"—outdoor sites where they could conduct long-term studies of plant communities. Although their first attempts to secure study sites were unsuccessful, New Deal environmental projects provided members of the newly established Ecological Society of America with a justification for nature reservations. By comparing protected nature reservations with managed areas, ecologists contended, they would be able to assess whether federal interventions were effective in reversing environmental problems like soil erosion and overgrazing. In doing so, ecologists recast nature reservations as "check-areas" or "baselines"—experimental controls—and framed ecological theory as a tool for national recovery.

The Rise of Scientific Ecology

The German zoologist Ernst Haeckel coined the term "ökologie" in 1866, referencing the Greek *oikos,* meaning "household" or "place to live." In support of Darwin's theory of natural selection, published just a few years earlier, Haeckel contended that his colleagues, who up until that point had focused on organismal physiology and morphology, should turn their attention to the relationship between a species and its environment: "all those complex interrelations referred to by Darwin as the conditions of the struggle for existence."[8] Henry Cowles first encountered Haeckel's term in a graduate botany course at the newly founded University of Chicago in 1896, in a book that thrilled him. Danish botanist Eugenius Warming's *Plantesamfund* (Plant Community) articulated a vision for a new science, "oecological plant geography," that would classify plant community types and explain their existence. Warming wrote, "In countries far apart there are to be found communities identical in type, but entirely different in *floristic* composition. Meadows in North America and Europe, or the tropical forest in Africa and in the East Indies, may show the same general physiognomy, the same kinds of constituent growth-forms, and the same types of natural community, though of course their species are entirely different." Warming theorized that community types like meadows and tropical forests were determined by physical variables such as temperature, rainfall, wind, and soil nutrients. He was not so much interested in the particular species present as the shared traits of those species. Why did plants growing in deserts across the globe from one another look similar? Why were certain regions dominated by grasses, and others by tall trees?[9]

Inspired by *Plantesamfund,* Cowles decided to use Warming's idea of the community type to frame his graduate work on the flora of Lake Michigan's southern shore. While exploring the Indiana Sand Dunes, Cowles had noticed that only grass species grew close to the lake, whereas cottonwood, juniper, and pine trees thrived farther away from the water. Applying Warming's framework, Cowles described the shoreline as a "disturbed community" and the forest as a "climax community." He hypothesized that these two types of plant communities resulted from the actions of distinct physical forces: wind, in the case of the dune grasses, and temperature and rainfall in the case of the forest community. In retrospect, Cowles's resulting 1899 paper can be considered one of the first ecological studies in North America.[10]

Cowles was soon joined by other botanists who sought to understand how environmental conditions determined the distribution of plant communities at a regional scale. Like Cowles, Frederic and Edith Clements first encountered the concept of ecological community through the work of European botanists. Frederic Clements was born to homesteaders in Lincoln, Nebraska, in 1874. Edith Schwartz was born the same year in Albany, New York, the daughter of an Omaha meat packer. The two met at the University of Nebraska, where each earned a doctorate in botany, Frederic in 1898 and Edith in 1904. In an informal seminar with the influential botanist Charles Bessey, they read *Deutschlands Pflanzengeographie,* published in 1896 by Carl Georg Oscar Drude of the Dresden Botanical Garden.[11] Building on the work of Alexander von Humboldt and other European naturalists, Drude contended that the distribution of plant communities was determined by geological structures and by regional climates, both of which changed gradually. If plants were adapted to their environments, and environments changed slowly, Drude contended, it stood to reason that plant communities also replaced one another gradually over time.[12]

Frederic Clements taught at the University of Nebraska from 1897 to 1907, and then at the University of Minnesota until 1917, when he was recruited by the Carnegie Institute in Washington, DC. Although Edith's name did not appear on his publications, her ideas and her labor informed them; she was a self-described collaborator, driver, mechanic, secretary, photographer, and illustrator. Meanwhile, Henry Cowles worked in the botany department at the University of Chicago until his retirement in 1934, bringing classes of students on excursions to sites such as Flathead Lake, Montana; the Florida Everglades; and Sitka, Alaska.[13] Many of the students he mentored, including Charles C. Adams, Paul Bigelow Sears, and Victor Ernest Shelford, would later hold leadership positions in the Ecological Society of America and promote the idea of ecological restoration.

Cowles's innovation was a space-for-time substitution: just as modern geologists could infer the order of past physical events from the layers of a sedimentary rock, he reasoned, ecologists could reconstruct the succession of plant communities through time by studying the distribution of communities at a given site. As Cowles walked from the grassy shore of Lake Michigan to the dense forest, he imagined he walked into the future. The Clementses, meanwhile, focused on uncovering the climatic factors that governed distribution. The plant community, in Frederic's words, was a "complex organism" that was adapted to its environment, and therefore bore "the unmistakable impress of the climate." Whether interpreting sand dunes, mountain scree slopes, river bottoms, or the recently erupted Krakatoa, ecologists emphasized the biotic dynamism of seemingly static landscapes.[14] Central to these early ecological studies was the question of succession: how plant communities emerged over time, from bare ground to complex assemblages of seemingly interconnected species.

Conferences and expeditions like the International Phytogeographic Excursion of 1911, a monthlong trip across the English countryside organized by British botanist Arthur Tansley, helped solidify a transnational network of botanists who were beginning to pursue ecological plant geography, or community ecology, as it is now called.[15] After Tansley and his colleagues founded the British Ecological Society in 1913, Robert Wolcott, a zoologist at the University of Nebraska, and Victor Shelford, one of Cowles's first graduate students at the University of Chicago, discussed the possibility of founding a similar group to coordinate botanical expeditions to sites in the midwestern United States. When Shelford mentioned the idea to Cowles, Cowles replied, "Why could it not be national?" The following December, at the 1915 meeting of the American Association for the Advancement of Science in Columbus, Ohio, a group of approximately fifty botanists and zoologists voted to form the Ecological Society of America (ESA). By 1917, the ESA had recruited 307 members and had elected Shelford its first president.[16]

Early adopters of the label "ecologist" came from backgrounds previously defined by taxa (botany, zoology, anthropology) or by habitat (limnology, forestry) and were united by the common goal of understanding how environments shaped species distributions. Cowles, Clements, and other theorists of plant succession were joined by zoologists, among them Joseph Grinnell, of the Berkeley Museum of Vertebrate Zoology, and Warder Clyde Allee of the University of Chicago. By the 1930s, students could train with self-proclaimed ecologists at a handful of universities, including Illinois, Chicago, Cincinnati, Nebraska, and Minnesota. In large part, early ecologists continued to study only plants or only animals, seeking to identify the environmental conditions that explained their geographic distributions.[17]

Nature Reservations

Soon after the founding of the Ecological Society of America, ecologists began to lobby their home institutions, as well as the Forest Service (founded in 1905) and the National Park Service (founded in 1916), for the creation of field sites dedicated exclusively to ecological study. The reorganization of universities at the turn of the twentieth century had led to a proliferation of biological research infrastructure. In addition to indoor laboratories and museums, biologists began to work in vivaria, marine laboratories, agricultural experiment stations, and forestry stations. Vivaria were buildings for keeping living animals on university campuses. Marine stations, the first of which was the Stazione Zoologica Anton Dohrn in Naples, Italy, founded in 1872, brought zoologists closer to their ocean-dwelling subjects. At inland field stations, the first of which were established in the early 1800s in France, Germany, and England, agricultural scientists and foresters developed and demonstrated new techniques for plant cultivation. By 1917, American universities, private foundations, and the Department of Agriculture had established dozens of experimental stations for agriculture and forestry across the country.[18]

The effort to secure research sites for the exclusive use of ecologists was led by Victor Shelford and Emma Lucy Braun. After teaching at public schools for a number of years, Victor Shelford wrote a doctoral dissertation under Henry Cowles, and in 1914, he accepted a professorship at the University of Illinois.[19] That same year, E. Lucy Braun became the second woman to graduate with a doctorate from the University of Cincinnati (her sister, Annette, was the first). In 1917, Braun founded the Cincinnati chapter of the Wild Flower Preservation Society and began to edit *Wild Flower,* which later became the society's official publication. Despite her flouting of gender norms—she was widely considered "blunt and ungracious" and "an intellectual snob"—Braun became a nationally recognized expert on deciduous forest ecology. After working for a decade as an instructor, she was made a professor at the University of Cincinnati in 1923.[20]

Like members of the American Bison Society, many members of the Ecological Society of America believed that rapid environmental change was a regrettable but necessary step in American colonialism. The development of wild places was "the necessary price which we have had to pay for our advance beyond savagery," ecologist Francis Sumner wrote in the journal *Science* in 1921.[21] But wartime efforts to increase agricultural and industrial production heightened the perception that the number of "natural areas" was diminishing. In order to have sites to study in the future, Sumner concluded, ecologists would need to place tracts of land "representing every type of physiography and of plant association" into permanent reservation. Whereas game restorationists had sought to rehabilitate large mammalian species of economic and recreational value, ecologists sought to reserve sites that contained plant communities of interest to scientists.[22]

Increasingly, ecologists viewed "natural areas" or "wild areas" as threatened not only by settler expansion, but also by federal conservation management. To organize ecologists' efforts to secure field sites, Victor Shelford founded the ESA's Committee on the Preservation of Natural Conditions for Ecological Study in 1917. This committee would eventually become an independent organization, The Nature Conservancy—today one of the largest and most influential environmental NGOs in the world, and a leader in ecological restoration (as later chapters reveal).[23] Within the span of a few years, the preservation committee grew from seventeen to seventy members. Its main contention was that existing federal lands, including those administered by the Forest Service and the National Park Service, were not suitable for ecological study because federal managers were altering "natural" conditions. Charles C. Adams complained in the *Scientific Monthly* that the Forest Service was harvesting timber, permitting cattle and sheep grazing, and constructing roads in places that might otherwise be studied by ecologists. Shelford described how, even in national parks, federal agents were introducing game and killing predators, a "pampering of herbivores" that left ecologists with few sites in which to study "original nature."[24]

An ideal ecological study site, the preservation committee believed, would not be actively managed. It would be a site where, in Shelford's words, "ordinary 'conservation' views are reversed." Rather than seeking to control organisms, managers would refrain from intervening so that organisms were "allowed free play."[25] Ecologists called these sites "nature reservations" or "nature reserves" to distinguish them from the experimental stations of forestry and the agricultural sciences. Whereas game restorationists had sought to reserve federal lands from private settlement, ecologists imagined reservations to be places protected from federal intervention.

After surveying ESA members, Victor Shelford and E. Lucy Braun compiled a list of nearly one thousand potential sites for nature reservations in the United States, Canada, and Central America. Shelford published the list in 1926 as the 761-page *Naturalist's Guide to the Americas,* a brown book embossed with a single American bison. While Shelford announced the preservation committee in scientific journals, Braun delivered public lectures with titles such as "A National Monument of Every Type of Native Vegetation" and "Our Present Emergency." Paralleling the language Hornaday had used to justify bison restoration, Braun argued that it was a "sacred duty" to preserve "museum specimens" of natural vegetation.[26]

As it turned out, the ESA's preservation committee was better at gathering information on potential nature reserves than at actually securing research sites. By 1926, their efforts had led to the establishment of only one nature reservation, Glacier Bay National Monument in Alaska.[27] Frustrated but determined, Braun and Shelford continued to write letters and pamphlets about the importance of nature reserves into the early 1930s. Wildlife reservations were not nature reserves, Shelford argued, but rather "farms" for game animals and birds.[28]

It would take the droughts of the 1930s, with their reshaping of the material and political landscape, to galvanize the nature reservation movement. In response to Roosevelt's New Deal environmental programs, ecologists began to reframe nature reservations as "experimental controls" against which to test the effects of federal environmental recovery projects. In doing so, they recast their theory of ecological succession as a tool for recovery and positioned themselves as experts on "restoration."

"Destruction of Natural Conditions"

In 1932, Franklin Delano Roosevelt was elected on the promise of "a new deal for the American people." Between 1929 and 1933, unemployment in the United States had surged from 4 percent to 25 percent. Thousands of banks closed, and millions of families lost their entire savings. The suite of legislation that Roosevelt signed in his first hundred days in office lastingly expanded the role of the federal government in the nation's economy and the welfare of its citizens.[29] Less obviously, but as importantly, the New Deal transformed physical environments across the nation. During the New Deal years, the federal government acquired more than twenty million acres of land, increasing its holdings by 15 percent. It reshaped these lands through initiatives like the Civilian Conservation Corps and the Resettlement Administration, and through the many people these programs hired.

The Civilian Conservation Corps (CCC) transformed environments at an unprecedented scale, as historians Sarah Phillips and Neil Maher have each detailed. A New Deal work relief program, the CCC was supervised by the Departments of Labor, War, Agriculture, and the Interior. For one dollar per day, CCC workers, generally unmarried men aged eighteen to twenty-five, performed manual labor for federal and state agencies. Between April 1933, when CCC enrollees first began working, and 1942, when Congress terminated the program, the CCC employed more than three million men, a number that dwarfed the United States' World War I military mobilization. At the behest of the Forest Service and the National Park Service, CCC workers built roads and fire lookout towers, dammed streams, drained wetlands, and constructed picnic grounds. During its existence, the CCC developed eight hundred new state parks, constructed ten thousand

2.1 **Civilian Conservation Corps workers felling trees to create a fire line in the Siskiyou National Forest, Oregon, for the U.S. Forest Service, 1936.** Courtesy Gerald W. Williams Collection, Special Collections and Archives Research Center, Oregon State University Libraries

reservoirs, erected one million miles of fence, stocked rivers with one million fish, and eradicated almost four hundred thousand predators. In total, CCC projects altered more than 118 million acres, an area larger than the state of California.[30]

The stage was set for disciplinary conflict when Roosevelt's administration turned to foresters, not ecologists, to advise the federal response to drought and economic depression. Ecology was still a young discipline; in 1933, few ESA members referred to themselves as ecologists, identifying instead as botanists, zoologists, or entomologists. In contrast, the discipline of forestry science was flourishing. Gifford Pinchot had founded the Society of American Foresters in 1900, and by the 1930s, a generation of U.S. Forest Service leadership had been recruited from the Biltmore Forest School (established in 1898), the College of Forestry at Cornell (1898), and the Yale Forest School (1900). Since its founding, the ESA had attempted to attract foresters and convert them to ecology, but not always successfully. A 1916 letter among the society's founding members cheerfully noted that increases in dues in other professional societies had foresters "flocking right in." But Raphael Zon, the first director of the Forest Service's Experiment Station program, complained of the Ecological Society of America's "boosting" of a new discipline that "produced only generalities."[31]

Eager to promote their science, ecologists were quick to critique CCC projects and the foresters who oversaw them. Addressing the 1935 meeting of the Ecological Society of America, Walter P. Taylor argued that ecologists were better qualified to respond to the Dust Bowl than the "laymen" overseeing the CCC, who were busy applying "certain obvious remedies to cure the ills of exploitation" as though they were "trying to cure a headache with headache powders and persisting in a round of night life." In Taylor's view, the "artificial measures" employed by the CCC, including predator control and the erection of erosion-control dams, were nothing but superficial remedies for environmental damage. Only a trained ecologist, Taylor concluded, could "picture accurately the land as it ought to look" and "advise safely what should be done."[32]

Besides arguing that CCC projects failed to address the underlying causes of soil erosion, ecologists also criticized the federal government for prioritizing public recreation over the maintenance of "natural conditions." In 1936, the National Park Service employed nearly four hundred landscape architects but only twenty-seven biologists, leading Newton Drury, the head of the Save-the-Redwoods-League, to deride the Park Service as a "glorified playground commission." In a speech that year, E. Lucy Braun disparaged "the often mis-directed activities of the CCC," which were "putting the stamp of man's interference on every natural area they invade." Wallace Grange, who collaborated with the now-famous restoration ecologist Aldo Leopold, contended that "it would be a great deal better to have the government buy up a dump-yard somewhere and put all the [CCC] boys to work on it." The ESA leadership, in turn, wrote to congressional representatives to warn them of the CCC's "destruction of natural conditions in state parks and forests."[33]

Ecologists also accused the CCC of shortsightedness, arguing that federal policy could only be improved through a heightened ability to forecast future climatic conditions. To determine whether drought in the West was cyclical, they argued, ecologists would need to begin gathering records of rainfall, stream flow, snow cover, and tree rings. Any periodicity would become evident only after a large amount of evidence had been collected over many years. Frederic Clements speculated that this new power of ecological prediction would result in an overhaul of America's social and

2.2 Artistic representation of shelterbelt spacing, 1934. Strips are about 100 feet wide, running north and south one mile apart. U.S. Forest Service

political systems through land classification surveys, de-settlement, and resettlement, "something far more intelligent than the original trial-and-error settling-up of the country." To achieve a "proper ecological synthesis," Clements argued, which was imperative for "rehabilitation and restoration," professional ecologists would have to obtain "comprehensive and objective measures" of the effects of federal management on soil, plants, and animals.[34]

Tensions between ecologists and foresters peaked in 1934 when Roosevelt announced the Prairie States Forestry Project, also known as the Shelter-Belt Program. Through this program, the Works Progress Administration, with the assistance of the Forest Service and the CCC, planned to plant a belt of trees from the Canadian border to the Texas Panhandle, roughly following the ninety-ninth meridian. The administration contended that this belt would reduce dust storms and decrease evaporation across the Great Plains, and it placed Raphael Zon of the Forest Service in charge of the program. Within its first year of operation, the Shelter-Belt Program received a one-million-dollar allocation from emergency relief funds and began planting. Initially the government leased farmland from its owners, but in subsequent years Congress demanded that owners donate the land. The CCC planted trees in parallel rows about 140 feet wide and up to one mile long. By 1936, they had planted 23.7 million trees.[35]

As soon as the Shelter-Belt Program was announced, ecologists argued that it was doomed to fail. Grasses, not trees, grew naturally on the Great Plains, they argued, and the program's failure would be an embarrassment to the nation. The Shelter-Belt Program, in Clements's view, violated "every canon of dynamic ecology." Ecologist Paul B. Sears criticized the Forest Service for implementing federal policy uncritically rather than trying to influence it. Ellsworth Huntington, a professor of geography at Yale and an active ESA member, wrote that the most that could be expected from the program was "an extravagant percentage of deaths among the trees," and that the Forest Service should instead situate its projects where "nature may do most of the work."[36] The arguments hinged on what a tree was. To the Forest Service, a tree was a transposable erosion prevention tool. To ecologists, a tree was an organism that needed the correct environmental conditions to survive.

When Royal S. Kellogg disputed the merits of the Shelter-Belt Program with Raphael Zon in the *New York Times*, the debate moved beyond the confines of professional journals. One of Zon's staff wrote a poem (to be sung to the tune of "Clementine") describing the confrontation:

Kellogg lives in New York City,
Far away from drought and wind.
Broadway dandies never fancy
Any need for shelter belts.

When he hears about the project
He sits down and writes the TIMES
In a letter: "I know better.
They don't need a shelter belt."

"Planted trees will die on prairie,
Eighteen inches not enough.
Suffocation, radiation
Kill the trees in shelter belt."

"Zon should move to New York City,
Live with me on old Broadway
Here it's cozy, here it's rozy
And forget his shelter belt."

What will happen in the future,
Zon and Kellogg only know.
Will they conquer or debunker?
I predict a shelter belt!

Chorus:
Thousand miles of living fences,
Pinus, Ulmus, Fraxinus,
Ponderosa, Resinosa,
Soon will grow in shelter belt.[37]

Ultimately, the Shelter-Belt Program proceeded despite ecologists' critiques, and by 1942, CCC workers had planted 220 million trees across a total of 18,600 miles. Many of these shelterbelts still persist in the landscape. In a twist of fate, the 1970s ecologists began to study how birds and mammals used these "man-made islands" for shelter.[38]

"Check-Areas" and Experimental Controls

Responding to the Shelter-Belt Program, ecologists argued that it would be impossible to assess the outcomes of New Deal land management without "check-areas" to compare with managed sites. The need for such sites was obvious, Shelford contended: whereas an agricultural scientist had two types of field sites—"one which he fertilizes and one which he does not fertilize"—ecologists could rarely access an "undisturbed" area.[39] Only by the study of "nature unmodified by man," E. Lucy Braun contended, could ecologists generate knowledge to apply "to the best uses in land utilization, in agriculture, in grazing problems, and in forestry."[40]

And yet, the idea of an ecological "control site" was far from obvious. The idea of an experimental "control group," an unmanipulated group that is compared to a manipulated group, had emerged only in the late 1800s in the fields of psychiatry and physiology. The intention behind this experimental design was that all variables would be identical between the two groups except for the one being tested. The pursuit of controlled experiments led biologists to standardize "model organisms," breeding mice, for example, that were genetically identical.[41] But scientists who studied groups—economists, psychologists, and ecologists—could not standardize their objects of study. Groups treated exactly alike by the experimenter would still vary because individual people were never exactly alike or because two forests never contained exactly the same species or experienced exactly the same environmental conditions. This conundrum inspired the development of statistical concepts that allowed for the characterization of group performances in terms of central tendency (e.g., mean, median, mode).[42]

One of the main promoters of experimental control groups was R. A. Fisher, author of the widely read *Statistical Methods for Research Workers* (1925) and *The Design of Experiments* (1935). Fisher was employed as a statistician at one of the first agricultural research stations in the world, the Rothamsted Experimental Station, about 25 miles north of London. Fisher's methods were soon applied to colonial experiments on tea in India, cotton in Sudan, and oil palm in West Africa.[43] By the 1930s, statisticians had popularized methods for comparing the mean responses of two or more groups. An agricultural scientist could, for example, compare whether the mean weight of grain produced in five fertilized plots was different from the mean weight of grain produced in five unfertilized plots. Variation among plots, which had once thwarted the experimenter, became the domain of the statistician.

Prior to the 1930s, ecologists had conducted descriptive surveys of multiple sites within a region, identifying and characterizing communities, groups of species that occurred together under common environmental conditions. Ecologists observed what was present at a site and relied on their personal knowledge of other sites to make comparisons. After years of traveling across Ohio, for example, E. Lucy Braun divided the flora of the Cincinnati region into six communities—upland forest, flood plain forest, mesophytic forest, ponds and swamps, upland meadows, and prairie meadow—based on her impression of the most common combinations of plants.[44] But such studies were understood to be descriptive, not experimental, and were not based on comparison with any "control" area.[45] Indeed, until the popularization of statistical methods for comparing group averages, ecologists generally believed that it was impossible to manipulate outdoor conditions in a controlled fashion.[46] Shelford wrote with envy of the "great satisfaction" of laboratory physiological experiments, in which experimenters could vary a single factor and set up an experiment "with its parallel check." Ecologists could not conduct such experiments, Shelford contended: "The difficulties of working with nature, with all the multiple factors involved, are really very great, the solution of problems comes slowly, and they have to be approached from a viewpoint different from that of the mechanistic physiologist."[47] Others maintained that any attempt to manipulate a physical variable outdoors would have an unintended effect on another variable—irrigating or shading a large area would also change temperature, for example. Those ecologists who did promote experimental study, including William Ganong and Victor Shelford, conducted this work exclusively in the laboratory. At the Carnegie Institution's Desert Laboratory in Tucson, Arizona, for example, Forrest Shreve placed cactus seedlings in a freezer to see how long they could survive, comparing them to unfrozen controls.[48]

During the Dust Bowl, however, American ecologists began to adopt the experimental design of agricultural scientists, and in a 1934 paper in the *Journal of Ecology*, Frederic Clements announced a new form of ecological study, distinct from descriptive and successional studies: experimental studies that took place outdoors.[49] With experimental field studies, ecologists manipulated the number or type of species in a plot and compared it to a control plot. By constructing fences to exclude herbivorous rodents and livestock, Clements suggested, ecologists could isolate climate as the only factor influencing the composition of a plant community. Alternatively, an ecologist could subject some plots to burning, clipping, or sowing and leave others as "controls." The environmental and political conditions of drought motivated ecologists to import and modify methods from agricultural science.

Ecology's shift from descriptive fieldwork to experimental fieldwork is exemplified by the career of University of Nebraska ecologist John Weaver, a student of Frederic Clements. By 1933, the hilltop prairie outside of Lincoln, Nebraska, where Weaver worked as a professor, had lost almost half of its plant cover to drought. In that year, Weaver and his students compared "prairie relicts" to cultivated fields of corn, wheat, clover, and alfalfa. Using "sunshine recorders" (photographic paper) and thermometers, they determined that 65 percent of sunlight reached the ground in cornfields, but only 25 percent reached the ground in prairie relicts, and that the temperature at the soil surface of a cornfield was 136 degrees, versus 98 degrees in a prairie. Using an "interceptometer"—an iron can set downslope from their research plots—they compared the movement of water through native prairie vegetation, pasture vegetation, and bare ground. They determined that prairie vegetation held soil in place with deep root systems and that fallen plant material formed miniature dams that held the water. The solution to the "national menace" of soil erosion, Weaver concluded, was not the engineering of check-dams and other soil erosion prevention technologies, but the restoration of "the stabilizing influence" of prairie plant communities.[50]

As ecologists began to do experimental fieldwork, they reinforced their arguments that ecological expertise was essential to a program of national recovery from the Dust Bowl. Weaver and Clements began the 1938 update of their 1928 textbook, *Plant Ecology*, by arguing that "the importance of

2.3 A photograph from October 1941 of fenced area (left) and unfenced area (right) in the Coconino National Forest, Arizona. The fence was installed in August 1938 to compare a grazed (managed) area to an area protected from grazers, a "check-area." U.S. Forest Service Research Data Archive

field experiments cannot be overestimated." Methods of experimenting on plant communities included removing vegetation through burning, flooding, or digging; fencing an area against grazers; watering a plot or cutting a ditch to divert rainfall away from a plot; and adding fertilizers. Such interventions, Weaver and Clements contended, permitted "experimental study of the processes of natural recovery or artificial regeneration."[51]

Managed or "artificially regenerated" areas, Weaver and Clements continued, could be compared to control or "relict" plots. Relicts were plant communities that had been "continuously protected against disturbance" and had therefore survived "general disturbance by man" whether by clearing, fire, or overgrazing. Relicts could be found along railways and highway rights of way, in cemeteries and schoolyards, and in remote valleys and hills. In a 1934 article, Clements described the "peculiar advantages" offered to North American ecologists by the continent's colonial history, in which the plant communities had experienced a "wave-like migration of population to the westward," resulting in a "graduated record" of the changes wrought by settler colonialism. By using relict sites "as a standard of comparison," he claimed, historical environmental change could be "evaluated objectively."[52] Indian reservations contained the best relicts, he contended, followed by some national parks. Particularly valuable to ecological study, he argued, were the game reservations such as the one in Wichita, Oklahoma—the one set up by the American Bison Society.[53]

The idea of the relict plant community as experimental control, or check-area, supplied ecologists with a new justification for the acquisition of land for exclusive ecological study. Check-areas, in ecologist Herbert Hanson's words, would serve as "natural yardsticks to measure man's land management by."[54] This emerging consensus among ecologists coincided with a federal effort to reincorporate agricultural land into the public domain. Between 1930 and 1940, approximately 3.5 million people moved out of the Plains states, and the federal government reclaimed millions of acres, many of which it had distributed only a few years prior via the Homestead Act. ESA members argued that these new federal holdings should be reserved as controls, or check-areas—nature reservations—to compare with federally managed sites.

Accelerating Nature's Recovery

One of the areas that the ESA's preservation committee hoped to acquire was a prairie tract in southwestern South Dakota. Through the early 1930s, Shelford had tried to convince the National Park Service to establish an ecological research station there by emphasizing how few grassland research sites existed compared with the dozens of forest research sites across the country. But the National Park Service instead focused on the establishment of parks near population centers, including the Great Smoky Mountains (1934), the Olympic Peninsula in Washington (1938), and the Florida Everglades (1947). Then, around 1938, the federal government acquired the land in question through the Resettlement Administration.

Roosevelt established the Resettlement Administration by executive order in 1935, charging it with the resettlement of "destitute or low-income families" to government-planned communities. But the Resettlement Administration's plans to move 650,000 farmers from one hundred million acres proved unpopular: many in Congress viewed it as socialistic and as a threat to the tenant farming system. The Resettlement Administration was ultimately granted only enough resources to relocate a few thousand people, but in its two years of existence, it acquired more than nine million acres of "exhausted" farmland for the federal government.[55]

Recognizing that the area they wanted to reserve for ecological study now belonged to the federal government, the preservation committee passed a resolution calling for the establishment of grassland reserves on land that had been part of the Great Sioux Reservation from 1868 to 1877. Agreeing with the Roosevelt administration that it was "necessary to withdraw land from cultivation and remove settlers from poor land," the preservation committee argued that nature reserves could play a vital role in the "proper development of the science of ecology," a science that would be invaluable to federal environmental management.[56] Ecology, they argued, promised to reveal information that could reverse environmental damage and prevent future disasters.

Ecologists' justifications for nature reservations were notably different from those of wilderness preservationists.[57] Nature reservations would be sites reserved for the purpose of scientific study, and not for their beauty or for their contrast to the modern, mechanized world. Clements, Braun, and others emphasized that ecologists saw scientific worth in grasslands, wetlands, and other places that the general public found unremarkable. Neither did the preservation committee align with foresters and game managers. As Shelford wrote in 1930, "Our interests are not exactly the same as those of the recreation and scenic groups since we need reservation of grassland, desert, and scrub in which they would not be interested. [. . .] Our interests are not the same as those practical organizations such as sportsmen, foresters, etc., as they frequently wish to dispose of certain species and introduce others."[58]

Although the land proposed for the Great Plains National Monument was far from untouched, Shelford and the preservation committee suggested that ecologists could reassemble ecological communities, and that with time these communities would revert to their natural state. Since much of the landscape had been overgrazed, the preservation committee suggested that ecologists could replant prairie species and reintroduce small numbers of "the more conspicuous mammals now exterminated," including bison, elk, and antelope. Even the wolf could be restored—a revolutionary proposal at the time, and a controversial practice still today—if, that is, "a tight fence could be constructed around the project." Shelford believed that it would take twenty-five years of the right kind of stewardship to get the proposed area "into a pristine condition."[59] Thus, despite its commitment to studying relicts and unmanaged sites—areas in which nature was "allowed free play"—the preservation committee also believed that expert ecological restoration could produce such sites.[60] Wildness was not forever banished from a place by human activity, and the right kind of management could invite its return. Once the "pristine condition" of the Great Plains National

Monument was restored, ecologists believed they could discover how nature sustained itself, and thereby, how the nation could sustain nature.

The Great Plains National Monument proposal went nowhere in Congress, and to the chagrin of the preservation committee, some of the area was actually placed under Forest Service management with passage of the 1937 Bankhead-Jones Farm Tenant Act. Consequently, most Ecological Society of America members considered the preservation committee to be a failure.[61] The initial results of the committee's efforts to establish nature reservations were trivial, and only one national park would be established in the Great Plains before 1950, Badlands National Monument in South Dakota. Yet, as will become evident as our discussion proceeds, the preservation committee ultimately had an enormous impact on global environmental management. The Great Plains Monument would never materialize, but the idea of a nature reservation would endure, reshaping material environments across the world. Today more than 240,000 protected areas worldwide encompass approximately 15 percent of the Earth's ice-free land area.[62] Ecologists have recently called for 30 percent or even 50 percent of Earth to be formally protected. Proponents argue that nature reserves are the only solution to catastrophic biodiversity loss in a warming world.[63]

Embedded in the early history of ecological restoration was an important change in the practice of ecological fieldwork, and an important debate over the prospect of managing wildness. Check-areas were the precursors to the "control sites" that are ubiquitous in ecological research today, and they moved experimental systems from the laboratory to the landscape.[64] In a 1935 article, "Experimental Ecology in the Public Service," Frederic Clements asked ecologists to consider that large national projects represented an opportunity to carry out ecological experiments "on a scale and for a period never before possible." Eventually, Clements concluded, ecologists would be able to restore damaged lands using the same tools "by which nature reclothes bare areas." Once scientists better understood the process of ecological succession, he argued, they would be able to retard or accelerate it, to "deflect it in any one of several possible directions," so that they could control a plant community "or at least shape it to the desired purpose."[65] By studying nature reservations, ecologists could learn how to mimic and perhaps even accelerate nature's recovery.

But while ESA members agreed that ecology should inform federal policy and management, they differed in their political commitments and imaginations. The Clementses, for example, believed that restored farmland could be resettled. (Bitingly, Edith later reflected that the Great Plains Drouth Committee had asked Frederic whether midwestern farmers should be "unsettled, resettled, subsidized, taught how to farm, or be painlessly chloroformed.")[66] Other ecologists rejected the nation's "laissez-faire policy of uncontrolled settlement" in favor of centralized land planning. Shelford believed that which lands would be farmed, forested, or protected ought to be decided by "disinterested experts, with sole reference to the higher welfare of the public and of posterity" and not by "the accidents of private ownership."[67]

Crucially, ecologists also disagreed about the extent to which humans could accelerate ecological recovery. Walter Taylor contended that while "the city can be rebuilt with relative ease and speed," ecological communities often could "be restored only with extreme difficulty, if at all; and such restoration requires a very long time."[68] Notably, Taylor was referring to the need to prevent forest fires; it would be another few decades before ecologists would embrace fire as a natural and necessary ecological force. Similarly, Frederic Clements and John Weaver speculated that "natural recovery" was "probably the most certain and rapid" and "the most inexpensive" of all restoration methods.[69] Others maintained that managers could accelerate restoration through active intervention, like restocking game animals or replanting an area. Shelford believed that by using "remedial measures" like replanting native species, ecologists would be able to slowly restore the biotic community to the point at which it was able "to take its course." Homer Shantz speculated that in the future ecologists would be able to use "knowledge of the natural trends of succession" to "work with nature to bring about desired results."[70]

The concept of ecological restoration thus emerged in a time of dramatic environmental, governmental, and scientific change. In response to New Deal environmental projects, ecologists reformulated their arguments for acquiring nature reservations, moving from the fairly abstract claim that unmanaged lands were essential to the discipline's advancement to the more urgent claim that they would serve as controls in a grand national experiment. Ecologists speculated that ecological restoration was both possible and manageable. Contesting the influence of foresters in national policy and land management, they claimed that the key to reversing environmental damage lay in their own disciplinary knowledge, while the efforts of the CCC were only a temporary fix—"headache powders," as Taylor had called them. Despite some disagreement about how restoration would proceed and how long it might take, ecologists shared the view that wildness was not a property inherent to species, but rather the condition of being unmanaged. They shared the goal of restoring "natural conditions"—not so much for the sake of individual species, but as a control against which interventionist environmental management could be assessed.

Although the ESA's preservation committee failed to acquire many experimental spaces for ecologists, its successor, The Nature Conservancy, would become a major landholder in the 1970s, eventually dedicating millions of acres to ecological restoration. Meanwhile, ecologists and botanists pursued a third type of experimental space: naturalistic gardens. Alongside game reservations and nature reserves, naturalistic gardens were spaces in which ecologists would experiment with restoring wild species and species wildness.

3

An Outdoor Laboratory

They began with a wildflower boycott. Elizabeth Britton, cofounder of the New York Botanical Garden and a renowned expert on mosses and ferns, created the Wild Flower Preservation Society in 1901 in hopes of protecting her botanizing sites just outside Manhattan.[1] She modeled the WFPS after the extremely successful Audubon movement, which coalesced when a group of wealthy Boston women pledged to boycott hats decorated with bird feathers.[2] Like feathers, wild flowers were fashionable Victorian accessories, adorning hats as well as dining room tables. An 1887 magazine article described an elaborate centerpiece made of native wildflowers: a pyramid of jack-in-the-pulpits, violets, and red maple twigs with a border of skunk-cabbage leaves. Citing the success of the Audubon societies, early WFPS members argued that if wealthy women refused to purchase wildflowers from pushcart vendors, it would put an end to wildflower harvesting in the countryside, thereby protecting the "victims of the massacre exposed for sale in our city streets."[3]

Like many women's clubs, the WFPS aimed to define moral sensibilities during a time when women's roles in society were highly contested. In addition to boycotting, wildflower advocates argued, women could reform their own behavior in the countryside. "Let us for a moment consider the cruel waste that is going on in the region of Colorado Springs," WFPS member Mary Perle Anderson wrote. "On certain days in the week special trains run 'flower-trips' which are largely patronized by tourists. They recklessly

3.1 A wildflower excursion on the Colorado Midland Railway, 1915. Mayall Photograph Collection, © Pikes Peak Library District, 102-5599

pull up and tear up the flowers, and return with great armfuls and basketfuls, and in their ungoverned enthusiasm, they often deck the cars and festoon the engine with them!"[4] This ungoverned enthusiasm, Perle and other WFPS members contended, was leading to local declines in Christmas greens, ferns, laurels, dogwoods, mayflowers, and other delicate and interesting species. "Weddings, by the way," Elizabeth Britton wrote in the *New York Times,* "are a new menace to our native plants."[5]

The WFPS initially aimed to influence the behavior of Anglo-Saxon women, but as the society expanded, it moved toward policing the behavior of so-called new immigrants to the United States—especially children. A 1904 article in *Plant World* decried the "hordes" that trampled wildflowers in the Hudson Valley, arguing that "fresh air and other charitable societies have unconsciously aided in this destructive work."[6] Increasingly, the WFPS strove to assimilate immigrant children through education. It was through schools, Jean Broadhurst wrote in *Plant World,* that women "reach, control, and elevate the masses brought into our country daily."[7] Local chapters developed and distributed educational pamphlets, lantern slides, and children's books, and presented plays and pageants, shaping a generation's relationship to nature.

The WFPS expanded rapidly during World War I, establishing chapters in Baltimore, Philadelphia, Chicago, Cincinnati, Milwaukee, and Washington, DC. Its roster included a number of the individuals that would popularize the science of ecology, including Henry Cowles, Charles Bessey, Edith Roberts, and E. Lucy Braun. But although the WFPS shared the goal of preserving scientific study sites with the Ecological Society of America's Committee on the Preservation of Natural Conditions, they based their appeals in aesthetic arguments. Together these organizations solidified national concern for native plant species, laying the groundwork for ecological restoration.

During the 1920s, the WFPS increasingly focused on a new threat to wildflowers: automobilists. Between 1908 and 1925, the cost of a Ford Model T plummeted from $850 to less than $300. With it, the number of Americans who owned cars increased from fewer than five hundred thousand to more than eight million. Road expansion became one of the main concerns of wilderness preservationists, who argued that the government should preserve large expanses of roadless lands.[8] While wilderness preservationists worried that automobiles would destroy opportunities for peaceful recreation, WFPS members worried that they would destroy native plant species. Automobiles made the countryside outside of the city—and the private country estates of urban naturalists—available to many. Native plants, Elizabeth Britton maintained, had "found a powerful new enemy in the automobile."[9] John Harshberger, a University of Pennsylvania ecologist and coiner of the term "ethnobotany," argued in 1923 that automobiles threatened plants and animals with extinction because they could "reach the previously almost inaccessible parts of the world." Today roadkill is considered a major problem in conservation biology, but in the first decades of the automobile age, naturalists worried far more about the people who drove cars than about the cars they drove. Pennsylvania botanist Mira Dock complained about the "anarchists in automobiles" who plundered roadside shrubs and wildflowers. Editorial cartoons published in newspapers and magazines across the country depicted motorists denuding landscapes with their supposed love of nature, trampling flowers and stripping trees bare.[10]

Soon Wild Flower Preservation Society members were split over whether to focus their efforts on immigrants or on automobilists. The schism ended only when Percy Ricker engineered a hostile takeover. Since before World

War I, Elizabeth Britton had cultivated a decentralized society in which local chapters handled their own finances, planned their own events, and focused on their local regions. The membership comprised mostly women, with a few male botanists holding leadership roles by invitation. In 1924, U.S. Department of Agriculture botanist Percy Ricker, president of the Washington, DC, chapter, began conspiring to take over the WFPS from Britton, arguing that automobiles had made wildflower preservation a national issue that needed to be addressed by a centralized national organization. Further, Ricker claimed that, under the leadership of women, wildflower preservation had become a "sentimental" subject and that "professional botanists"—meaning male botanists—had become "disgusted with the over-zealous efforts of individuals and organizations wishing to forbid all flower picking."[11]

Unlike some sciences, botany had been considered a suitable study for women as early as the nineteenth century.[12] Of members listed in the first directory of American botanists, approximately 11 percent were women; by 1890, women made up almost half of the Torrey Botanical Society.[13] Along with the professionalization of science at the turn of the century came the anxiety that, as an 1887 article in *Science* put it, botany was not "a manly study"—that it was "merely one of the ornamental branches, suitable enough for young ladies and effeminate youths, but not adapted for able-bodied and vigorous-brained young men."[14] Indeed, the view that botany was feminine, and thereby less reasoned, helps explain the efforts of male ecologists to standardize their fieldwork. Historian Robert Kohler maintains that early ecologists imported quantitative and experimental methods from laboratory sciences such as physiology in order to distinguish themselves from "amateur naturalists." Kohler attributes ecologists' professional anxieties to the material constraints of working outdoors, arguing that, unlike laboratories, where access was restricted to experts, the field could be accessed by anyone, including tourists.[15] But in distinguishing between laboratory and landscape, Kohler does not consider how the language of "amateurism" was gendered.[16] The field could be accessed by anyone, including women. Often the link was explicit, as in this quote from botanist Willard Clute in 1908:

> Too long concerned with leisurely jaunts about the country in search of "specimens" and the pulling of flowers to pieces in order to "analyze" them, botany has come to be viewed by many with a sort of amused contempt which has found expression in the quip that "botany, like croquet, is a fitting pursuit for elderly ladies and ministers." The boy who elects botany, or the man who teaches it, is regarded by the community as having a feminine streak somewhere in his make-up, no matter what other good qualities he may possess.[17]

By deriding the work of female WFPS leaders as "sentimental," Ricker sought to redefine wildflower preservation as a masculine endeavor. In doing so, he channeled a broader postwar campaign to characterize women's professional and suffrage groups as radical and dangerous. Britton recognized Ricker's challenge to her leadership and attempted to negotiate with him, but in a December 1924 meeting in Cincinnati that Britton could not attend, Ricker led a vote to reorganize the society with himself as president and John Harshberger and Henry Cowles as vice presidents.[18] Britton was given a seat on the new executive committee, but resented Ricker. In the margin of a letter from Ricker, in which he described the need of "ladies with such radical ideas" to listen to a "proper presentation of the subject [of wild flower preservation]," Britton scrawled, "A characteristically sarcastic and rude letter. A real sample of his character!"[19]

Unlike Britton, Ricker did not believe that plant species could be protected through appeals to individual behavior; he argued that wildflower decline was due to road building and real estate development, causes over which, in his view, "little or no control is possible."[20] Unlike animals, which were mobile and whose bodies were regarded as property of the state, plants belonged to the owner of the land on which they grew.[21] Ricker's WFPS lobbied unsuccessfully for the passage of a federal law by which certain native plants might not be transported in or out of state without a certificate of ownership, referencing the Lacey Act of 1900, which restricted interstate transportation of game and birds.[22] When this failed, the WFPS planned to purchase land and assume ownership of wildflower populations. In the *Scientific Monthly*, Ricker explained that the WFPS planned to raise funds for establishing "protected sanctuaries," where native plants would increase in abundance under "expert supervision" (the supervision of male ecologists). Once they did, the experts would distribute seeds or seedlings to other protected areas.[23] They would, in other words, couple land preservation with species restoration—a strategy that paralleled the American Bison Society's game reservations and the Ecological Society of America's nature reserves.

Into the 1930s, wildflower advocates increasingly sought to establish permanent sanctuaries that would be owned by private organizations, universities, or governmental agencies and would be managed to promote native plant species. Today one such site—the University of Wisconsin–Madison Arboretum's Curtis Prairie—is often celebrated as the first major ecological restoration site in the United States; one of its designers, Aldo Leopold, is frequently cited as the sole inventor of ecological restoration, a prophet of modern environmentalist thought.[24] From 1933 until 1948, when he died battling a neighbor's brush fire, Leopold oversaw cutting-edge work at the arboretum that synthesized ideas from the burgeoning fields of ecology, game management, and naturalistic gardening. The Wisconsin Arboretum's work on prairie restoration continues to this day; in the 1980s, the founders of the Society for Ecological Restoration claimed the arboretum as the first ecological restoration project in the United States.[25]

But the Wisconsin Arboretum was neither the only nor the first effort to restore and study native plant communities—it was just the best publicized. Nor does Leopold deserve sole credit for inventing ecological restoration. Rather, he was embedded in a large network of scientists and land managers working to restore native plants, including Eloise Butler, Edith Roberts, Elsa Rehmann, and other women botanists and ecologists with roots in the WFPS. Their work provided the discipline of ecology with botanical gardens and "outdoor laboratories" in which to train ecologists and conduct experiments, spaces designed to be "wild" assemblages of native species. They established networks of knowledge and material support that became central to ecological restoration, via the nurseries these women set up for the propagation of native plants. To understand what Leopold and his colleagues were doing at the Wisconsin Arboretum, we must first consider the work of these wildflower restorationists.

Botanical Gardens and the Aesthetics of Wildness

The fact that botany was considered a suitable study for women created opportunities for Eloise Butler. Butler was born on a farm in Appleton, Maine, in 1851, where an aunt who lived with the family taught Eloise and her sister, Cora, about the local plant species. At the age of twenty-three, after teaching high school in Maine and Indiana, Butler decided to move to the rapidly growing city of Minneapolis, Minnesota, a place of sawmills and flourmills. Although she continued working as a teacher, teaching was never her passion: she wrote in an unpublished autobiography that "no other career than teaching was thought of for a studious girl," and, more pointedly, "In my next incarnation I shall not be a teacher." So when, in the summer of 1881, the University of Minnesota offered its first Summer School of Science, Butler jumped at the opportunity. Charles E. Bessey, then professor of natural history at Iowa State College of Agriculture at Ames, was one of her instructors.[26]

Over the next decade Butler developed a reputation for her algal collecting and observations, and by the 1890s she was a nationally known phycologist. Her sister, Cora Pease, was a well-known naturalist in Malden, Massachusetts, and Pease connected Butler to a community of women scientists that included Elizabeth Britton and Ellen Swallow Richards, the first female instructor at the Massachusetts Institute of Technology and inventor of home economics and human ecology. Butler and Pease, self-avowed "bog-trotters," often traveled together on collecting expeditions.[27] Among her "loves and likings," Butler included "beauty of line," "old ruins," and "the scent of conifers, white violets, twin flower"—and thirteen other plants. Among her "hates," she listed "the scraping of chairs," "the odor of carbolic acid," and "toothpick shoe heels."[28]

In 1907, Butler organized local science teachers to petition the Minneapolis Park Board for space to establish a botanical garden "to show plants as living things and their adaptations to their environment, to display in miniature the rich and varied flora of Minnesota, and to teach the principles of forestry." Rather than fight to preserve wildflowers in sanctuaries, she would endeavor to reintroduce them. Like the first botanical gardens, which arose in Italian universities in the sixteenth century, Butler's project had a pedagogical aim: it would serve as a living reference collection for ecologists. With the rise of taxonomic botany in the eighteenth century, European botanical gardens had organized their displays as living representations of scientific understanding, with evolutionarily related species grouped together. As the era of scientific ecology began, community displays began to replace taxonomic ones. In Germany, for example, Oscar Drude opened the Dresden Botanic Garden in 1893, with distinct arrangements of alpine plants, wetland plants, and forest plants.[29] The archives do not contain a record of what inspired Butler, but it is likely that she encountered Drude and other German botanists in her studies. Her botanical garden was organized like theirs.

Butler was also influenced by the naturalistic gardening movement, and she favored native species. At the turn of the twentieth century, American horticulturalists were divided between Victorian cosmopolitanism and an emergent, austere design aesthetic. Acclimatization societies favored cosmopolitanism and sought to import useful or beautiful species from other continents. At their urging, many species today considered "invasive"—Norway maple, European starlings, and brown trout, for example—were introduced to North America in the name of environmental improvement.[30] Against this spirit of biotic cosmopolitanism, a growing number of naturalistic landscape architects began to advocate for an aesthetic unique to the United States and distinct from European garden design. These designers opposed nonnative plants on aesthetic grounds; they dismissed Old World plants as showy and florid and rejected the straight lines of the Renaissance garden.[31] Moreover, naturalistic gardeners argued that native plants were better adapted to local environmental conditions, and therefore required less maintenance. Frank Albert Waugh, a professor of landscape architecture at Massachusetts Agricultural College (now the University of Massachusetts), described the naturalistic style as "intelligently letting alone a natural landscape." A naturalistic garden would *appear* as though it was untended.[32]

Butler's Wild Botanic Garden, the first public wildflower garden in the United States, opened on April 27, 1907, on three acres of tamarack swamp and hillside donated by the Minneapolis Park Board. A local newspaper article titled "Shy Wild Flowers to Be Given Hospice" announced that the garden would "supplement the text books for students." Rather than travel long distances to show students relict plant communities, botanists would now be able to teach from an organized, easy to reach location.[33] The express goal of the Wild Botanic Garden was to "avoid all appearance of artificial treatment." The word "appearance" was key here. The garden was a highly managed place, but it was designed to appear wild. Once transplanted, plants would be allowed to grow, in Butler's words, "according to their own sweet will and not as humans might wish them to grow." Elsewhere, she wrote, "My wild garden is run on the political principle of laissez-faire." Seedlings would not be watered once they were well-rooted, and they would not be fertilized. But they were not entirely unmanaged—Butler and her colleagues thinned specimens when they were too prolific, and they uprooted nonnative plants like dandelion, burdock, and Canada thistle. For Butler, the aesthetics of wildness and nativity were closely connected. She critiqued "foreign plants" for looking "formal and stiff," much as "impaled butterflies do in a museum case." And she critiqued Minneapolis suburbanites for planting lawns of "monotonous, songless tameness." Butler purported to prefer the "fragility, delicacy and artless grace of the wildings."[34]

From the beginning, Butler and her collaborators viewed the garden not just as a pedagogical tool, but also as an experimental space for botanists and ecologists. In an article in the *American Botanist,* Cora Pease reported that the garden was designed "to study at firsthand the problems of ecology and forestry." Butler maintained meticulous records of specimen origins as well as annual survival rates and flowering dates. She also developed methods for propagating hundreds of native plant species, many of which were difficult to rear.[35] Orchids were "uncertain, coy and hard to please." She designed a special water tank to keep *Viola lanceolata* seedlings moist. To grow saxifrage, she moved a limestone slab from the cliffs of the St. Croix River in Wisconsin to the garden.[36] The garden would become a seed and cutting source for amateur and professional ecologists across the country, and other restorationists would learn from her propagation techniques. When Butler began the Wild Botanic Garden, few nurseries sold wildflowers, and little information was available on how to propagate nonhorticultural plants. Federal tree nurseries, the first of which was established in the Nebraska Sand Hills in 1902, catered to the Forest Service. Wild Flower Preservation Society member Mary Perle had predicted in 1904 that "a new industry, the raising of wild flowers on their native soil, will certainly arise in the near future," and indeed such an industry arose from the network of botanists and ecologists with whom Butler was connected.[37]

Unlike restoration ecologists in the late twentieth century, who would become concerned with hyperlocal genotypes, early wildflower restorationists defined nativity regionally. On a trip to Massachusetts to visit Pease, Butler collected species including buttonbush, epilobium, and dwarf ginseng. In Providence, Rhode Island, she collected acorns, and at Boston's Arnold Arboretum, morning glory. During an expedition for squirrel corn on the Big Island of Lake Minnetonka, she and Pease dug an access hole under nine-foot-high chicken-wire fence: "We kilted our skirts and, weighted with impediments, trudged through the wet grass some three miles across the country," to emerge "dusty and triumphant!" When Butler was in a train accident in Ontario in 1908, she disembarked and used a broken penknife to collect willowherb. Other plants were sent to Minneapolis by correspondents, such as Fannie Mahood Heath, the "flower woman" of North Dakota, a self-taught botanist who grew hundreds of wildflowers on her farm in Grand Forks. Butler once traded one hundred maidenhair ferns with a woman in Bemidji, Minnesota, for one clump of the rare ram's head lady's slipper. By the end of her life, Butler had planted 710 species at the Wild Botanic Garden, which joined the more than four hundred species already growing on the site.[38]

Butler curated the Wild Botanic Garden from its founding in 1907 until her death in 1933. Wearing brown overalls and high-laced black leather boots, armed with a broken-off machete and wearing a park watchman's star pinned to her chest, she would run out the "spooners" who might trample her plants.[39] Tellingly, the Wild Botanic Garden was renamed the Native Plant Reserve in 1929. Two years later, Butler tried to convince the University of Minnesota to assume supervision of the garden, "an excellent field for the study of and experiments in ecology," but without success. Nevertheless, her approach to naturalistic gardening, her propagation methods, and the offspring of her specimens would be essential to future restoration ecologists. You can still visit the garden today.

3.2 Eloise Butler in a bog adjacent to the Wild Botanic Garden, from which she sourced plants, in 1911. Hennepin County Library, Minneapolis Newspaper Photograph Collection, P05706

American Plants for American Gardens

Whereas Butler designed the Wild Botanic Garden primarily as a reference collection of sorts, a resource for botanists and ecologists, Edith Roberts was the first ecologist to design a botanical garden for the express purpose of researching how to restore native plants. Edith Roberts was born to farmers in Rollinsford, New Hampshire, in 1881—thirty years later than Butler. After attending Smith College, Roberts received a doctorate in botany from the University of Chicago in 1915, where she worked with Henry Cowles. Following a brief stint as an associate professor at Mount Holyoke College from 1915 to 1917, Roberts worked as a field representative for the Department of Agriculture during World War I, traveling through the forty-eight states and advising women who were managing farms while men were away in the service. In 1919, she was hired as a professor of botany at Vassar College.[40] Thus Roberts was more firmly ensconced than Butler in formal networks of academic teaching and research.

As soon as she arrived at Vassar, Roberts began developing the Ecological Laboratory. Her goals were twofold: first, to establish an on-campus facility in which to train students in the new discipline of ecology, and second, to test whether native plants could be reestablished on degraded lands. Beginning in 1921, Roberts and her students cleared more than four acres of grasses and poison ivy on the Vassar campus. By 1924 they had grown enough plants to begin transplanting, with at least ten students working on the project. "It was hoped that it might prove that native plants of the county could be used to reclaim waste land," Roberts reflected in 1933, "and could blend into an attractive landscape picture." On the cleared land they planted around six hundred native species collected from around the country, arranging them into thirty plant communities that Roberts and her students had identified in the area, including the "open field association," and the "bog association."[41]

Roberts's garden-laboratory, as distinct from Butler's, began with a primary emphasis on scientific experimentation. At the Vassar Ecological Laboratory, Roberts aimed to stimulate interest in ecology, as well as to acquire "ecological data" and thereby identify the factors that promoted certain assemblages of native plants. She saw the Ecological Laboratory as a place in which ecologists would generate knowledge about the cultivation of native plant species. Like the Minneapolis Wild Botanic Garden, the Vassar Ecological Laboratory stimulated new research on the propagation of native plants, many of which were difficult to germinate.

The garden-laboratory's institutional affiliation makes it much easier to demonstrate the reach of its influence. Ecologist Henry Cowles, whose daughter attended Vassar, visited the Ecological Laboratory in 1922. From 1923 to 1926, the Garden Club of America Conservation Committee awarded scholarships to students at Vassar College, the University of Chicago, the University of Washington, and Ohio State University to study the propagation of native plants. When Roberts and her students studied artificial methods of propagating native trees, shrubs, and flowers, the Wild Flower Preservation Society and Garden Club of America disseminated their results to a broad audience. Inez Haring, for example, worked to establish some seventy-five mosses in the greenhouse to be transplanted to the garden, and Opal Davis discovered that the two-year period of dormancy required to germinate dogwood seeds could be simulated by a ninety-day period in a refrigerator. A number of Roberts's students went on to complete graduate research on native plant cultivation in universities across the country, and they brought with them the techniques they had learned at the Ecological Laboratory.[42]

3.3 **Vassar College Ecological Laboratory site before *(above)* and after *(opposite)* ecological restoration.** Republished with permission of John Wiley & Sons from Edith Roberts, "The Development of an Out-of-Door Botanical Laboratory for Experimental Ecology," *Ecology* 14 (1933): 163–223, figures 30 and 31.

Work at the Ecological Laboratory also led Roberts to collaborate with Elsa Rehmann, a landscape architect who began teaching part time at Vassar in 1923. Rehmann was born in Newark, New Jersey, in 1886. After attending Wells College from 1904 to 1906, with the intention of becoming a writer, she transferred to Barnard College, where she studied architectural history and geology. Upon graduating in 1908, Rehmann attended the Lowthorpe School of Landscape Architecture for Women in Groton, Massachusetts. Founded in 1901, the Lowthorpe School offered courses in surveying, engineering, entomology, forestry, and soil science. Rehmann next apprenticed with the Hudson County Park System, at which point she began writing articles for periodicals like *Country Life* and *Better Homes and Gardens*. In 1919 she established her own practice, working out of her home to design estate gardens in New Jersey, New York, Pennsylvania, and Delaware.[43]

The book on naturalistic gardening that Roberts and Rehmann copublished in 1929, *American Plants for American Gardens,* would influence a generation of landscape architects and also shape the aesthetics of American restoration ecology. *American Plants for American Gardens* listed plants by the environmental conditions that promoted their growth, rather than by taxonomy or by geography. It also encouraged the cultivation of native plant species. The book was written for owners of private estates, but the authors also hoped to reach, in Rehmann's words, "all those interested in national, state, and county parks," including real estate subdividers, city foresters, and engineers designing roadway construction.[44]

Notably, Roberts and Rehmann synthesized trends in landscape architecture and ecology. The job of the landscape architect or ecologist, as they saw it, was to establish the environmental conditions in which native plants could thrive. Gardeners could clear dried underbrush and diseased specimens, and they could thin forest cover. They could plant new trees and herbs. In the book, Roberts and Rehmann explained that some species, like milkweeds and arbutus, thrived in sunny spots. Others, like bellworts, required shade. The carefully observed landscape presented "opportunities for the most delightful naturalizing," a sometimes painstaking process. The ideal gardener, for example, would make sure to plant species of different heights and arrange them in a pleasing manner. "It requires no little art to leave the woods absolutely natural and seemingly untouched," they wrote, "and yet, nature can be aided."[45]

Drawing on both their field experience and on theories of ecological succession, Roberts and Rehmann argued that the "re-creation of natural scenes" was a slow process, one that could be understood and anticipated by studying what happens to an abandoned farm within a generation or two, "when nature is left to run its own course." Once-cultivated fields were soon covered by shrubs and herbs, they explained, which only later were followed by birches and then pines. By understanding the "fundamental principles upon which the indigenous vegetation is established," they concluded, a designer or ecologist could re-create a natural landscape.[46] Along a similar vein, John Harshberger wrote that in the future, ecologists might "reproduce nature so closely by the use of native plants that our fellow men are deceived and believe that they look upon a wild growth when in fact it is artificial."[47] This restoration, like that proposed by ecologists in response to widespread soil erosion, was premised on giving plant communities time and space to grow free from human-caused disturbance. Restoration was to be achieved by intervening to establish the conditions by which natural processes could take over. A wild aesthetic was achievable through constrained human intervention.

"Game Farms Itself"

Aldo Leopold's path to founding a naturalistic garden at the Wisconsin Arboretum did not run through the study of botany, but rather through his work for the U.S. Forest Service. Leopold was born in Burlington, Iowa, in 1887. His father was a traveling salesman who sold barbed wire and roller skates to those moving west. By the time Leopold was a teenager, he was hunting and keeping detailed notes of bird arrivals and departures. He decided he wanted to train to be a forester when he learned of Yale University's new forestry school, and from 1905 to 1909, he studied timber management and forest law there. It was his early dream to be supervisor of his own forest reserve, and after graduating in 1909, Leopold boarded the

Atchison, Topeka, and Santa Fe Railroad at Fort Madison, Iowa, bound for Arizona. From 1909 to 1924, he worked for the U.S. Forest Service in the region that became Arizona and New Mexico, where his main duties involved suppressing wildfires, managing tourists and hunters, and, eventually, killing wild predators.[48]

Likely inspired by Hornaday's 1913 book *Our Vanishing Wild Life,* Leopold was one of the first Forest Service employees to argue that the agency should actively manage game species, and he became a vocal advocate for the establishment of game reservations on national forests.[49] But the Forest Service did not have jurisdiction over game animals—states did—and further, the Forest Service had no statutory mandate to manage wildlife on its lands when Leopold started his career. Indeed, the role and purpose of the Forest Service was still under debate, as only fairly recently had federal policy shifted from disposing of public lands to restricting the public's use of them. Two congressional acts in the 1890s began the establishment of the national forests, and the idea that the administrative state would manage these lands in perpetuity emerged gradually and haltingly.[50]

During the time that Leopold worked as a forest ranger, predator control would become the most widespread, the best resourced, and the most intensive mode of federal species management. Its ecological effects are still apparent today. Predator control in national forests began in earnest in 1914, when the Forest Service announced that it would cooperate with the Bureau of Biological Survey to protect migratory and insectivorous birds, linking the two agencies in species management.[51] The Biological Survey—which in 1940 would be incorporated into the newly organized Fish and Wildlife Service—had replaced the Division of Economic Ornithology and Mammalogy in the Department of Agriculture in 1905. Its main charge was to reduce populations of animals "injurious" to agriculture, and predator control would become one of the main functions of the Biological Survey.[52]

Although federal management of public land was in some ways new in the 1910s, with new institutions recently established for that purpose, predator control had been a government pursuit for centuries. State governments had encouraged settlers to kill predators, offering bounties for animals like bears, wolves, and mountain lions, and after 1860, the federal government distributed strychnine to farmers across the country, who laced carcasses with it to poison predators. In 1915, Congress instructed the Biological Survey to hire trappers to kill predators directly, leading some naturalists to call it "the Bureau of Destruction and Extermination." At the time, the accepted method of increasing the populations of game species like deer and elk was to kill anything that preyed on them, and killing predators was also thought to protect livestock.

After World War I, the Bureau of Biological Survey established the Eradication Methods Laboratory in Albuquerque, New Mexico (later moved to

Denver, Colorado) with the goal of testing war gases for their effectiveness as animal population control. Wildlife managers argued that poisons were more efficient and more humane than guns and steel traps. One of the most important and controversial of the poisons the Biological Survey developed to kill wildlife was Compound 1080, or sodium fluoroacetate, a colorless salt and metabolic toxin. The Biological Survey used Compound 1080 against coyotes and wolves until 1972, when President Nixon banned the practice.[53] Wildlife management as a profession and a scientific discipline thus has a long history as an endeavor to destroy wild species, not to restore them.

Federal wildlife eradication efforts were stunningly successful in the early part of the twentieth century, and, indeed, much of the work of contemporary restoration ecologists attempts to undo the results of these early federal interventions. Between 1916 and 1933, the Biological Survey killed 458 bears, 6,141 bobcats, 54,629 coyotes, 148 mountain lions, and 33 wolves in the state of Oregon alone.[54] Federal and state environmental agencies eradicated wolves from everywhere but northeastern Minnesota and Michigan.[55] And although the Fish and Wildlife Service moved toward restoring wildlife in the 1970s, as later chapters reveal, the Wildlife Services branch of the U.S. Department of Agriculture continues to cull animals deemed economically damaging. In 2018, it killed more than 2.6 million animals, including prairie dogs, wolves, and short-eared owls.[56]

After the Forest Service's 1914 agreement to cooperate in predator control, Leopold's district became a leader in exterminating bears, wolves, and mountain lions.[57] In 1920, Leopold triumphantly reported a 90 percent reduction in the wolf population of New Mexico, from three hundred to thirty.[58] But when World War I shifted the Forest Service's priorities toward food production, Leopold began brainstorming ways to further augment game populations beyond killing predators and regulating hunting. Restrictive hunting laws had been "of little effect," he later wrote, and scientists and sportsmen were beginning to realize that wildlife management required "a deliberate and purposeful manipulation of the factors determining productivity,—the same kind of manipulation as is employed in forestry and agriculture." Indeed, Leopold and his students would come to see game management as the frontier of modern agriculture. Game was "a wild crop," they argued, and restoring it would involve establishing breeding stock, controlling natural enemies, and augmenting food and cover plants.[59]

What was novel about Leopold's proposal is that he argued for the need to control not only the factors that "limited productivity" of game species—hunting, predation, disease, and so forth—but also those that increased productivity, such as food, water, coverts (a thicket in which an animal can hide), and other habitat features like salt licks for herbivores

3.4 Five mountain lions killed in three days by government hunter Cleve Miller and party, contracted by U.S. Bureau of Biological Survey, Apache National Forest, Arizona, 1922. Denver Public Library Special Collections

and hibernation places for bears. Drawing on ecologists' recent studies of succession, which indicated how species communities were understood to change over time, Leopold envisioned the potential for game managers to design productive systems that required little intervention. His was not a proposal to abandon predator control in all situations, but rather to supplement it with another form of control: a constructive one. "Game management," he wrote, "does not consist of farming game. It consists of so regulating the natural factors of productivity that game farms itself."[60]

From a new position at the U.S. Forest Products Laboratory near Madison, Wisconsin, Leopold completed his now celebrated *Report on a Game Survey of the North Central States* for the Sporting Arms and Ammunition Manufacturers' Institute in 1931. In this text, he elaborated on his argument that game species could be restored by planting food and cover species, rather than by killing predators or—and this was in the interest of the gun manufacturers—further regulating hunting. Restoration, in other words, promised an end to hunting restrictions. Planting food and cover crops to promote game was not about returning to a prior ecological

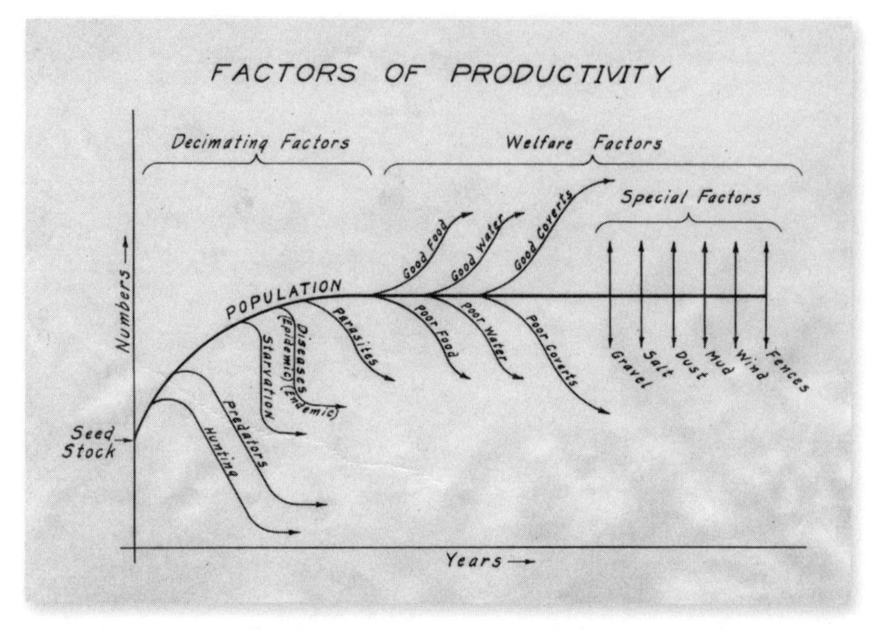

3.5 Figure from Aldo Leopold, draft of "Southwestern Game Fields" (1927). Leopold and his collaborators' innovation was to argue that wildlife scientists and managers could restore game species by manipulating "welfare factors" like vegetation as well as "decimating factors" like predators. Courtesy of the Aldo Leopold Foundation and University of Wisconsin–Madison Archives

baseline. As Barrington Moore noted in his review of the *Report on a Game Survey* in the journal *Ecology*, "The author does not advocate a return to the former conditions, but the restoration of brush in waste and odd corners."[61]

Leopold would later credit his friend and collaborator Herbert L. Stoddard with the idea that animals could be managed by manipulating plant species. Leopold first met Stoddard in 1928, when Stoddard was traveling across the country working on a quail inventory for the Biological Survey.[62] Stoddard was studying quail food requirements as well as predation on quail, and shortly thereafter he wrote to Leopold to assure him that "vast areas that now have few if any quail could be made productive by restoration of food, cover and so forth," and that "it should cost very little to let cover restore itself over a period of years on the average northern farm."[63] Another of Leopold's collaborators, Paul Errington, was also coming to the conclusion that winter food and cover were more important determinants of bobwhite quail population size than the presence or absence of predators.[64] Plant restoration, it seemed, could be an efficient way of producing more game. If people could learn how to establish the right plant species, the plants would manage the game.

The President's Committee on Wild Life Restoration

The *Report on a Game Survey* circulated widely and solidified Leopold's reputation as an expert on game management, and in 1933, the University of Wisconsin–Madison hired him for what the *New York Times* noted was the only "wild game chair" position in the country.[65] That year, he was also appointed to President Franklin Delano Roosevelt's new (and short-lived) Committee on Wild Life Restoration. In an October 1933 memorandum to President Roosevelt, Thomas H. Beck, editor of *Collier's* magazine and a friend of the president, suggested that the federal government use land recently acquired through the Resettlement Administration to augment game bird populations. (The Ecological Society of America's Preservation Committee was not the only group vying to control the use of reclaimed land.) As a founder of the More Game Birds in America Foundation, the precursor to Ducks Unlimited, Beck envisioned rearing birds in captivity and then stocking them in designated areas, as the American Bison Society had done with bison. Intrigued, Roosevelt created a three-member committee, initially inviting Beck; Jay "Ding" Darling, the political cartoonist and noted conservationist; and John Merriam of the Smithsonian Institution to fill its seats. When Merriam declined the invitation, the administration extended an invitation to Leopold.[66]

On the committee, Leopold and Ding Darling found themselves at odds with Beck. At the twentieth American Game Conference, held that January in New York, rumors abounded that Roosevelt's Bureau of the Budget was about to shutter the Biological Survey. Beck was in favor of this, as he considered the Biological Survey an ineffective and "semi-scientific" agency. Leopold and Darling, meanwhile, defended the role of the Biological Survey.[67] And while Beck advocated for captive breeding and restocking of game birds, Darling and Leopold made a novel proposal that was in keeping with Leopold's developing views about the capacity of nature to restore itself.[68] Darling later recounted, "Beck advocated the theory held by the 'More Game Birds' crowd—that the way to restore ducks was to hatch them in incubators and turn them loose into the flight lanes, in other words restocking by artificial methods. Leopold and I held to the principle that nature could do the job better than man and advocated restoring the environment necessary *to* migratory waterfowl, both in the nesting areas, the flight lanes, and the wintering grounds."[69]

The committee's report, delivered in February 1934, reflected Darling and Leopold's vision, rather than Beck's. It requested the federal purchase of some twelve million acres of submarginal farmland and allocated $25 million dollars "for restoration and improvement of the land acquired."[70] In a cartoon-covered letter to Roosevelt, Darling claimed the Biological Survey could make better use of retired agricultural land than anybody. He wrote, "Others just grow grass and trees on it. We grow grass, trees, marshes, lakes, ducks, geese, fur-bearers."[71] At the end of the letter he drew a family of ducks dressed in tattered clothing asking why they, too, have not benefited from the New Deal. While two baby ducks cry, the father duck says, "Redistribution of wealth, eh? Where do we come in?" The mother duck retorts, "Yeah! How about subsistance [*sic*] homesteads for us?"[72] Darling imagined the nascent welfare state encompassing bird species.

The Biological Survey was aware of the existential threat posed by the President's Committee on Wild Life Restoration. The committee had proposed that the Biological Survey be renamed the Wild Life Service and that it hire a commissioner of restoration—a plan that the Biological Survey critiqued as "extravagant" and "incompatible with good administration." In a memorandum to the president, the Biological Survey countered that it had been working for decades to restore game species by killing predators and enforcing hunting laws. This work had been in cooperation with many federal agencies, including the Forest Service, the Bureau of Reclamation, and the Bureau of Indian Affairs. Further, the Biological Survey noted, ownership and control of all nonmigratory species was vested in the states. Besides, the Biological Survey concluded, many lands purchased by the Resettlement Administration would be unsuited to game production: "Game and other animals must be looked upon as a crop, the production of which goes back to soil fertility; quantity production, therefore, cannot be expected from poor lands." It concluded, "With necessary funds, the Survey can do the work as promptly as any emergency organization and can do it better."[73]

Although Congress appropriated only a fraction of the funds that the President's Committee on Wild Life Restoration requested, its report informed the subsequent Duck Stamp Act of 1934 and the Federal Aid in Wildlife Restoration Act of 1937. Both acts created mechanisms for the federal government to fund wildlife management, which until that point had fallen to clubs like the Audubon Society, the Boone and Crockett Club, and the American Bison Society. Under the Duck Stamp Act of 1934, any person who hunted ducks, geese, swans, or brant had to carry a $1 Duck Stamp, the proceeds of which went into a special treasury account that could be used for the acquisition of waterfowl habitat. The Federal Aid in Wildlife

Restoration Act of 1937, also known as the Pittman-Robertson Act, redirected the proceeds of an excise tax on firearms and ammunition from the U.S. Treasury to the secretary of the interior. States could then apply to use the funds for wildlife research, surveys, artificial breeding, or the acquisition of land. Crucially, the act maintained state jurisdiction over wildlife while creating a role for federal wildlife biologists, who would vet the states' wildlife management proposals.[74]

Darling continued to shape the Biological Survey when Roosevelt appointed him its director in July 1934 (a position Leopold would later decline). During his eighteen-month tenure, Darling intensified federal game law enforcement and also helped institutionalize ecological principles and practices at the federal level by establishing a wildlife research program in cooperation with the states.[75] Referencing Victor Shelford's recent article on check-areas in *Ecology*, the program required each cooperating state to set up unmanaged "natural areas" as "checks or controls" for comparison with managed areas.[76] Thus began the Cooperative Fish and Wildlife Research Units program, now administered by the U.S. Geological Survey. When Ira Gabrielson succeeded Darling as director of the Biological Survey, the agency's expansions into experimental wildlife management continued. In 1936, Roosevelt established the first national wildlife experiment station with the objectives of providing "optimum conditions for wildlife production" and to see how "deliberate changes in food and cover" would affect wildlife production.[77] The station was established on 2,670 acres of "submarginal farms retired from agriculture" acquired by the Resettlement Administration on Patuxent River in Maryland. It would become the Patuxent Wildlife Research Center, the main research center for captive breeding to fulfill the requirements of the Endangered Species Act of 1973.

Federal expansion into game restoration was eagerly supported by arms manufacturers and by hunters, as it promised a means to curtail hunting restrictions. As early as 1911, the Winchester Repeating Arms Company had offered William Temple Hornaday at least $10,000 a year for wildlife-protection efforts—on the condition that he abandon his campaign against automatic and pump shotguns. Hornaday turned them down, but the company didn't abandon its efforts. Gun manufacturers, including Winchester, the Remington Repeating Arms Company, and E. I. du Pont de Nemours Powder Company, continued to seek allies among game restorationists, and in 1911 they created the American Game Protection and Propagation Association, which would go on to lobby for and fund restoration projects across the country.[78] The 1934 report of the President's Committee was endorsed by the American Game Association, the North American Game Breeders Association, the Camp Fire Club of America, and local fish and game clubs like the Brattleboro Rifle Club and the Illinois Sportsmen's League.[79] Federal natural resources managers, in turn, concluded that it was easier to design an ecologically productive area than it was to enforce hunting regulations. In a 1939 leaflet, the Biological Survey described the nation's move away from restrictive hunting laws and toward game restoration, noting, "It is much easier to prevent all shooting or trapping on a number of sanctuaries than it is to maintain supervision over the personal activities of a large number of gunners in such a way as to compel each of them to obey every requirement of a complex code."[80]

The U.S. Fish and Wildlife Service (FWS) was created in 1940 by combining the Bureau of Biological Survey and the Commerce Department's Bureau of Fisheries, which had been created in 1871 to investigate declining numbers of "the most valuable food fishes of the coast and the lakes of the United States."[81] The new agency was housed in the Department of the Interior, but the bureau's origins in the Department of Agriculture left a deep imprint on the FWS. For decades after reorganization, the FWS would focus on commercially valuable species, framing them as crops to be managed—both through predator control and through habitat manipulation. With the 1940 reorganization, the FWS was given responsibility for managing 193 bird and game refuges, including the wildlife reservations utilized by the American Bison Society.[82] The FWS would go on to shape wildlife management practices in its National Wildlife Refuge System and across federal and state lands.

The Wisconsin Arboretum

Leopold, for one, was not a fan of FDR and was not in favor of government expansion. He believed restoration would be most efficient and effective if pursued by private citizens. He argued to Thomas Beck, for example, that the government should encourage agricultural colleges to teach farmers how to raise game birds, and not take on the task itself.[83] One of the resulting recommendations of the President's Committee on Wild Life Restoration was that subsistence farm homes be established on all the wildlife restoration areas acquired, the farmers to serve as caretakers under the direction of trained district supervisors.[84] Instead of restocking degraded areas with birds, Leopold suggested, why not pay the same sum to the farmer to fence off areas for birds and feed them? How, Leopold asked, could land in the public domain be kept without a "huge expansion of federal machinery," or given away "without the certainty of misuse"? The answer, he believed, was to foster a care for wildlife in private landowners.[85]

He wrote, to himself, "The basic problem is to *induce the private landowner to conserve on his own land,* and no conceivable millions or billions for public land purchase can alter that fact."[86]

This was the vision of restoration Leopold put forth in a series of leaflets titled Wildlife Conservation on the Farm, and it is the ethic that would suffuse his famous collection of essays, *A Sand County Almanac.* Leopold and his collaborators set out to demonstrate that farmers could produce wildlife as well as domestic life, a claim consistent with the aesthetics of cultivated wildness that Butler, Roberts, and Rehmann all advocated. But Leopold did not entirely share their preference for native species. Farmers, Leopold argued, had the opportunity to conserve plants such as ragweed and foxtail (an introduced grass), ones "on which game, fur, and feather depend for food."[87] They could also attract species like pheasant, prairie chicken, quail, and squirrels by putting out offerings of corn, soybeans, buckwheat, or other grains. Leopold and his colleagues wrote of the need to provide rations to wild animals in times of hardship and drought—to provide wire baskets with ears of corn, or scatter grain under briars.[88] And finally, farmers could build coverts for game by constructing piles of felled oaks or by leaving grape tangles and evergreens. "It is useless to plant birds on a farm without good cover," Leopold wrote, "New fences may bring more birds than new laws."[89] Unlike members of the American Bison Society or the Wild Flower Preservation Society, Leopold did not seek the institutionalization of ecological care. He imagined, rather, that individual landowners could cultivate wildness on private land.

Leopold's service on the President's Committee coincided with an unusual opportunity to create the type of wildlife restoration area he envisioned. In 1932, the University of Wisconsin–Madison purchased a swampy 430-acre property on the edge of Lake Wingra with the plan of establishing an arboretum there. When the university hired Leopold in 1933, he came on board as the new arboretum's first research director. William Longenecker, a landscape architect, was its executive director, and the two immediately clashed. Longenecker believed the arboretum should plant ornamental trees that would inspire Wisconsinites to beautify their backyards. He would organize these trees by taxonomic group. Leopold, though, was uninterested in what he derided as "a collection of imported trees." Instead, he argued that they try something "new and different" by using the grounds for wildlife research. In a report to the university president, he explained that nobody had ever tested whether game populations could be promoted by restoring vegetation, and that, if managed as an experimental site, the arboretum could provide insight on how to encourage the success of any animal—whether a game species or not.[90]

Neither Leopold nor Longenecker would compromise on his vision for the arboretum, and in 1934, the university gave each responsibility for half of the property. At the dedication ceremony for the University of Wisconsin Arboretum and Wild Life Refuge that June, before a crowd of two hundred in a heavy-beamed barn, Leopold described his vision for the grounds, including a new site-specific take on restoration. Unlike a typical arboretum, designed to exhibit apples, lilacs, roses, and the like, Leopold explained, the University of Wisconsin would attempt to reconstruct "a sample of old Wisconsin." He asked the audience to imagine the surrounding landscape as it would have appeared in 1840: orchard-like stands of oaks, interspersed with shrubs and prairie flowers and populated with sharp-tailed grouse, partridges, elk, and deer. The arboretum researchers would try to re-create that lost landscape.[91]

In justifying his plans for the arboretum, Leopold stressed its relevance to Dust Bowl recovery. To reconstruct Wisconsin's past landscape would be to illuminate those changes that threatened "to undermine the future capacity of the soil to support our civilization." Familiar with ecologists' growing advocacy of check-areas, he argued that the arboretum could furnish a nearly wild comparison to used land. Echoing the language of the Ecological Society of America's preservation committee, he suggested that a reconstructed prairie sample would "serve as a bench mark" for future land management. Familiar with Edith Roberts's work, he wrote to the university president that the purpose of the arboretum was "to become an outdoor classroom and research laboratory" where "long-time experiments can be undertaken without risk of disruption or interference."[92]

Leopold's plans for the arboretum were delayed as the university appealed to the Biological Survey and the Forest Service for funding, to no avail. But then, in 1935, the university was assigned a unit of Civilian Conservation Corps (CCC) workers, and Leopold was able to pursue his vision. Leopold placed Theodore Sperry, a graduate of the University of Illinois botany department, in charge of managing the project, and that summer, Camp Madison began to plant prairie vegetation in a former horse pasture. Under Sperry and Leopold's direction, CCC workers cut out chunks of sod with long-handled shovels from "prairie remnants," such as untilled ground in cemeteries and railroad rights-of-way, and trucked them to the arboretum. To make space for the remnants, the workers tore up common species like quackgrass and goldenrod, and poisonous ones like prairie larkspur and sundial lupine. In their place they planted icons of the prairie: blazing star, big bluestem, purple coneflower, hairy grama, prairie tickseed, rattlesnake master, cut-leaf violet, and bur oak.[93]

3.6 Civilian Conservation Corps workers water a recently planted pine tree at the University of Wisconsin Arboretum, 1936. University of Wisconsin–Madison Archives Collections

By 1938, Camp Madison workers had moved 186,000 yards of dirt, planted one hundred thousand trees and nine acres of prairie plants, and dug fourteen acres of lagoons. Sperry, Leopold, and their CCC crew had successfully transplanted forty-nine species of prairie flowers and grasses. In *Parks & Recreation* magazine, Paul Riis announced, "Faculty and students in natural history have, in the ecological garden, a practical outdoor laboratory for botany, zoology, entomology, limnology, game management and landscape gardening." A 1939 newspaper announced that the Wisconsin Arboretum was working to restore prairie species that had been brought to the brink of extinction by the settler's plow, in an article titled, "Clod by Clod, Historical Prairie Returns to Madison's Yard."[94]

"Controlled Wild Culture"

Nonnative species are today considered one of the main threats to native species, but this view extends back to only the 1980s. Prior to that, most environmental managers and ecologists expressed concern about nonnative species only if they proved injurious to agriculture.[95] In the early twentieth century, privileged white women, followed by recent immigrants and middle-class automobilists, were considered the main threats to native plants. In focusing on native species like flowering dogwood and blazing star, early plant restorationists made two arguments. The aesthetic argument claimed for native species a form of biological belonging: naturalistic gardeners held that native and uncultivated species were more visually harmonious with American landscapes. The second argument lay at the intersection of evolution and management: restorationists argued that native species were well adapted to their environments and therefore required little maintenance. Left alone, they would thrive. Members of the Wild Flower Preservation Society, like members of the American Bison Society, believed that a reintroduced native population would soon restore itself.

Aldo Leopold's plenary address at a joint meeting of the Ecological Society of America and the Society of American Foresters, in 1939, merged the values of wildness and nativity with an explicit call for ecological study. Leopold had long been a member of the Society of American Foresters, but this meeting solidified his visibility within the ESA, and his presentation, "A Biotic View of Land," makes clear that he intended to build bridges among the disciplines: what united the "researches at Vassar and Wisconsin for methods of managing wildflowers" and the ESA's "campaign for natural areas," Leopold argued, was that they sought to "preserve samples of original biota as standards against which to measure the effects of violence."[96]

Unlike evolutionary changes, which were "slow and local," Leopold argued, recent environmental changes were of an "unprecedented violence, rapidity, and scope." Whereas in the 1910s it had seemed possible to save native plants simply by convincing people not to pick them and allowing them to regrow in place, by the 1930s restorationists had begun to doubt nature's capacity to regenerate without human intervention. The problem confronting ecologists, Leopold contended, was that they would not know what to expect of "healthy land" unless they were able to access "a wild area for comparison with sick ones." A science of restoring ecological health would require "a base datum of normality, a picture of how healthy land maintains itself."[97]

The ecological gardens in Minneapolis, Vassar College, and the University of Wisconsin were joined by others. After moving from Tucson, Arizona, to Santa Barbara, California, in 1925, Frederic and Edith Clements established "combined botanical and experimental gardens" there.[98] In 1935, the Works Progress Administration funded the creation of the Lee Park Wild Flower and Bird Sanctuary in Petersburg, Virginia. Groups of unemployed women, mostly Black, worked together to build more than ten miles of paths and transplant more than 365,000 plants into the reserve.[99] Meanwhile, Elizabeth Prescott led a prairie restoration project at the Wisconsin Prison for Women.[100] At the University of Illinois, Shelford and his colleagues worked to restore a prairie remnant through the 1940s.[101] At the University of Wisconsin Arboretum, John Curtis, Henry Greene, and others continued to experiment with tallgrass prairie restoration.[102] These projects laid the groundwork at local scales for practices that today are global in scope.

As the methods of wildflower propagation reached experimental areas at colleges and universities—where ecologists researched how to control succession—they were normalized as acceptable ways of studying wild species and, crucially, as elements that could be used to rebuild wild spaces beyond garden and university walls. By the 1940s, ecologists and game managers had converged on the idea that plants could be reestablished on formerly cultivated land. Whereas the American Bison Society had worked to reintroduce a single species to federal lands, ecologists and naturalistic gardeners began to plant desired species on private lands, either for their aesthetic value or to provide habitat for other species. Importantly, they argued that this strategy could be applied to commercially valuable species and noncommercial species alike. The Wild Flower Preservation Society often noted that, unlike with the game conservation movement, the species they sought to protect had little economic significance.[103] The method of "controlled wild culture or 'management,'" Leopold argued in his famous essay "The Conservation Ethic," could be "applied not only to quail and trout, but to *any living thing* from bloodroots to Bell's vireos." Planting—providing shelter and food—was an effective and efficient alternative to removing predators. Leopold concluded, "A rare bird or flower need remain no rarer than the people willing to venture their skill in *building it a habitat*."[104]

Elizabeth Britton had argued as early as 1904 that those concerned with the fate of birds should support the preservation of remaining wildflowers.[105] What was novel in Leopold's formulation was not the idea of ecological interconnectedness, but the idea that habitat could be built, rather than regrown. Into the 1940s, other academic ecologists and wildlife managers would experiment with assembling entire ecological communities, building off the work of Leopold, Edith Roberts, Eloise Butler, and their collaborators. They sought to manipulate plant communities as an alternative to killing predators, and as an alternative to artificial propagation.

In recommending "controlled wild culture," Leopold struggled to articulate what level of human intervention was appropriate, and much of his later writing dealt with the place of technology in land management. In a 1940 essay, he described the Wisconsin Arboretum as a "synthetic prairie" that "will always be synthetic" and that was costing taxpayers twenty times as much as what it would have cost to buy and protect a prairie remnant.[106] Elsewhere he suggested that "very intensive management" of fish and game "artificializes it." He criticized new outdoor recreation "gadgets" like factory-built duck calls that "dangle from neck and belt" and "fill the auto-trunk," arguing that they eroded the "atavistic" value of hunting and fishing, yet he valued his rifle and binoculars, admitting, "I do not pretend to know what is moderation, or where the line is between legitimate and illegitimate gadgets."[107] The same technologies, Leopold recognized, could be used for environmental destruction and for environmental restoration. Leopold's famous collection of essays, *A Sand County Almanac,* published posthumously in 1949, ended with a call for attentive restraint: "We shall hardly relinquish the shovel, which after all has many good points, but we are in need of gentler and more objective criteria for its successful use."[108]

And yet ecologists and restorationists were already moving to embrace what quite a vastly more powerful and violent technology could offer them. On July 16, 1945, the United States Army detonated "The Gadget," the first atomic bomb, in New Mexico's Jornada del Muerto desert, not far from where Aldo Leopold had begun his career. Technologies closely associated with the weapon would rapidly transform ecological theory and research, as the weapons themselves transformed landscapes and lives.

PART II

Recovery, 1945–1970

4

Atoms for Ecology

On August 18, 1943, Lauren Donaldson received an urgent telegram from the Office of Scientific Research and Development while on his way to a fisheries management conference in British Columbia. The telegram requested his presence in Washington, DC, to discuss a sensitive matter. It gave no further explanation.[1] Unbeknownst to Donaldson at the time, it was not the OSRD that had sent the telegram, but the Manhattan Engineer District (MED). The federal government had recently acquired land in eastern Washington State on which to produce plutonium for atomic weapons. Designs for the Hanford Engineer Works called for pumps to channel 30,000 gallons of water per minute through each of three reactors. This water would come from the Columbia River, and it would return to the river warm and radioactive. Eager to ensure that this effluent would not damage the Columbia's valuable fisheries, the Manhattan Engineer District requested that Donaldson, an expert on salmon and trout physiology, study whether radiation harmed fish. In DC, the supposed OSRD officials asked Donaldson to lead a grant titled "Investigation of the Use of X-rays in the Treatment of Fungoid Infections in Salmonid Fishes" and to rename his University of Washington laboratory the Applied Fisheries Laboratory. Both titles concealed the project's true objective: to study whether nuclear reactor effluent killed salmon.[2]

That fall, Donaldson and his research assistants began exposing salmon eggs, embryos, and fingerlings to X-rays in their Seattle laboratory. The design of their initial experiments reflected the belief, widespread at the time, that the major biological hazard of radiation was prolonged exposure to external sources. The Applied Fisheries Laboratory (AFL) took the results of their early experiments to be reassuring. At 100 rads (a unit of absorbed radiation dose), the fingerlings appeared normal. At 250 rads, they were noticeably thinner. Above 500 rads, the fingerlings quickly died. Because Manhattan Engineer District officials anticipated radiation levels no higher than 100 rads in the Columbia River, the AFL concluded that fish were killed only by "unusually high" doses of ionizing radiation.[3]

Donaldson's atomic work, as it turns out, was only just beginning. The federal government invested heavily in scientific research during World War II, reconfiguring the relationships among the military, universities, and corporations. Like thousands of biologists across the country, Donaldson would take advantage of funding opportunities through the U.S. Atomic Energy Commission (AEC), established in 1946 and charged with continuing atomic weapons production while developing peacetime uses of atomic energy. Postwar federal sponsorship reoriented the goals of many sciences, including physics and oceanography, and expanded science's influence in national and foreign policy.[4] Ecology was no exception. As we will see, Donaldson aligned his fisheries research to meet the AEC's goals, and he was enthusiastic about partnering with corporations like General Mills on projects to restore wild salmon runs by establishing industrial-scale aquaculture operations and "farming the sea." His career exemplifies the shifting postwar relationship between academic ecology and species management.

Along with new funding opportunities, the Atomic Age provided scientists with new tools: radioisotopes. Radioisotopes entered environments through the production and detonation of atomic weapons; between July 1945 and 1992, the United States detonated 1,134 nuclear devices.[5] The AEC also sent tens of thousands of shipments of laboratory-produced radioisotopes to American scientists.[6] Biologists would use radioisotopes to track molecules as they circulated and transformed in cells, organisms, and, ultimately, ecosystems. As historian Angela Creager details, radioisotopes transformed biomedicine, reorienting it toward a molecular, process-oriented vision of life.[7] They also played a key role in the shift from single-species restoration to ecosystem restoration.

Over the course of Donaldson's career, atomic technologies transformed both the methods for studying wild species and the methods for cultivating them. Donaldson's early involvement with the AEC gave him access to the radioactive environments and materials from which ecosystem theory emerged. Yet Donaldson—like many biologists of his generation—remained focused on managing one species at a time. The promise of radiation for restoration, he maintained, was that it would help biologists "accelerate" the evolution of particular wild species. Donaldson hoped to breed new strains of trout and salmon that could thrive in a rapidly changing world. In the history of ecological restoration, fish would remain their own complicated category, slipping between the categories of wild and domesticated, their caretakers' visions for restoration never quite aligning with those of restorationists working on land.

Building Better Salmon

As early as the 1850s, naturalists blamed water-powered industry and agriculture for a decline in the number of trout, salmon, and shad runs in the eastern United States: as they saw it, farmers had cleared forests, leading to erosion and the subsequent silting of streams, while factories had dammed and polluted rivers.[8] In 1872, the federal government established a system of national fish hatcheries "to restore wasted waters to their primitive or more than primitive fruitfulness" and "to extend the geographical range of the more important food-fishes, such as shad, salmon and trout, by naturalizing them in new waters."[9] Fish culture was not without its difficulties, however: hatchery fish often died before they could be released into streams, whether from malnutrition or disease. But those working to introduce "food-fishes" to new bodies of water persisted. Fisheries biology emerged as its own scientific discipline in the early 1900s around efforts to improve hatchery yields, borrowing from zoology and the nascent field of ecology.[10]

By the time Donaldson began a master's program at the University of Washington School of Fisheries, a nationwide network of research centers connected fisheries biologists and managers.[11] The UW School of Fisheries emphasized research on salmon and trout, closely related species that migrate from the ocean to freshwater to spawn. Salmon and trout were (and still are) important commercial and cultural resources in the Pacific Northwest, species that were understood to be threatened by overfishing, damming, and urbanization.[12] While completing his master's thesis, Donaldson became interested in rearing salmonids that could flourish in a rapidly changing world.

In 1932, Donaldson decided to remain at the School of Fisheries to complete a PhD. That year he learned of the work of George Charles Embody, a Cornell University zoologist, who reported that in only three generations he had selectively bred brook trout that were resistant to a common hatchery disease, furunculosis.[13] Embody's publications inspired Donaldson and his friend Clarence Pautzke, a recent biology graduate (and a future commissioner of the U.S. Fish and Wildlife Service), to begin selectively breeding trout for West Coast waters. Together Donaldson and Pautzke collected rainbow trout and steelhead trout from Washington streams and hatcheries. They began interbreeding the rainbows and the steelheads, watching for individuals that were larger or produced more eggs than their progenitors.[14]

Like Aldo Leopold and federal game managers, fisheries biologists looked to agricultural science for concepts and methods. Through selective breeding, they would attempt to "improve" wild species. Later in life, Donaldson would explain that he modeled his efforts to breed trout and salmon on "modern beef production," in which carefully bred cattle were raised in controlled areas and were fed pills that "defeated diet deficiencies." He recounted, "I've always looked at the shape of fish in terms of what we're trying to accomplish. But there isn't any pattern for the ideal shape of a fish. So I go to the fair and look at the beef cattle, the Aberdeen angus and the hefherds [*sic*], and the big blocks of potential meat. And then I go back and look at the fish and think 'what would happen if we changed the fish? If we made them heavy?'"[15]

Between 1932 and his retirement from the School of Fisheries in 1973, Donaldson indeed made fish heavy. The fish Donaldson and Pautzke caught in Washington streams in 1932 reached sexual maturity in four years and weighed 1.5 pounds at maturity. In 1955, after twenty-three years of selective breeding, Donaldson's trout breed reached sexual maturity in two years and weighed 4.1 pounds at maturity.[16] To accomplish this, Donaldson and his collaborators employed traditional methods of selective breeding, choosing to mate only the fish with the most rapid growth rates. Donaldson argued that by improving salmonid bodies, scientists could produce fish that would thrive even in industrialized waters; in a 1961 presentation he explained that, in the West, "the great rivers that were the ideal spawning and rearing areas for Chinook salmon were used more and more by industry," and that in addition to the "impressive array of rehabilitation measures" that state and federal agencies employed to protect salmon, such as regulating fishing and constructing fish ladders at dams, his laboratory was pursuing "yet another area of effort—that of selective breeding."[17] By breeding better salmon and trout and then releasing them, Donaldson and his collaborators hoped to take pressure off wild fisheries. They imaged a future of biotic abundance, one in which fishing laws and regulations were unnecessary. They imagined it was easier to change animal bodies than it was to change human behavior.

The MED had originally contracted Donaldson to determine whether fish were harmed by radiation exposure. Through the AEC research network,

4.1 Lauren Donaldson (center) transporting hatchery-reared salmon to the Yakima River in Washington, c. 1935. University of Washington Libraries, Special Collections, UW35916

however, Donaldson soon learned of efforts to use radiation to induce beneficial mutations in species. On learning of the potential for radiation to "accelerate" genetic mutation, Donaldson reframed his laboratory's research and began to explore radiation's potential as a tool for species management.

The AFL's work in this area was novel, but not unique. Interest in "mutation breeding" exploded around 1955, as President Eisenhower's Atoms for Peace campaign expanded into an international program with the support of the United Nations.[18] During this period, to counter public objections to nuclear proliferation, the federal government emphasized peacetime applications of nuclear technology, promising cheaper energy and powerful medicines. The AEC also began to invest heavily in research on breeding applications for atomic technologies. Most of this research focused on agricultural species. At Brookhaven National Laboratory, for example, biologists announced that continuous irradiation with cobalt-60 produced a 17,000-fold increase in the rate of mutation in corn, and the *New York Times* reported that radiation offered "the possibility of speeding up the creation of new varieties of valuable food plants."[19] Amid the widespread attention to agricultural species, though, a few researchers attempted to mutate species of conservation concern. The geneticist W. Ralph Singleton, for example, began exposing American chestnut trees to radiation in the hope of inducing a mutation that might make trees resistant to chestnut blight.[20] Donaldson's AEC-supported research fit into both categories. Salmon and trout were commercially valuable food species *and* of concern to conservationists, and as Donaldson worked to speed evolution and create robust strains, he selected for traits that would make salmonids suitable for industrial-scale production.

In 1958, Donaldson's laboratory began exposing Chinook salmon eggs to radioactive cobalt-60 at the campus hatchery. Many of the salmon displayed abnormalities as they developed, including fused vertebrae. But some were larger and seemingly more robust than the nonirradiated controls.[21] The researchers grew the salmon in captivity and, in May 1961, released 21,217 control fish and 22,273 irradiated fish into Portage Bay to migrate to the ocean. To see whether any of the fish returned to their spawning grounds—the University of Washington hatchery—they marked the salmon by notching their fins. In 1962, two times more irradiated fish returned than control fish. Donaldson concluded that "low levels of irradiation" actually had a *beneficial* effect on salmon, increasing either their ability to survive in the wild or their ability to find their original spawning grounds.[22] In a letter to a colleague at Los Alamos Scientific Laboratory in 1963, he wrote, "We are having interesting experiences with our very low exposure to developing salmon embryos and have about reached the conclusion that a half roentgen/day during the developmental period of 100 days, plus or minus, is very beneficial. I am sure this will cause some of the critics to really get up on their hind legs and scream, but such is the way of living things."[23]

Donaldson had reason to anticipate critics. Public concern about nuclear fallout was at its peak, and Donaldson, in addition to working for the AEC in Seattle, worked at the Pacific Proving Grounds, where the United States had detonated more than one hundred nuclear weapons. Donaldson had recently found himself at the center of the controversy around *Mondo Cane,* an Italian documentary that included images of Bikini Atoll, where Americans had been detonating atomic weapons since 1946. The film featured images of thousands of unhatched seabird eggs and tortoises dying of exhaustion before finding their way to the sea; it claimed that atomic technologies had perverted nature, inverting animals' basic instincts and destroying their fertility.[24] The film's premiere prompted a flurry of correspondence among media outlets, the National Academy of Sciences, and the AEC, who called on Donaldson to serve as their spokesperson.[25] In one letter, the director of the National Academy of Sciences Pacific Science Board wrote to Donaldson, "It seems to me that we are dealing with some clever Soviet propaganda which was trying to embarrass us into not carrying out further tests in the Pacific islands."[26] Responding to a *New York Times* reporter, Donaldson toed a fine line. "We have yet to discover in or near the proving ground any biological aberration or peculiarity that is definitely ascribable to the effects of radioactivity," he began, "but this is not to say that some such effects have not taken place, for radioactivity, of course, is capable of producing biological change and we always are aware that such possibilities must be followed as far as instruments and human intelligence will allow."[27]

Through the 1960s, Donaldson actively defended the value of nuclear technologies for ecological research and management. Like a number of AEC-funded scientists, he was working toward an ecological intervention that historians have barely explored, one that holds lessons for those debating some of today's most controversial restoration proposals, including novel ecosystems and de-extinction.[28] Rather than protecting particular places or populations, Donaldson sought to repopulate species by genetically altering them. Hoping to restore salmon's ability to survive in a damaged world, Donaldson changed salmon themselves.

"A More Exact Science"

Donaldson and his colleagues' successful selective breeding of salmon and trout would reshape the biotic world. So, too, would their work on artificial fish diets. Donaldson had set aside this line of research in 1943, when the MED contracted him to study the effects of Hanford Works effluent on salmon. But he came back to it circuitously, in the 1950s, through the AFL's work at the Pacific Proving Grounds. Curiously, it is the convergence of Donaldson's nutrient studies with atomic warfare that made bioaccumulation and nutrient cycles visible to ecologists.[29]

In the weeks after the United States bombed Hiroshima and Nagasaki, the U.S. Senate and Joint Chiefs of Staff entertained proposals to test atomic weapons against naval warships, a set of plans they codenamed Operation Crossroads.[30] Widespread radiation poisoning in Hiroshima and Nagasaki made officials wary about conducting further atomic tests in the continental United States, and so the Joint Chiefs of Staff decided that Operation Crossroads would be conducted "overseas." (The United States had already detonated The Gadget, an implosion-design plutonium device, in New Mexico's Jornada del Muerto desert three weeks before bombing Hiroshima, and they began detonations in Nevada in 1951.) From a short list of Pacific islands, including the Caroline Islands and the Galapagos Islands, the Joint Chiefs selected Bikini Atoll, a C-shaped coral atoll surrounding a deep central lagoon, 2,500 miles southwest of Honolulu. To justify the forced resettlement of 167 Bikini Islanders, the navy argued that Bikini was unsuitable for human habitation because it produced little food.[31] And yet, American lobbyists soon voiced their concern that weapons testing might damage commercially valuable Pacific fisheries. Although the U.S. Fish and Wildlife Service testified that the fisheries resources at Bikini were "negligible," the MED hastily convened a conference to discuss biological monitoring at the test site.[32] There it was decided that Lauren Donaldson would lead the Bikini Radiobiological Survey.[33]

Donaldson and fellow AFL members reached Bikini on June 13, 1946, eighteen days before the first scheduled detonation, test Able. Also arriving at the site were approximately 250 naval vessels and 150 aircraft, as well as 200 pigs, 204 goats, 60 guinea pigs, 5,000 rats, and 200 mice that were slated to be bombed. Donaldson led a team tasked with determining the effects of the detonations on wild marine fauna. Like most scientists at the time, the AFL team expected the expansive Pacific Ocean to quickly dilute and disperse any fission products from the blasts. Their first assignment was to collect "control material" to compare with any organisms collected after test Able. Over an area of almost 250 square miles, they hurriedly gathered as many specimens as they could. They killed smaller fish by poisoning tide pools with derris root and caught larger fish by hook and line. By hand they picked algae, coral, clams, and sea cucumbers from reefs at low tide. As of two days before Able, the AFL had collected 1,926 "control" specimens.[34]

On July 1, 1946, at approximately 9 a.m. Bikini time, the B-29 aircraft *Dave's Dream* dropped an atomic bomb on a battleship stationed in Bikini lagoon. It was a choreographed international event. From thousands of applicants, Joint Task Force One had selected hundreds of reporters, scientists, and diplomats to witness the detonation. Its sound was broadcast into homes, bars, hotel lobbies, and offices. Residents of Sydney, Australia, awaited a tidal wave. And yet most witnesses found test Able an anticlimax. The bomb burst as planned some 500 feet over its target, but approximately 1,500 feet to the west of it. By 2:30 p.m. the next day, the navy had declared

4.2 **Test Baker as seen from Bikini Atoll, July 25, 1946.** University of Washington Libraries, Special Collections, DON0032

Bikini lagoon safe for reentry, and Donaldson and his crew were unable to find any dead or injured fish.

Unlike Able, however, test Baker was spectacular. On July 25, the Joint Task Force detonated an atomic bomb ninety feet below the surface of Bikini lagoon. Within seconds, a hollow column containing some ten million tons of water rose to a height of more than one mile. In his notes, Donaldson wrote, "The one July 1 was awe-inspiring and in many ways beautiful, but the one today just frightened the very daylights out of one."[35] After Baker, Donaldson's crew had no problem finding dead fish. They visited collection points from one of the *Haven*'s whaleboats, and when beach landings were necessary, they used rubber rafts. They placed small fish whole into Geiger counters, first reducing larger fish to ash in laboratory ovens. By the end of summer, the AFL had processed 1,021 specimens in the field and had sent thousands more ahead of them to Seattle for analysis.[36]

The MED did not anticipate a return trip to Bikini, but a few months later, the Joint Chiefs of Staff announced the Bikini Scientific Resurvey as the concluding phase of Operation Crossroads. Naval officials hoped to inspect the target vessels a year after they were sunk in Baker, and the newly formed Atomic Energy Commission asked physicists, geologists, and biologists to participate.[37] Unlike Operation Crossroads, the resurvey would bring only 700 people to Bikini, including some twenty biologists. Over

4.3 Scientists wading in reef around Namu Island to net poisoned fish, 1947. University of Washington Libraries, Special Collections, DON0365

six weeks in the summer of 1947, Donaldson's team collected 5,883 specimens from the lagoon. By eye, the specimens appeared normal: "The usual patterns of life seemed unaltered, and there were no specimens of freaks or cancers or evidence of mutations in Bikini's living system," one AFL associate wrote. Resurvey physicists did record high levels of radioactivity in a layer of mud at the lagoon bottom, but the radioactivity seemed to be confined to a five-foot-deep layer, and on July 25, the navy information office reported, "Sun-tanned sailors and scientists observed the anniversary of the world's first underwater atomic bomb explosion today by going swimming in the clear blue-green 84 degree warm waters of Bikini lagoon."[38]

This might have been the end of the story, if not for the actions of hydroids. Hydroids are a life stage of the hydrozoans, a class of small aquatic predators related to jellyfish, which attach themselves to rocks and other substrates. While at anchor, sailors on the resurvey's transport vessel, the

USS *Chilton,* used large wooden crates to support smaller picket boats that were floated in the water. Over the month of July, hydroids and other fouling organisms grew on the crates, and when they were pulled up, an AFL member decided on a whim to run a Geiger counter over them. To everyone's astonishment, the radioactivity of the hydroids was about a thousand times that of the lagoon water.[39] The AFL team speculated that perhaps radioisotopes were still circulating in the lagoon and that the hydroids, somehow, were concentrating them.[40]

For the time being, though, the AFL could not pursue the matter of the radioactive hydroids, and when the team returned to Seattle after the resurvey, they were uncertain that they would ever visit Bikini again. Donaldson hoped to, however. He wrote to AEC officials, arguing that continuing fieldwork could help determine when Bikini Islanders could return home. As it happened, Donaldson would return to the Marshall Islands, though with no pretense of aiding dispossessed residents.[41] On July 22, 1947, the AEC announced that it would be establishing a permanent proving grounds at the newly established Trust Territory of the Pacific Islands, an administrative unit encompassing two thousand islands spread over three million square miles.[42] Then, in April and May 1948, the United States detonated three atomic weapons at Enewetak Atoll, 190 miles west of Bikini. That July, the AEC sent twelve AFL researchers to Bikini and Enewetak. At Bikini, they attempted to repeat the hydroid incident by anchoring twelve pieces of scrap lumber in the lagoon. Hydroids attached themselves to the wood, and once again they were highly radioactive. The AFL also recorded radioactivity in coral samples collected upwind of the detonation site. With these findings, AFL scientists began to speculate that species, not water currents, were primarily responsible for transporting radioisotopes.

Working for the AEC connected Donaldson and his labmates to a large network of physicists and biologists sharing classified methods. Until the summer of 1948, the AFL had quantified radioactivity in its biological samples with Geiger counters. At this juncture, however, they began experimenting with another method of measuring radioactivity—radioautography—developed by AEC-funded physiologists at Berkeley and Chicago.[43] To produce a radioautograph in a lab, a researcher placed a slice of an organ on a photographic plate. Emissions from a radioactive sample would produce a brighter or darker image, depending on how much radiation reacted with the plate's substrate.[44] When AFL members began to place fish organs and small whole fish on photographic plates, they were astonished at what they saw. In dazzling white, the film revealed the previously invisible phenomenon of internal tissue contamination.

In a high-security talk delivered at UCLA in 1948, Donaldson announced that the AFL team had found evidence of "selective absorption" and "concentration" of radioactive materials by all kinds of living forms, from algae to crabs to fish.[45] Rather than being distributed evenly across an organism's body, radioactivity seemed to be concentrated in the digestive system. Feeding in the lagoon, Bikini's biota had ingested products of the explosions, radioisotopes of elements necessary to life, such as phosphorus and calcium; some species, feeding on others, had concentrated these radioisotopes in their bodies. This was a new way for ecologists to perceive ecological interconnectedness. Rather than observing the moment of interaction—recording a predator eating its prey, for example—ecologists could infer interconnection from shared radioactivity. Food webs would give way to trophic levels and nutrient flow diagrams. Atomic weapons had made connections among species visible. It was a development that, like nuclear proliferation, would profoundly shape the future of the global environment.

In a 1954 report to the AEC, Donaldson extolled the "unparalleled scientific experiments" at the Pacific Proving Grounds—that is, the detonations of atomic weapons—that had provided ecologists with a new tool: radioisotopes. The radioactive residue of fission bombs, and then fusion bombs, had cycled through Bikini's and Enewetak's lagoons, enabling

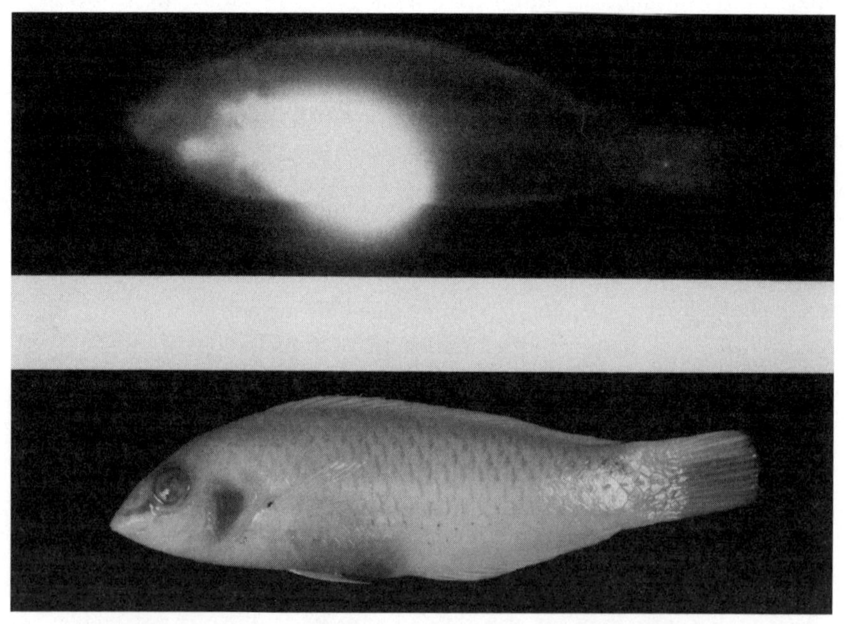

4.4 Radioautograph of wrasse collected by University of Washington biologists at the Pacific Proving Grounds. A radioactive sample placed against photographic film produced a brighter or darker image, depending on how much radiation reacted with the film. University of Washington Libraries, Special Collections, UW35914

scientists to visualize relationships among species in a "natural environment" and to make ecology "a more exact science," in Donaldson's words. And because radioisotopes "did not interfere with normal metabolic processes," he wrote, they were an ideal, noninvasive observational tool.[46]

That same year, an Alaskan territorial delegate asked Donaldson whether the data he had collected for the AEC at the Pacific Proving Grounds on trace elements might help managers to improve salmon fisheries. Donaldson replied that yes, radiobiology provided the tools to measure which minerals were deficient in the environment and, accordingly, which minerals managers could add to streams to improve salmon yields. Donaldson believed it was inevitable that damming and logging would destroy natural salmon runs, but that the runs could be restored with "mineral regeneration" and then populated with genetically improved salmon.[47]

In 1951, and again in 1954, Donaldson proposed studies to the AEC to follow cycles of "essential food elements" in the Pacific Northwest, an environment that he figured to be similar to the Pacific atolls in its "nutrient limitation." The data would be valuable not only to the military, Donaldson argued, but also to scientists working to restore fisheries. The AEC rejected Donaldson's first two mineral regeneration proposals, but it approved his third request in 1957. With a $20,000 AEC grant and land donated by the State of Washington Department of Game, Donaldson launched the Fern Lake Trace Mineral Metabolism Project.[48]

Through his work on the movement of radioisotopes through the Bikini lagoon community, Donaldson came to depict salmon as vessels that exchanged nutrients between sea and land. As he put it, when salmon returned to their birthplaces to spawn and die, they transferred energy "earned in the ocean" to freshwater rivers and lakes, depositing "valuable" minerals in the terrestrial environment. This was an early articulation of the ecosystem concept, although Donaldson did not use that term. In one seminar he explained: "And one must realize that in this whole Northwest area [. . .] life was possible really only because [. . .] the salmon went to the sea and gathered the minerals, many of them trace minerals, the 16-plus elements needed for life. They carried them up the hill. [. . .] We know, for example, that the western red cedar won't grow unless there's calcium present. Well, how does the calcium get in these calcium-deficient areas? Well, it came up with the salmon."[49]

With the Fern Lake project, Donaldson attempted to jump-start such a system by supplementing "nutrient deficient" waters with the elements necessary to sustain aquatic life. Fern Lake, on the rainy Kitsap Peninsula, had been carved out of volcanic rock by glaciers, making it "mineral deficient," and there was no evidence that it had ever sustained salmon runs.

Donaldson believed this made it a good site to study whether "artificial fertilization" could create a habitat that supported salmon, and he divided the project into three stages. In the first stage, the AFL inventoried the lake to establish a baseline for subsequent surveys. In the second stage, they "fertilized" the lake by introducing radioactively tagged elements, including phosphorus, calcium, and potassium. In the third stage, they documented the physical and chemical effects of trace element supplementation on algal growth and, later, on introduced salmon.[50] One reporter described the lake as "a huge test tube" in which to study "the area's mineral deficiencies which limit growth of fish."[51] But the Fern Lake project was riddled with failures.[52] When the AFL constructed a new outlet, beavers blocked it. Stickleback and yellow perch preyed on the introduced salmon. Neither steelhead, nor steelhead-rainbow hybrids, nor sockeye salmon seemed to thrive in the new environment of Fern Lake. These difficulties, however, did not deter Donaldson and other ecologists from continuing to use radioisotopes to study "community metabolism" in the field with the aim of restoring fish populations—or introducing them—through nutrient supplementation.

Domesticating the Sea

Donaldson's work with trace elements and fish nutrition involved him in discussions of human nutrition as well. In March 1954, three months after Eisenhower's "Atoms for Peace" speech, the United States detonated its first thermonuclear weapon, Castle Bravo, at Bikini Atoll. The bomb had over a thousand times the destructive force of the atomic bomb dropped on Hiroshima, and its fallout contaminated the inhabited Rongelap Atoll as well as the Japanese tuna fishing boat *Daigo Fukuryū Maru.* The boat's crewmembers, suffering from burns, headaches, nausea, and bleeding from the gums, were diagnosed with acute radiation syndrome upon their return to Tokyo, and seven months later, one crewmember died. The AEC, eager to avoid international criticism while still safeguarding details about the hydrogen bomb, enlisted Donaldson as a scientific ambassador to Japan and as a consultant on the diets of the Rongelap Islanders. AEC officials were aware that the Japanese investigations of the incident would "turn up some very interesting and rather exciting material," and that, for this reason, Donaldson's work had become, in the words of W. R. Boss, "of more importance than ever."[53]

From 1954 to 1973 Donaldson traveled to Japan six times to take tuna samples and meet with Japanese government officials. He contended that fish aboard the *Fukuryū Maru,* which Japanese scientists had found to be highly contaminated, were only externally coated with radioactive "ash," and that the edible parts of the fish were well within acceptable limits for consumption by humans.[54] Meanwhile, AEC officials downplayed the threat of radiation poisoning to Rongelap residents, at times mocking their concerns. A 1959 AEC report stated that Rongelap residents "seemed to try to blame everything on the radiation accident. They claimed they have been feeling weak and not up to standard since their return to Rongelap. [. . .] They were assured that the symptoms described were not due to radiation but were due to fish poisoning." The report also noted that residents inquired, "Why do we say they cannot eat the coconut crabs and yet allow them to feed them to the pigs and then eat the pigs? As you can imagine, the answer to this question was not easy to explain, and I had to go over this again and again with them." With a note in pencil in the margin, an AEC official commented, "Primitive logic is the worst kind."[55] The concerns of Rongelap Islanders were, of course, well founded, and consistent with the ecosystem theory that the AEC, in other contexts, embraced. The United States had catastrophically polluted their home. The rate of miscarriage among women who returned was more than twice that of women who had never been exposed to such high levels of radiation.[56]

Donaldson's work on trace nutrients also connected him to companies researching American diets. In the fall of 1963, the president of General Mills, Edwin W. Rawlings, wrote to Donaldson asking if he would be a consultant for the General Mills Isolated Proteins Division as "an expert in the field of marine life as a protein source." Food-grade soy protein had recently become commercially available, and General Mills, which owned Betty Crocker, Pillsbury, Cheerios, Bisquick, and Wheaties, among other brands, and which sponsored *The Lone Ranger* and *The Bullwinkle Show* on television, had a vested interest in keeping up with food industry trends. Donaldson had ties to Minnesota, where General Mills was based, Rawlings reminded him, and the project presented "possibilities in the field of conservation." Donaldson took Rawlings up on the offer.[57]

One of Donaldson's first assignments for General Mills was to procure information on fisheries with surplus stocks that could be "harnessed as sources for commercial protein." But such aggregated information was difficult to come by in 1963. When Donaldson wrote to the chairman of the International Whaling Commission to ask for information on worldwide whale harvesting rates, the chairman replied that exact numbers were available only for Norway, where an average of 6,300 tons of whale meat were landed per season, 45 percent for human consumption, the rest for animal food. Despite the paucity of concrete information, General Mills was pleased with Donaldson's fact-finding, and in subsequent months, Rawlings sent Donaldson complimentary premarket samples of WONDRA flour and Bacon Bits.[58]

Donaldson soon befriended Rawlings through a mutual interest in trout fishing. In an invitation to join General Mills employees on a company retreat, Rawlings wrote to Donaldson that his "tour of duty at WONDRA Isle" in Ontario would include "piscatorial pursuits, feeding the inner man, detailed recital of past military and civil personal accomplishments, round-table discussions, drinking in the beauties of the surroundings with special guests Jack Daniels and Jim Beam."[59] Through his connection to Rawlings, Donaldson was copied on more and more General Mills correspondence, some of which concerned the Pacific atolls with which Donaldson was so familiar. In 1965, for example, a representative of Central and South Pacific Pan American World Airways alerted Rawlings to a recent congressional plan to enact "economic and social development of the Trust Territories." The air service contractor for the Trust Territories, Pan Am was keenly interested in developing tourism infrastructure, with the aim of selling seats on commercial flights. Describing the venture to an editor at *The Reader's Digest,* the director of Pan Am wrote: "Stone-age peoples, the descendants of and the confused product of mixed racial stocks and cultures from Malaya, Asia, Spain, Germany, Japan and the U.S.A. could offer wonderful new and exciting cultural-collecting tourism opportunities to today's sophisticated multi-destination travelers."[60] General Mills should also be interested in the fate of this legislation, Pan Am argued, because "the warm, clean, tropical seas which surround the 58,000 atolls and islets could serve as 'anchored factory bases' from which to cultivate the ocean areas for all forms of protein which the expanding populations of the Earth require." The atolls—which in the 1950s the United States had portrayed as barren of food in order to justify their destruction—were now being celebrated as the future "protein 'bread basket' of the entire world's population."[61]

General Mills was indeed interested in the idea of cultivating a tropical protein basket, and Donaldson, whose conception of ecological restoration was so closely connected with the improvement of food species, began brainstorming. In follow-up meetings, Donaldson emphasized that food from the ocean was an excellent resource because it contained "a perfect distribution of the 14 elements of diet which are so important to human beings." He suggested that hake, never regarded as a table fish by the Russians, Japanese, or Americans, was packed full of concentrated protein as well as cobalt, iodine, copper, zinc, and other trace elements that were "sometimes hard to procure in controlled or budgeted amounts." Donaldson also noted that his Radiation Biology Laboratory (formerly the

AFL) had excellent data on the locations of areas around the Pacific atolls where General Mills might establish "ocean ranches"—pens of fish reared for consumption. Through selective breeding, he explained, he had created trout "which could live in impossible conditions and thrive."[62]

Donaldson's and General Mills' interests in trace elements matched up with broader trends in the field of nutritional research at the time. Recent debate in the Food and Drug Administration had centered on whether there was a mineral imbalance or shortage of minerals in the typical American diet. To address the question directly, the U.S. Department of Agriculture launched a project to analyze whether there were differences in the mineral constituents of wheat grown in different locations in the United States. The position of General Mills was that "minerals, particularly the so-called 'trace' minerals, are on the threshold of assuming a prominence in nutritional studies at least equal to that attained by vitamins and proteins to date," and that a study by Donaldson of the mineral needs of fish, the results of which were translatable to humans, "could be interesting to [Donaldson] and profitable to General Mills."[63] Donaldson's interest in hatcheries and nutrient cycling coincided with General Mills' commercial interests, and so, in addition to selective breeding and nutrient supplementation, Donaldson began to pursue commercial "ocean ranching" as a method of salmon restoration.

At a 1968 meeting, Domesticating the Sea, hosted by the Hawaiian Sugar Technologists, Donaldson argued that ocean ranching promised to aid species in the struggle for existence by "healing" and "protecting" natural fish runs, while at the same time feeding an expanding human population, so that "in a few generations, the result would be a balanced, healthy mixture of farm, pasture, ranch, park, and wilderness, capable of feeding and providing a living for generations of man yet to come."[64] Donaldson thus extended his vision for fish stocking beyond state government to an international and industrial scale, where the goal was the recovery of "wild fish" through the establishment of separate ocean ranches where people would cultivate protein. He imagined that ocean ranching—what we are more likely to call fish farming or aquaculture today—would protect certain environments even as it heavily utilized others. Donaldson reasoned that once wild fish populations were restored, nutrient cycles that had been broken by industrialization would be revitalized. An AEC promotional pamphlet of 1968 featured Donaldson's irradiated salmon, concluding that, across the nation, university faculty, federal and state conservationists, and fish and wildlife personnel were "beginning to take advantage of the nuclear age."[65] Donaldson's vision of ecological restoration was thus compatible, if not synonymous, with American capitalism and empire.

Academic ecologists and employees of environmental agencies were linked by the AEC, and Donaldson mobilized such connections very successfully in his effort to promote aquaculture in the United States and abroad. For example, in 1965 Donaldson proposed to General Mills that they establish a nonprofit organization (modeled after their Wheaties Sports Association) to pursue fisheries restoration projects. General Mills was interested in the publicity, and Rawlings, an avid fisherman, was keen to establish a Great Lakes stocking program. At Donaldson and Rawling's prompting, and with General Mills' backing, the Minnesota Department of Natural Resources began stocking Donaldson trout.[66] Soon Donaldson trout (also known as "hurry-up trout" and "super trout") were reared in Alaska, California, Colorado, Idaho, Michigan, Minnesota, Montana, Oregon, Canada, China, Germany, Ireland, Japan, Korea, New Zealand, Norway, and Sweden.[67] Today rainbow trout, native to a narrow band along the northern Pacific coasts, are now stocked almost everywhere in the world.[68] If you have eaten a trout, it very well may have been a Donaldson trout.

Radiation and Restoration

From the vantage of the twenty-first century, the connections between U.S. atomic colonialism, Bacon Bits, aquaculture, and ecological restoration are no longer obvious. But they are key to understanding just how recent the stark conceptual divide is between wild and agricultural species. Leopold and Donaldson were only two of the many twentieth-century biologists who looked to agriculture as a model for fish and wildlife management, and who sought to breed wild species to make them more productive, and thus more abundant. Donaldson believed that the development of large-scale commercial aquaculture would save wild fish from overexploitation. And he believed that by managing evolution, he would create salmon that could thrive in a rapidly industrializing world.

In hindsight, this has not been the case. Around the world, wild fisheries are in crisis. At least two-thirds of the world's fisheries are overfished.[69] Meanwhile, annual global food fish consumption has increased since 1961 at a rate almost twice that of annual human population growth, and today, as in Donaldson's times, proponents of commercial aquaculture cast it as both a solution to world hunger and a solution to a worldwide fisheries crisis. The Food and Agriculture Organization estimates that, as of 2018, aquaculture accounted for 52 percent of fish consumed by people, an all-time high.[70] Aquaculture is known to cause pesticide and nutrient pollution—excrement and uneaten food add nitrogen and phosphorus to the surrounding waters, leading to oxygen-starved areas or dead zones.[71] It is also an increasing source of greenhouse gas emissions.[72]

Meanwhile, salmonid stocking remains a pervasive environmental management practice across the world. In New York state, for example, each year the Department of Environmental Conservation stocks around 2.3 million brook, brown, and rainbow trout in almost three hundred lakes and ponds and roughly 3,100 miles of streams.[73] These stocked fish compete with wild fish, and they hybridize with them, so that it is perhaps anachronistic to call any fishery "wild." They also alter food webs and nutrient cycling. It is worth questioning the way in which fish species continue to be managed as agricultural ones. Contemporary restoration ecologists decry species introductions and would generally balk if the New York Department of Environmental Conservation were stocking forests with millions of Eurasian red squirrels per year, say. And yet relatively few fisheries scientists study the impact of stocked fish on native fish, or advocate for the end of stocking.[74] Because fish do not fit neatly into the history of ecological restoration, it is tempting to write them out. But they serve as a reminder that conceptual categories matter—that plants and animals are brought into being, encouraged, ignored, or killed based on whether they are considered to be wild or domesticated, eaten or not eaten, native or nonnative, charismatic or overlooked, predator or prey, fowl or fish.

Donaldson's fieldwork also exemplifies how U.S. nuclear colonialism put scientists in the position of searching for anthropogenic ecological changes, and provided biologists with new tools, like radiotracers, that would spur several decades of growth in the theory and practice of ecological restoration. Despite being first contracted by the MED to check whether radioactive reactor effluent harmed fish species, Donaldson himself never marked nuclear technologies—even weapons—as definitive ecological threats. In this sense he was like many scientists of the 1950s and 1960s, including ecologists, who saw in nuclear technology a promising and powerful tool for basic and applied research. Indeed, through the 1950s, many scientists viewed atomic technologies as powerful and even redemptive tools. A *Time* magazine feature on the UN International Conference on Peaceful Uses of Atomic Energy explained that "in the atom lies not just menace but hope, a new start, a new future. [. . .] The atom can ultimately move mountain ranges, drain seas, irrigate entire deserts, transmute poverty into plenty, misery into mercy."[75] Many ecologists held that atomic technologies would enhance scientific understanding of nature's structure and, therefore, human ability to manage and restore the environment.

Prior to the 1950s, the accepted method of "improving" or "rehabilitating" streams and lakes was to poison "trashfish"—species that might compete with or eat sport fish—and then to stock with hatchery-reared fish.[76] Through the work of Donaldson and his contemporaries, fisheries managers began to seek to promote productivity by manipulating nutrient levels as well as fish genomes. This approach prefigured two important trends of later ecological restoration. The first is off-site mitigation: Donaldson argued that by enabling salmon to live in new sites, scientists could mitigate environmental destruction elsewhere. Off-site mitigation would become a widespread practice of ecological restoration in the 1990s and remains so today. Second, the effort to design a functional system to support farmed salmon prefigured ecosystem theory.

The AEC remained the main funder of ecology until the early 1970s, when it was eclipsed by the newly founded National Science Foundation. With AEC funding and access both to weapons test sites and radioisotopes, ecologists solidified the ecosystem concept with atomic fieldwork, setting the stage for ecological restoration's eventual transition from a single-species approach to an ecosystem approach. And, as the Cold War intensified in the 1960s, the science of ecological recovery would take on a new significance, as ecologists planned for the recovery of wild species after Doomsday.

5

The Specter of Irreversible Change

"A nuclear war would not end when blast and fire subside," cautioned the activist newsletter *Nuclear Information* in 1963. The blasts and fires would initiate a cascade of ecological changes. What would this postwar world look like? "How would the living things that cannot be sheltered—native plants, crops, wild and domestic animals—fare after a nuclear war?" The answer to this pressing question was far from clear, according to Robert Wurtz, an author for *Nuclear Information* and a scientist with the St. Louis Citizen's Committee for Nuclear Information. World War III would be a type of war that had never been waged, and data on the ecological effects of nuclear weapons, such as those data collected by the Applied Fisheries Laboratory, were rare—"particularly for wild species." Anticipating the effects of nuclear weapons on wild plants and animals would require a new type of science, Wurtz concluded, one focused on the ecology of post-disaster recovery.[1]

By the time Wurtz wrote "War and the Living Environment," public concern over global-scale nuclear disaster had peaked. The 215 aboveground and underwater nuclear weapons detonated by the United States between 1945 and 1962 had utterly transformed the physical and biological environments of the Marshall Islands and the American Southwest. At the same time, nuclear weapons development utterly transformed the professional networks, experimental practices, and theories of the environmental sciences. When fisheries biologist Lauren Donaldson was hired in 1943 by the Manhattan Engineer District to study whether radioactive effluent from Hanford Works harmed Columbia River fisheries, most scientists considered nuclear contamination to be a localized, contained threat. This view shifted as scientists demonstrated that nuclear fallout traveled through the upper stratosphere, through ocean currents, and through food chains. By the time of the Castle Bravo detonation in 1954 (the largest U.S. nuclear blast to this day, with one thousand times the power of the Hiroshima bomb), scientists and the public had begun to conceptualize radioactive fallout as a regional threat to human health—and perhaps even a global one.[2]

Accompanying this concern over the health risks posed by nuclear fallout was the even blunter fear of nuclear annihilation. In 1950, the United States had 299 nuclear weapons in its stockpile. By 1960, it had 18,638. And by 1965, it had 31,139.[3] As the United States and the Soviet Union massively increased both the power and the range of their nuclear weaponry, it became possible to conceive of a catastrophic, global-scale war. In the early 1960s, the U.S. Atomic Energy Commission (AEC) funded studies that attempted to predict the economic and ecological consequences of such a war.[4] Along with military planners, sociologists, and even science fiction writers, ecologists were tasked by the U.S. government with envisioning a world shaped by the immense destructive power of nuclear weapons. In so doing, they did not picture the outcome of World War III as the total annihilation of life on earth; there would have been no point to such an exercise. Instead, ecologists pictured biotic and economic recovery *after* World War III and considered how the government could hasten that recovery—how they could pursue ecological restoration.

In planning for apocalypse, ecologists and military strategists revisited studies of past environmental disasters, including the American Dust Bowl. For their Doomsday imaginings, they drew substantially on ecological succession theory, expanding the category of "environmental disturbance" beyond windstorms, fires, and floods to include nuclear bombs—and, ultimately, other human actions like deforestation and pollution. Crucially, ecologists also drew on an increasingly popular concept from physiology: homeostasis, or the process by which organisms maintain stable conditions, such as how warm-blooded animals maintain a constant body temperature. Ecosystem homeostasis—the mechanisms through which ecosystems repaired damage and restored stability—became a central topic of inquiry for Doomsday ecologists. The idea of homeostasis presumes that an entity—whether a body, a missile, or an ecosystem—experiences perpetual "disturbances" to which it organically reacts so that, paradoxically, it does not seem to change. Atomic ecologists of the 1960s would thus embed the idea of perpetual threat into the very idea of an ecosystem.

The rise of the ecosystem science represented a major transformation in the theory and practice of ecological restoration. Succession theory dominated ecological thought prior to the 1960s, and under succession theory, human activities were not widely considered to be drivers of permanent ecological damage; ecologists believed that human-wrought changes, with the important exception of species extinctions, were reversible. Emblematic of this view is the now-classic 1864 book, *Man and Nature*, by American scholar-diplomat George Perkins Marsh. Marsh wrote that "natural arrangements, once disturbed by man," would be "restored" once "he retires from the field, and leaves free scope to spontaneous recuperative energies."[5] A similar view prevailed through the 1950s, bolstered by studies of farmland abandoned in New England and, during the Dust Bowl, in the Midwest, in which ecologists observed rapid reforestation.[6] Indeed, participants in the influential 1955 Man's Role in Changing the Face of the Earth symposium at Princeton University concluded that nature would recover from the damage caused by intensive cropping, grazing, and lumbering if people simply desisted from the damaging action. Ecological communities, one participant argued, were self-healing: they had the "re-creative power" to "reconstitute themselves when the cause of disturbance disappears."[7]

However, the terrifying prospect of Doomsday, and the experiments that ecologists conducted in response, challenged the view that nature could heal itself. Seeking to simulate nuclear war, ecologists began to destroy ecosystems intentionally. They irradiated forests and poisoned entire islands, trying to measure how intentionally stressed biotic communities responded. Through these studies, ecologists came to define the "ecosystem" as an ecological unit that was vulnerable to *permanent* human-caused change. Further, they began to distinguish "ecosystem function" from "ecosystem structure," and hypothesized that more diverse ecosystems were more resilient to disturbance. These now-familiar theories were assembled on a bedrock of destruction. Ecologists laid waste to forests and dispersed poisons, all to challenge the view that nature could heal itself. Restoration then, in unexpected ways, emerged from the ruin of the Cold War arms race and birthed the specter of irreversible human-induced change.

Radioecology and Ecosystem Theory

Ecosystem theory emerged at the intersection of ecologists' interest in food webs and the AEC's interest in managing its growing amount of nuclear waste. By 1957, highly radioactive liquid from the nation's reactors amounted to sixty-two million gallons, most stored in underground tanks at Hanford Works. The AEC diluted low-level wastes and released them into streams or placed them in earthen pits to seep into the soil. They then charged ecologists with studying whether these "dilute and disperse" methods affected local flora and fauna.[8] When, for instance, the AEC drained a settling basin at Oak Ridge National Laboratory in Tennessee in 1955, exposing bare soil containing strontium-90, cesium-137, and other radioisotopes, it tasked ecologist Stanley Auerbach with determining whether plants and animals colonizing the former lakebed became radioactive.[9]

Increasing AEC patronage transformed ecology from an esoteric discipline to an internationally recognized branch of biology. The AEC Division of Biology and Medicine founded a new national ecology program in Washington, DC, in 1955 and appointed John N. Wolfe, a plant ecologist at Ohio State University, as its director. Wolfe quickly developed radioecological programs at approximately fifty universities across the country, as well as at AEC facilities like the Oak Ridge National Laboratory and the Savannah River Plant in Aiken, South Carolina.[10] As radioecologists joined biology departments across the country, more and more people came to know of the discipline.

The ecosystem paradigm, which continues to dominate environmental management today, emerged from these new radioecological programs and their atomic fieldwork. Understanding its history is key to understanding subsequent developments in ecological restoration. In telling the history of ecosystem theory, scholars have typically emphasized the influence of thermodynamics and other concepts from physics on its emergence.[11] But the field of physiology shaped ecosystem science as much, if not more, than physics did. Many scientists found their way to ecology through the field of organismal physiology, including brothers Eugene Odum and Howard "Tom" Odum, authors of the influential textbook *Fundamentals of Ecology*, which played a central role in popularizing the ecosystem idea.

While banding birds for the Ohio Fish and Wildlife Commission in Cleveland, Ohio, Eugene Odum befriended ornithologist S. Charles Kendeigh, a former student of Victor Shelford (and, later, a cofounder of The Nature Conservancy), and followed him to Urbana-Champaign to enter the biology PhD program there. In his first doctoral experiments, Eugene used piezoelectric crystals to build a "cardio-vibrometer" that recorded the heart rates of birds while they sat in their nests. His goal was to study the heart rates of birds "at rest" in the field rather than the heart rates of "disturbed" birds in a laboratory. Summarizing his results, Eugene emphasized that ecologists should be most concerned with the function of an organism *in the field* and *as a whole*—unlike physiologists concerned with the functions of discrete organs in laboratory settings—and he treated heart rate as a "physiology-of-the-whole indicator."[12] Species, he maintained, could only be understood as functional components of their surroundings, and this principle would push to him to study species in the field and, eventually, ecosystems.

Eugene found an important collaborator in his brother, Tom Odum, eleven years his junior. Tom entered Yale University's graduate program in zoology in 1947, intending to study bird physiology like his brother. But during his first semester, he was charmed by limnologist G. Evelyn Hutchinson and his "great diversity of abilities and knowledge."[13] For his PhD research, Tom explored the circulation of strontium in oceans, more than a decade before the fallout of radioactive strontium-90 catalyzed a national controversy over atmospheric nuclear weapons testing. As Tom worked on his dissertation, he wrote to Eugene frequently, sharing his lecture notes, news from Hutchinson's laboratory, and gossip about other young ecologists.[14] These updates informally linked Eugene to one of the leading zoology programs in the country and would shape his future work.

Tom, and by extension Eugene, were deeply influenced by G. Evelyn Hutchinson's work on homeostasis. They were not alone; many ecologists became interested in the idea of organic self-regulation during the 1940s, when the idea dominated physiological research.[15] The term homeostasis was popularized by Harvard physiologist Walter Cannon in his influential 1932 book, *The Wisdom of the Body,* which described how the human body maintains a stable temperature and other vital conditions.[16] The idea of homeostasis was also important to MIT mathematician Norbert Wiener's cybernetic theory. Cyberneticists analogized the organismal brain to a self-guided missile that gathered information and used that information to correct its own path en route. Homeostasis as understood in the fields of cybernetics and physiology was a process of self-preservation.[17] Through Doomsday studies, ecologists in the 1960s would come to think of homeostasis as a set of mechanisms for repairing damage and restoring stability across a system of interrelated individuals, species, and materials: an ecosystem.

In 1949, while Tom was studying for his preliminary exams, Eugene, now a faculty member at the University of Georgia, asked him if he'd consider coauthoring a general ecology textbook. Eugene believed the discipline's popular textbooks, Charles Elton's 1927 *Animal Ecology,* and John E. Weaver and Frederick Clements's 1929 *Plant Ecology,* did not capture the discipline's emerging emphasis on how plants, animals, and the physical environment *interacted with* one another.[18] In the end, Tom declined official coauthorship, but he wrote chapters on the topics that the Hutchinson laboratory had best trained him in: population biology, biogeochemistry, and cybernetic theory.[19]

The first edition of Odum's *Fundamentals of Ecology* was published in 1953 and sold a few thousand copies—a respectable number, considering the size of the discipline at the time, but hardly a triumph. The textbook's influence grew over time, however—and, crucially, the Odums kept it up to date. In *Fundamentals,* Eugene split ecology into four divisions of increasingly complex interrelations: species ecology, population ecology, community ecology, and ecosystem ecology. He defined an ecosystem as "any entity or natural unit that includes living and nonliving parts interacting to produce a stable system in which the exchange of materials between the living and nonliving parts follows circular paths." Although he was not the first to define ecosystem—the term was coined by British ecologist Arthur Tansley in 1935—*Fundamentals of Ecology* popularized it.[20] Notably, Eugene defined an ecosystem as an inherently stable entity, and one that maintained *its own* stability.

For the second edition of *Fundamentals,* published in 1959, Eugene added a chapter on the new subdiscipline of radioecology, drawing on his own experiences working for the AEC. These experiences began when the AEC announced plans to establish a plutonium production plant along the Savannah River, outside of Augusta, Georgia, and invited University of Georgia faculty to compete for grants to conduct preinstallation biological surveys of the site.[21] Two years later, in 1953, the AEC awarded Eugene and Tom a grant to visit the Pacific Proving Grounds for six weeks to conduct "community metabolism" studies.[22] And so, on the evening of June 24, 1954, Eugene and Tom deplaned onto the sands of Enewetak Atoll with 155 pounds of scientific equipment and a desire to advance ecological theory. For a month they snorkeled in the coral reefs of the atoll, taking samples and analyzing them in a new biological laboratory that also boasted a library, organized sports, and nightly movies. At night, in the Back N' Atom Bar, the Odums mingled with AEC officers and other scientists from across the United States.[23]

Members of the Applied Fisheries Laboratory working at Enewetak at the same time were interested in the distribution of radioactivity in the atoll environment: How long did it persist, and where? The Odums, meanwhile, wanted to understand how nutrients circulated among interconnected organisms. Taking advantage of visiting Enewetak mere months after the Castle Bravo blast, they produced radioautographs of coral heads that visualized the symbiotic relationship between coral polyps and their resident algae. Using methods from Hutchinson's laboratory, they measured oxygen and nutrients in the water column at different points along the reef and used these measurements to estimate the respiration and nutrient cycling of the ecological community as a whole. After the trip, the Odums quickly wrote up their results and published them.

In their now-classic article, the Odums argued that Enewetak's reef community, with its close symbiotic relationships between coral and algae, was an excellent example of a homeostatic system. Echoing the language of AEC officials, they described the reef as an "isolated system" in a "rather constant environment" that had been "little affected by nuclear explosions."[24] The atolls, however, were neither isolated nor pristine. Rather, they were globally connected, geopolitically central, and radically transformed by military and scientific activity. The beaches were littered with barges, steel cable, scrap metal, beer and sake bottles, and abandoned furniture.[25] For Operation Ivy (1952) alone, the United States had transported seventy-five million gallons of fresh water, one million meals, 89,968 square feet of tent, and three million board feet of lumber to the Marshall Islands.[26]

Nor were the atolls uninhabited. As the rest of the world worried about the possibility of atomic violence, Marshall Islanders lived it. Between 1946 and 1962, the AEC conducted 105 atmospheric and underwater nuclear weapons tests at the Pacific Proving Grounds, releasing the equivalent power of more than seven thousand Hiroshima bombs.[27] Marshall Islanders suffered forced relocations, destruction of ancestral lands, and radiation sickness.[28] The materialization of ecosystems, then, entailed injustice and horror: the dispossession of Marshall Islanders; the radiation sickness of thousands of soldiers and civilians; and the creation of the massive U.S. nuclear complex, which by the end of the Cold War occupied more than 8,500 km^2 and the radioactive legacy of which will persist for at least ten thousand years.[29]

Like many ecologists, the Odums were not troubled by atomic colonialism. Rather, they celebrated the spaces and tools made available to them through nuclear weapons development and detonation. Eugene's Enewetak contract led to an invitation to the International Conference on the Peaceful Uses of Atomic Energy in Geneva, Switzerland, in August 1955. The Atomic Energy Act of 1954 had set declassification in motion, and a number of U.S. ecologists presented in Geneva, including members of the Applied Fisheries Laboratory, who, for the first time, presented their concept of bioaccumulation to a public audience.[30] Odum used the momentum from the Geneva conference to organize a radioecology symposium at the annual American Institute of Biological Sciences meeting in 1956. That year he also received a National Science Foundation fellowship to write the new chapter on radioecology for the second edition of *Fundamentals*.

The fellowship funded trips to visit AEC-funded laboratories at the University of California Los Angeles and the University of Washington's Applied Fisheries Laboratory. During his four months at UCLA, Eugene regularly joined biologists for three-hour train trips to Nevada to observe nuclear detonations. He wrote to his parents that the Nevadan desert struck him as "a good system for our type studies because it is, like the old fields and marshes, relatively uniform and simplified biologically."[31] Just as laboratory scientists sought to simplify and standardize model organisms, ecologists would seek to identify model ecosystems.[32] Deserts, islands, and abandoned agricultural fields particularly appealed to Eugene because they were thought to contain fewer species than mainland forests. Fewer species, he reasoned, would mean fewer variables to control for.

Eugene would carry these experiences with him as he sat down to write a new chapter on radioecology for *Fundamentals*. In it he argued that just as physiologists were using radiotracers to study human metabolism, ecologists were using radioisotopes to better understand "community metabolism," or the circulation of nutrients and energy among organisms.[33] He described studies from the Pacific Proving Grounds and the Savannah River Ecology Laboratory that explored how species mediated the distribution of radioactive substances in the environment.[34] Like Lauren Donaldson and other ecologists, Eugene viewed radioisotopes as tools rather than contaminants, as tracers for naturally occurring elements. In the new edition of the preface, he added, "Some of the things which we fear most in the future, radioactivity, for example, if intelligently studied, help solve the very problems they create."[35]

The advent of radioecology dramatically changed how ecologists demarcated their study sites. Instead of delimiting sites by visual observation ("The oak stand ends here") or by identifying "check-areas" or property lines, ecologists increasingly relied on the spread of radioisotopes to mark out a study's boundaries. Radioecology thus changed how ecological assemblages were conceived; instead of communities of species that shared an abiotic environment, ecologists now began to see ecosystems: groups of mutually dependent species, each performing a function necessary to the continued survival of them all. Ecologists analogized the boundaries of an ecosystem to the boundaries of an organism's body. And yet these boundaries, seemingly external to the experimentalist, were determined as much by experimental design and the human production of radioisotopes as by intrinsic properties of species and their interactions. Ecosystems may now seem natural and universal, but their emergence depended on the tools of a particular time and place.

In the first decade of radioecological studies, ecologists largely harnessed the fieldwork "opportunities" afforded by nuclear waste to study species interconnections, whether these were waste products from nuclear production (in the case of Hanford Works, the Savannah River Plant, and Oak Ridge National Laboratory) or fallout from detonations (in the case of the Pacific and Nevada Proving Grounds). This changed with the passage of the 1954 Atomic Energy Act, when the AEC began mass-producing radioisotopes and distributing them to American researchers. By 1955, the Oak Ridge National Laboratory had sent nearly 64,000 shipments of radioisotopes to scientists and physicians.[36] Increasingly, ecologists applied laboratory-manufactured radioisotopes directly to field sites of their choice. Ecologists were thus no longer constrained to AEC sites; now, they could order vials of radioisotopes to apply to field sites themselves, wherever they could obtain permission.

Laboratory-produced radioisotopes enabled ecologists who were already studying food webs through direct observation (recording which insects visited a particular plant, for instance) to study them without spending as much time in the field. Stanley Auerbach, for instance, injected measured doses of cesium-137 into tulip poplar trees in Tennessee. He and his collaborators later sampled insects and animals in the general vicinity to see if they contained cesium-137. If they did, they could assume that those species had fed on the radioactive poplar trees. In similar experiments at the Savannah River Ecology Laboratory, ecologists sprayed solutions of phosphorus-32 or iodine-131 onto meadow plants. They later collected arthropods, snails, crickets, ground beetles, and other species near these "hot quadrats." Any that were radioactive had eaten from the radio-tagged plants.[37] Such experiments changed the role of the ecologist in the field from observer to collector. In this way ecologists moved away from studying the effects of external radiation on individual organisms and toward using radioisotopes to study the circulation of elements among organisms and within ecosystems.

The period from 1954 to 1963 was a time of massive growth for ecological science. Radioisotopes, whether released by atomic detonations in unpredictable quantities or carefully deployed in designated experimental field sites, were the means by which ecologists visualized interconnections among species. During this time, most ecologists remained optimistic that the constructive uses of nuclear technologies, including ecosystem studies, outweighed their destructive potential. Frank Golley, the first director of the Savannah River Ecology Laboratory, would later reflect,

5.1 Eugene Odum working on radioecological studies at Oak Ridge Institute of Nuclear Studies in 1963. Courtesy of Hargrett Rare Book and Manuscript Library / University of Georgia Libraries

> Our connection to the AEC always surprised foreign ecologists when they understood the basis of the funding of our research. These scientists immediately saw the connection between the AEC and the military. But at the time, we were unconcerned about the production of nuclear bombs; our focus was on ecological research. I can't recall a serious discussion about that issue in those days. Later, as opposition to the Vietnam War became more general in society, the disadvantage of being associated with a military production facility became more apparent and began to affect our recruitment of faculty and students. But at the time, the AEC support allowed us to be on the cutting edge of ecological research.[38]

If nuclear technology made ecosystems manifest and the interconnections of their components visible, Odum's *Fundamentals of Ecology* made ecosystems popular. In 1969, annual sales of the textbook were approximately 6,200 copies. In 1971, 42,000 copies were sold, and *Fundamentals* was translated into twelve languages, becoming arguably the single most influential text in the history of ecology.[39] Robert Wurtz drew heavily on *Fundamentals of Ecology* when he asked, "How would the living things that cannot be sheltered—native plants, crops, wild and domestic animals—fare after a nuclear war?"

Attacking Ecosystems: Destruction as a Method of Study

While radioecology flourished, the United States and the Soviet Union raced to expand their nuclear stockpiles. The rapidly increasing power and range of nuclear weapons sparked fears of a catastrophic, global-scale war. Studies by the RAND Corporation, a think tank with roots in the War Department, estimated that a first Soviet attack would be aimed at fifty U.S. cities and result in ninety million casualties. The government responded by promoting the construction of fallout shelters in state buildings, offices, and schools. But what about wild species? Wurtz warned in 1963 that "shelter programs are largely irrelevant to the bioenvironmental consequences of nuclear war," that humans depend on a "complex of animal and plant communities" that would be "disrupted," if not obliterated, by nuclear war.

In 1961, the RAND Corporation urged the Pentagon to investigate the ecological dimensions of post–World War III recovery. Their report, "Ecological Problems and Post-War Recuperation," argued that following a nuclear attack, the two main ecological problems would be fire and nuclear fallout. The direct result of widespread fires would be the destruction of crops, timber, livestock, and wildlife. Their indirect result would be the destruction of ground cover, which might cause erosion and turn large areas into "dust bowls." Fortunately, the RAND authors argued, ecologists had generated "a wealth of information" that would be "pertinent to the problems of post-war recovery of devastated biotic environments," including studies of forests after fires, range management, and "dust-bowl recovery."[40] Thus RAND positioned ecologists as disaster experts, decades after ecologists themselves had attempted to influence federal management of the Dust Bowl disaster.

The RAND report maintained that much less was known about the ecological effects of radiation. At the time, most scientists, including ecologists, downplayed the differences between atomic weapons and nonatomic—so-called conventional—weapons. Nuclear detonations were notable for their heat and blast pressure, these scientists argued, not as sources of pollution.[41] In keeping with this logic, RAND reminded readers that "natural radiation" was "an integral part of the equilibrium of life." The effects of radiation in a post-attack environment, the report maintained, could be separated into "the lethal concentration of radioactive substances by plants and animals" and changes in the composition of ecological communities due to "differential radiosensitivity" among species. The first effect referred to the bioaccumulation studies that ecologists had been conducting for almost two decades, beginning with the Applied Fisheries Laboratory's work at the Pacific Proving Grounds. The second effect, radiosensitivity, was a "new consideration," and RAND recommended that the experimental effort be "enormously increased."

In turn, the AEC began to fund studies in which ecologists would purposefully damage "ecosystems"—rather than warships or model towns—to study whether they recovered, and if so, how. Ecologists began *simulating* nuclear attack, and so began a series of experiments that concretized the idea that ecosystems were material things that humans could damage or even destroy. The first experiment to simulate nuclear attack without an actual detonation occurred on the Brookhaven National Laboratory grounds on Long Island. Between November 1961 and April 1962, ecologist George Woodwell and his colleagues exposed a former agricultural field and an oak-pine forest to continuous gamma radiation from cesium-137 and cobalt-60 point-sources.[42] The AEC Office of Civil Defense supported this project, as its primary objective was to evaluate radioactive contamination at a scale that could result from nuclear war. Woodwell and his coauthor justified the experiment by arguing that it was important to study the effects of nuclear war in eastern deciduous forests near urban centers, because bombsites had been "restricted generally to deserts and tropical atolls with limited floras."[43] Through this fieldwork, Woodwell set out to compare the "radiosensitivities" of wild plant species. He found that *Senecio* species (a member of the daisy family) survived high levels of radiation, whereas pine species were the most "sensitive" to radiation. This was a new way of categorizing species—not by taxonomy or by what-ate-what, but by their *ability to withstand disturbance.*

Along with comparing the radiosensitivities of particular species, Woodwell also set out to characterize the vulnerability of entire ecosystems to radiation. He concluded that the oak-pine forest was much more sensitive to irradiation than the old-field ecosystem, the difference "spanning a factor of 5–10 in exposure necessary to produce equivalent change in structure."[44] Woodwell thus treated the whole field site as an ecological unit and categorized that unit's ability to maintain homeostasis under stress. Further, he and his collaborators observed that selective elimination of the most radiosensitive organisms reduced the species diversity in the ecosystem. The fact that disturbance by radiation resulted in "simplification of the ecosystem," they argued, was consistent with the effects of other types of "disturbances," such as fires and floods.[45]

Woodwell and his collaborators spent years documenting the recovery of experimentally irradiated sites. To plan for Doomsday, Woodwell wrote, scientists needed to examine the "fundamental unit" of ecology: "the eco-

5.2 An aerial photograph of the Brookhaven National Laboratory Irradiated Forest Experiment after six months of exposure to gamma radiation ranging in intensity from several thousand R / day near the center to about 60 R / day at the perimeter. George M. Woodwell, "Radiation and the Patterns of Nature," Brookhaven Lecture Series, March 24, 1965, Figure 9.

system, which has been defined most thoroughly in contemporary context by Eugene Odum and Howard [Tom] Odum."[46] In their studies at AEC proving grounds, ecologists had not been able to separate the effects of radiation from those of blast and fire. But in experimentally irradiated field sites, they would endeavor to document the specific effect of radiation both on individual species and on groups of species.[47]

Public concern about the human health effects of fallout reached a crescendo in the early 1960s, leading the United States, the Soviet Union, and the United Kingdom to sign the Partial Test Ban Treaty in 1963.[48] With the treaty, ecologists lost the ability to study new aboveground detonation sites. This constraint only increased the value of simulated attacks, which were now the only way to study comparative radiosensitivity among species. In January 1965, Tom Odum and a team of AEC-funded ecologists installed a 10,000-curie cesium-137 gamma source at the Luquillo Experimental

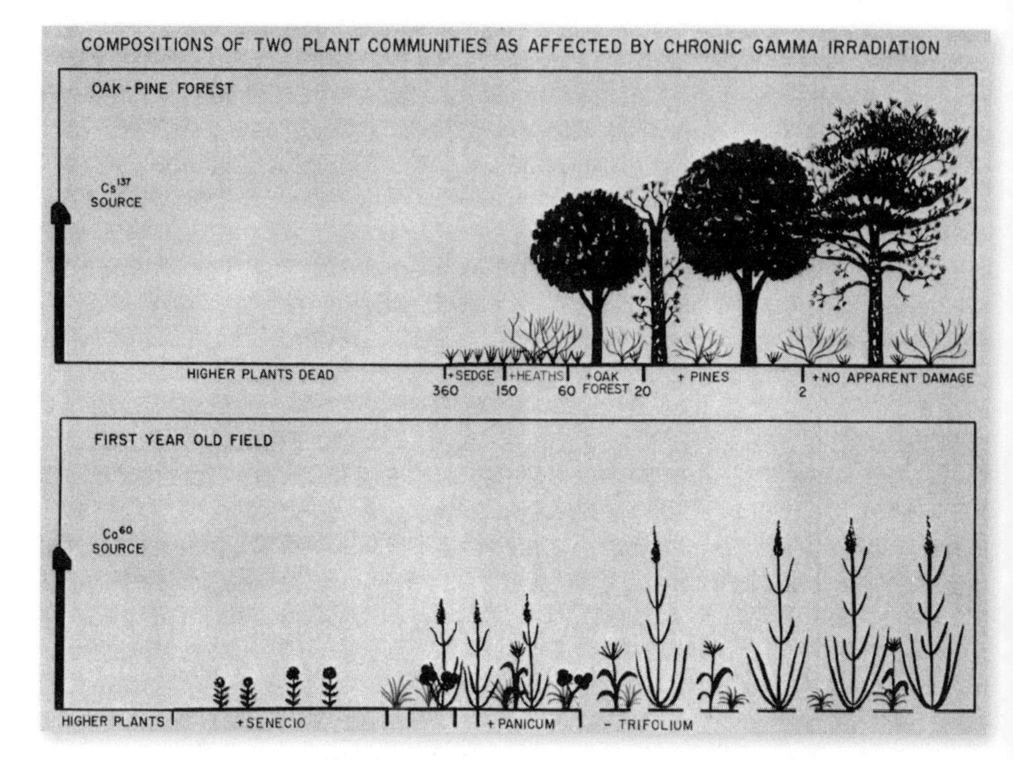

5.3 A figure depicting the species that survived six months of gamma irradiation in an oak-pine forest (top) and a former agricultural field (bottom) at Brookhaven National Laboratory. The species had different "radiosensitivities": the pines were more "sensitive" to radiation than the sedges and *Senecio,* which survived at greater levels of radiation exposure (closer to the radiation source, depicted at left). George M. Woodwell, "Effects of Ionizing Radiation on Ecosystems," Brookhaven National Laboratory (January 1963).

Forest in Puerto Rico.[49] The AEC supported the Luquillo irradiation study with more than $1 million; in addition to understanding ecological recovery after nuclear war, the AEC sought to anticipate the impacts of a proposed nuclear-excavated canal, the "Pan-Atomic Canal," which was supposed to replace the aging Panama Canal. The Luquillo Experimental Forest was actually a failed Civilian Conservation Corps project. The CCC had planted some four million seedlings on abandoned farmland, but most of the seedlings died; it was exactly the sort of result that galvanized ecologists against New Deal projects like the shelterbelts. Notably, Luquillo was also a site at which the Department of Defense tested Agent Orange and other "tactical herbicides" for use in the Vietnam War. Once a site was designated for ecological destruction, it was subjected to multiple layers of destruction.[50]

Tom Odum and his collaborators, eventually numbering in the hundreds, irradiated the Luquillo tropical forest site for three months and studied its "course of recovery" over the next six years. They compared the irradiated plot to two "control" plots: one denuded of all vegetation, and one that received no treatment. In Odum's rendering, the forest ecosystem was an agent: faced with the stress of irradiation, the forest "actively resisted loss of its complexity with such mechanisms as seedling release." In other words, he imagined the forest to be healing itself by sprouting new trees. He imagined early successional species as "wound healers in small damaged spots." The recovery of the "radiation-decimated area began from the forest floor with explosive new growth of seeds," Odum wrote. Two years later, the irradiated zone looked "like the scrubby growth in the Appalachian Mountains after the dominant chestnut was struck by disease." By 1969, bare rocks were re-covered with moss and *Cecropia* trees were 30 feet high. In Odum's assessment, the irradiated forest's "healing system" and "repair mechanisms" were similar to those of a human body.[51]

By 1970, following the RAND Corporation's recommendation that "studies of comparative radiosensitivity be enormously increased," ecologists had placed radiation sources in deciduous forests at Brookhaven, in a

5.4 Stanley Auerbach operates a hoist handling a cask of cesium-137-tagged sand in the post-nuclear-attack ecology study plots at Oak Ridge National Laboratory, Tennessee, c. 1968. Oak Ridge National Laboratory

tropical rain forest in Puerto Rico, in a desert in Nevada, and in old-fields at the Savannah River site and at Oak Ridge National Laboratory, where ecologists aimed to study the effect of irradiation on "wild, native species."[52] These were "basic" ecological studies with an eye toward post–World War III recovery. Succession theory had dominated ecology when settler colonists justified their actions with the idea of manifest destiny, the belief that a succession from Native American to white ownership was natural and inevitable. Suggestively, the idea of ecosystem homeostasis emerged when the United States perceived itself to be under perpetual threat.

In imagining, writing, and speaking about ecological recovery after World War III, ecologists situated themselves as experts on what military planners and citizens could expect of their surrounding biotic environment after nuclear attack. Drawing on the results of a forest irradiation experiment in Dawsonville, Georgia, for instance, ecologist Robert Platt speculated that after a summertime attack, people leaving their fallout shelters on the East Coast would be "pleasantly surprised to find that the familiar surroundings of field and woodland looked as they did before the explosion."[53] A few days after the attack, Platt continued, pine species would turn brown and die. Then, about a month later, oaks and hickories would lose their leaves, and they would not produce them the following year.[54] Though always conducted someplace in particular, ecologist's World War III simulations were oriented toward developing generalized, transposable strategies for the survival of American citizens. As Woodwell explained in his introduction to the 1965 book, *Ecological Effects of Nuclear War,* such work was meant simultaneously to define the "normal patterns of structure, function, and development characteristic of natural ecosystems" and to anticipate "the complex ecological problems involved in a nuclear holocaust."[55]

As destruction became a standard method of studying ecosystems, ecologists also began clear-cutting, burning, and applying biocides to their field sites. In a particularly dramatic example, in 1966 E. O. Wilson, an entomologist at Harvard, and one of his graduate students, Daniel Simberloff, chose six islands in the Florida Bay on which to kill every living animal. First, they censused the insects on each island. Then they tented entire islands and fumigated them with methyl bromide. After this "defaunation," Simberloff re-censused the insect communities. To make sure the recolonizing insects were arriving by "natural" means, and not on Simberloff himself, he immersed himself in Off! insect repellant between visits. The project was partially funded by the Defense Advanced Research Projects Agency (DARPA) of the Department of Defense.[56] In the write-up of their experimental results, Wilson and Simberloff noted the precedent for their experiment in ecologists' studies of field sites subjected to various "perturbations," including

5.5 "Defaunation" experiment in which E. O. Wilson and Daniel Simberloff fumigated seven islands in the Florida Bay with methyl bromide in 1966–1967. Republished with permission of John Wiley & Sons from Daniel S. Simberloff and Edward O. Wilson, "Experimental Zoogeography of Islands: Defaunation and Monitoring Techniques," *Ecology* 50 (1969): 267–278, Figure 8.

insecticides and fire.[57] Ecology's involvement with Doomsday planning not only made this experiment possible—it made it conceivable in the first place.

Ecologists' embrace of the "ecosystem disturbance" concept also shaped the well-known Hubbard Brook Ecosystem Study. The Hubbard Brook Experimental Forest in the White Mountains of New Hampshire was first established by the U.S. Forest Service in 1955 as a place to study flood control and hydrology. During its first eight years, scientists developed a network of stream-gauging stations and installed weather monitors. Then, in 1963, ecologists F. Herbert Bormann and Gene Likens and geologist Noye Johnson received a grant from the National Science Foundation to establish the Hubbard Brook Ecosystem Study. In November 1965, the research team clear-cut a 15.6-hectare area of forest and then routinely applied herbicides to prevent regrowth. After clear-cutting, the team observed increased erosion and increased streamwater concentrations for all the nutrients they studied. The researchers concluded that human-caused disturbance dramatically changed the flow of water, nutrients, and energy in the Hubbard Brook ecosystem.[58] By 1970, they had reframed the experimental site; instead of flood control, ecologists would study "the homeostatic capacity of the ecosystem to adjust to the cutting of vegetation and herbicide treatment."[59] Destruction as a method of study had crystallized the ecosystem as a homeostatic entity with functionally defined boundaries. Homeostasis was about that entity's reactivity and agency when faced with perpetual threats.

Disturbance beyond Repair

Doomsday simulations and other ecosystem destruction studies led ecologists to the idea that there is a threshold of damage beyond which an ecosystem can no longer repair itself—a point beyond which homeostasis breaks down. For decades, ecologists working on successional theory had imagined that ecological communities would repair themselves once the damaging action—hunting, say, or plowing—ended. Frederic Clements, for instance, had written, in 1935, "From the very nature of climax and succession, development is immediately resumed when the disturbing cause ceases, and in this fact lies the basic principle of all restoration or rehabilitation."[60] But through experiments designed specifically to damage ecosystems, it became conceivable that ecological "development" was not inevitable, and that ecosystems might cease to function entirely if sufficiently harmed by humans. George Woodwell wrote, in 1965, "Most natural ecosystems of temperate zones retain their capacity for regenerating the climax after a wide range of types and degrees of damage. Forests are usually self-regenerating units, even after clear cutting; abandoned fields revert to stable native vegetations through a series of developmental stages. [. . .] Destruction of the ecosystem, however, may reduce the potential of the site for supporting life for long periods, possibly for scores of years."[61]

In the context of planning for Doomsday, ecologists oriented themselves toward identifying the threshold of damage at which ecosystems would lose their ability to restore themselves. Writing for the Department of Defense, economist Robert Ayres reported that ecologists studying ecosystem homeostasis had discovered that ecosystems may be so damaged "that restoration can never be more than partial and incomplete." Once the "ecological balance is seriously disturbed," Ayres continued, "some species, no longer controlled by their natural enemies, may multiply enormously; others, deprived of their normal sources of food or otherwise affected by the total change in the system, may disappear."[62] Different species would thrive in a damaged world than those that had come before.[63] Likewise, a 1975 Office of Technology Assessment study for the U.S. Senate Committee on Foreign Relations suggested that, after a Soviet attack, it would be difficult or impossible to restore an ecosystem "to its pre-attack condition" because of "the possibility of irreversible ecological changes."[64]

Although many ecologists remained optimistic about the capacity for life to return (and human civilization to resume) mere weeks or months after the conclusion of a large-scale nuclear war, they began to imagine that a recovering ecological community might contain different species from that which had come before. It was not unreasonable to imagine, for example, the obliteration of certain species' entire populations. To make sense of "recovery" under such circumstances, ecologists began to distinguish "ecosystem structure"—the types and numbers of species in the ecosystem—from "ecosystem functions"—processes like nutrient cycling and water purification. In a military planning context where human survival was the concrete goal, ecosystem functioning was plainly more important than the restoration of particular species. If ecologists could not restore an exact ecosystem structure because some species had been obliterated, perhaps they could find other species to fill desired functional roles. Today the goal of many restoration projects is to restore ecosystem functioning, but the origin of this goal in Doomsday planning remains obscured.

Along with spurring research on ecosystem functioning, Doomsday ecology also entrenched the diversity-stability hypothesis, the basis of contemporary environmental management: the idea that species diversity confers resiliency, that ecosystems containing more species will vary less in response to disturbances. With the emergence of the diversity-stability hypothesis, biodiversity became a value in its own right, apart from the survival or demise of particular species.[65] Indeed, since the 1970s, biodiversity has become a key measure of an ecosystem's health and even a proxy for wildness.

Ecologist Robert MacArthur is typically credited with first articulating the diversity-stability hypothesis. A student of G. Evelyn Hutchinson, MacArthur published a 1955 paper in which he argued through equations that more diverse food webs were more stable: that more interactions among species in a food web made it more probable that the number of species would remain stable though time.[66] Similarly, in 1958, British ecologist Charles Elton argued at some length that more diverse ecological communities are also more stable, citing the occurrence of population outbreaks in communities simplified by human actions and the relative stability of species-rich tropical forests.[67] But while MacArthur and Elton were among those responsible for first articulating the diversity-stability hypothesis, ecologists' Doomsday simulations provided data that supported it. As ecologists strove to understand and theorize ecosystems' differential vulnerability to radiation, species diversity emerged as a compelling predictive variable; at the 1963 ESA symposium on nuclear war, Robert Platt invoked the diversity-stability hypothesis to argue that there was functional redundancy built into ecosystems, "replacement species" in the case that "certain species are removed by insect injury, extreme drought, ionizing radiation, or other stresses."[68] In justifying his tropical forest irradiation study, Tom Odum explained that the experiments would reveal whether, in tropical ecosystems, high levels of species diversity "provide more mechanisms for survival and recovery."[69] George Woodwell wrote that species diversity makes "the community as a whole resilient in the face of disaster," whether that disaster was "ionizing radiation" or "a gardener's hoe."[70]

Ecologists' World War III simulations established three interrelated principles that would gradually but considerably reshape the practice of ecological restoration: the idea of homeostatic thresholds; the distinction between ecosystem function and ecosystem structure; and the diversity-stability hypothesis. Together they suggested that if a damaged ecosystem could not repair itself because the homeostatic threshold had been crossed—because one or more species had been eliminated entirely, say—then perhaps humans could introduce replacement species, altering the ecosystem's structure to restore its functions. It followed, too, that if badly damaged ecosystems recovered in the direction of simpler communities, then ongoing human intervention might be necessary to maintain ecosystems. The establishment of these principles thus introduced a logic of maintenance to ecological restoration. Earlier in the century, most ecologists believed in the "spontaneous recuperative energies" of nature and a teleological process of succession that resulted in a "climax community." They imagined that ecological communities naturally progressed from simple to complex arrangement. By the 1970s, however, ecologists saw nature as a series of interlocking, self-regulating ecosystems. An ecosystem's ability to recover from disturbance, they newly imagined, stemmed from species diversity and from the complex interrelationships between species.

Both the succession model and the ecosystem model explicitly drew on and influenced metaphors of human society. In a 1936 paper titled "Succession, an Ecological Concept," for instance, sociologist Robert Park described the development of the United States as a progression from "primitive" to "cultured": from Native Americans, to "trappers" and "outlaws," to "frontier farmers," to "the men who eventually became the lawyers, politicians, and newspaper men of the booming settlements."[71] This was white supremacy as succession theory, and proponents used the ecological idea to naturalize and justify settler colonialism. What, then, did the ecosystem metaphor naturalize? Tom Odum wrote in 1973 that "forests, seas, cities, and countries survive that maximize their system's power for useful purposes."[72] The idea of the ecosystem encoded ideas about efficient production, division of labor and specialization, and maintenance of stability. Ecologists were explicit in offering the ecosystem as a new metaphor. "Simply stated," Eugene Odum wrote in 1976, "the problem is how to make an orderly transition from pioneer, rapid-growth civilizations to mature civilizations" in which feedback loops "guard against excesses that could destroy the system."[73] During a time when civil rights activists fought to disrupt the status quo in the pursuit of justice and equality for African Americans and women, the ecosystem, for the Odums and other ecologists, was about maintaining stability. It naturalized a different politics than succession theory, but a conservative politics, nonetheless.

Ecologists writing about ecosystem survival were thinking about the survival of American society in the context of the Cold War, too. This was most obviously true in the case of ecologists' Doomsday experiments. But it was true in studies of homeostasis and biodiversity as well. Tom Odum wrote, "Systems win and dominate that maximize their useful total power from all sources and flexibly distribute this power toward needs affecting survival."[74] Here, tellingly, Odum referred simultaneously to nonhuman ecosystems and to human societies. Whereas the ecologist of succession theory had been an observer, documenting change over time, the ecologist of ecosystem theory was an engineer, familiar with the components of the machine and able to repair them. The world of irreversible ecosystem damage was a world of chronic need for ecological expertise. Woodwell wrote in 1970 that if humans continued on their current trajectory, they would cause a move "away from a world that runs itself" to "one that requires constant tinkering to patch it up."[75]

"The Storm of Modern Change"

The popular environmental movement of the 1960s is often framed as a response to ecologists' discovery of nature's vulnerability.[76] Ecologists would ultimately marshal ecosystem theory to critique nuclear proliferation, the use of pesticides like DDT, biological warfare in Vietnam, and fossil fuel combustion. And yet, counterintuitive though it may seem, ecosystems are not the preexisting casualties of environmental degradation, but rather came into being with the intentional large-scale destruction of environments. During the Cold War, the AEC's interests in nuclear waste containment, ecologists' interest in community metabolism, and the specter of World War III formed a network of people, places, and experimental techniques that was crucial to the development and popularization of ecosystem ecology. Through experiments funded by the AEC and DARPA, ecologists constructed ecosystems as ecological units vulnerable to human-caused and irreversible change. Destruction became a research method.

What continued to haunt the discipline of ecology even after the end of the Cold War was not the blunt possibility of global annihilation, but the subtler specter of irreversible ecological change. In 1973, Robert Jenkins, the vice president for science at the newly revitalized The Nature Conservancy and former student of E. O. Wilson, and a colleague wrote that "man-engendered environmental changes" might be "of such magnitude and so rapid" as to "disrupt the internal cohesion" of an ecosystem. Some environmental modifications might even "become irreversible after passing an unknown threshold point."[77] Ecologists were converging on the view that humans, whether through nuclear weapons or through other large-scale "disturbances" like deforestation and pollution, might create systems in which ecosystems could no longer repair themselves.

In February 1965, President Lyndon B. Johnson declared in a special message to Congress that, in "the storm of modern change," the nation needed to pursue not only "classic conservation of protection and development," but also "a creative conservation of restoration and innovation."[78] The emergence of the idea of permanent ecological damage sharpened the paradoxes inherent in human efforts to repair nature by suggesting that, once intervention was necessary, it might have to be sustained indefinitely. It also made ecosystem functioning the target of some restoration projects, especially where it was difficult or impossible to restore particular assemblages of species. In the world that ecologists envisioned, human management of ecosystems would by necessity be ongoing, and without end.

When, in 1965, ecologists sought funding to participate in the International Biological Programme (IBP), a large collaborative research project modeled on the highly successful International Geophysical Year, they framed ecosystem science as a tool for human survival.[79] Testifying in front of the House Subcommittee for Science, Research and Development, ornithologist Dillon Ripley argued that humans were faced with the urgent problem of "achieving homeostasis" in ecosystems "before irreversible damage occurs."[80] With such rhetoric, ecologists successfully pitched the IBP as a mechanism for national ecological recovery, and in August 1970, the Senate dedicated $40 million to large-scale ecological projects. This was the beginning of "big ecology," and over the next five years, the IBP coordinated and funded ecological fieldwork at sites across the United States.

The IBP defined an ecosystem as the smallest unit to which environmental management could be applied if problems were "to be solved rather than moved."[81] Explicitly identifying ecosystems as the sites of solvable problems, the definition both assumed prior ecosystem damage and made management central to the ecosystem idea. Nuclear detonations—and detonation simulations—had reshaped material environments worldwide, both visibly and invisibly. So, too, did the ideas about ecological restoration that Doomsday planning made manifest. Perpetual ecological intervention would, in the 1970s, become the major work of the Fish and Wildlife Service, an organization that had for many decades been in the business of eradicating predators but that was transformed, by fits and starts, into the seat of U.S. restoration under the Endangered Species Act of 1973.

PART III

Regulation, 1970–2010

6

Extinct Is Forever

"This is the bottle for the Age of Ecology," proclaimed a 1970 ad for Coca-Cola in reusable glass bottles, "the answer to an ecologist's prayer."[1] That April, more than twenty million people participated in the first Earth Day, one of the largest public demonstrations in U.S. history. The previous year's environmental disasters, including a massive oil spill in Santa Barbara, California, and the Cuyahoga River fire in Cleveland, Ohio, had galvanized a national environmental movement. Congress considered a suite of environmental bills on topics ranging from water pollution to species extinction, and ecologist Barry Commoner wrote of the environment as "a huge enormously complex living machine," emphasizing that human survival depended on "the integrity and proper functioning of that machine."[2] Environmental activism and scientific ecology were converging into what journalists, politicians, and the Coca-Cola ad heralded as the "Age of Ecology."[3]

The Age of Ecology was also the Space Age, and both Coca-Cola employees and ecologists analogized the Earth and its living systems to mechanical ones. In addition to advertising glass bottles as appropriate for this "age," Coca-Cola employees collaborated with University of Georgia ecologists to produce and distribute a board game called Make Your Own World to four thousand elementary schools nationwide.[4] The premise of the game was that two spaceships were on a mission to Mars. Once they reached Mars, one spaceship broke down. The students had to figure out how to bring all the astronauts safely back to Earth in the "closed ecological system of a single cramped spaceship." The game was designed to teach students that the Earth's natural resources were limited, just like a spaceship's, and that only democratic (and not Communist) environmental planning could avert disaster. In an interview with CBS after the first Earth Day, Eugene Odum explained that, just like Apollo 13, the "earth spacecraft" could experience a "major breakdown in some one of its natural regenerating systems."[5] With the spaceship metaphor, Odum and other ecologists described environmental destruction—whether water pollution or species extinction—as akin to a mechanical systems failure. Ecological restoration, then, was vital maintenance. Ecologists were engineers.

In this context ecologists, activists, and legislators downplayed their prior commitments to the aesthetics of wildness. The 1970 *Sierra Club Handbook for Environment Activists,* for instance, urged its readers to "remember that we are not concerned here merely with the aesthetics of open space. [. . .] The issue is survival."[6] Survival was the intent of the 1973 Endangered Species Act, arguably the most powerful environmental law in the United States to date. A report to accompany the legislation argued that the human "ability to destroy, or almost destroy, all intelligent life on the planet became apparent only in this generation," and that if a species like the blue whale "were to disappear, it would not be possible to replace it—it would simply be gone. Irretrievably. Forever." If endangered species were to survive, the report continued, the federal government would have to "concern itself with more than a simple 'hands-off' attitude toward the animals and plants themselves."[7] In other words, it would have to pursue restoration in addition to conservation and preservation.

The Endangered Species Act was not simply a legislative triumph. Although its story is most often told as a legal history, tracing the decisions of politicians and the language of the law, the act was also the spur to, as much as the result of, new species management practices. This chapter takes a different approach to the Endangered Species Act: it begins by illuminating how the changing wildlife management practices of the U.S. Fish and Wildlife Service (FWS) shaped endangered species legislation, and then it turns to the burdens of Section 7 of the 1973 act on the FWS, which led to policies and practices that professionalized restoration ecology. As FWS ecologists scrambled to respond to their new regulatory obligations, they leaned heavily on captive breeding and on requiring restoration as "compensatory mitigation" for ecological damage caused by other federal agencies. But as ecological maintenance came to be seen as essential for the survival of ecosystems, some managers and theorists worried that maintenance could change the fundamental character of a species, even domesticate it. Was the reintroduction of captive-bred species "to the wild" a means of restoring the wild, or itself a threat to wildness? Did ecological intervention threaten wildness itself?

Revolutions in Predator Control

The transformation in how the FWS managed wild species during the "Age of Ecology" cannot be understated: the agency went from rampantly killing native predators, including endangered species, to captively breeding endangered wildlife, including predators, and reintroducing them to the wild. Considered in hindsight, this transformation appears dizzyingly swift. But in reality the process was halting, legally ambiguous, and deeply contentious. The goal of protecting endangered species did not fit easily into an agency that employed experts at endangering species. Ecologists who researched how to restore wild species found themselves working for the same agency as ecologists who researched how to efficiently kill wild species.

The Fish and Wildlife Service of the early 1960s maintained the approach of the former U.S. Biological Survey and followed an agricultural model of "game production." The bulk of this highly interventionist work involved partnering with state wildlife agencies to kill "injurious species," a term widely defined to include animals that might decrease the productivity of agriculture, forestry, and game through either predation or competition.[8] The FWS also directly managed the burgeoning wildlife refuge system. As of June 1964, the agency oversaw 297 wildlife refuges, including those game reservations first established by the American Bison Society. The refuges covered an aggregate of more than twenty-eight million acres.[9] On these lands, the FWS routinely embraced highly interventionist technologies to promote particular game species. By 1962, "for the principal purpose of increasing duck production," the FWS had partnered with Ducks Unlimited to construct more than 875 dams, 1,500 water control structures, and 150 miles of canals across the country. It killed beavers and muskrats to prevent interference with water control structures and eradicated snapping turtles and snakes to reduce predation on waterfowl nests.[10] Increasingly, the FWS used helicopters, floodlights, grenades, and other surplus army equipment to keep waterfowl within the borders of its refuges.[11] The FWS was similarly interventionist in its management of fish species. By 1960 the federal government was operating more than one hundred fish hatcheries and the states around five hundred. These combined to produce hundreds of millions of fish per year.[12]

Change was coming, however. Public disapproval of federal wildlife management was peaking, and as the Department of the Interior responded to public outcry and a series of scandals, it mandated that the FWS adopt new practices grounded in newly coalesced ecosystem theory. Indeed, in 1962, a public relations fiasco at Yellowstone National Park led the Department of the Interior to fundamentally reenvision the ecological interventions of both the National Park Service and the FWS.[13] In 1961–1962, park rangers responded to a (human-caused) overabundance of elk in Yellowstone National Park by shooting more than four thousand of them, using helicopters to herd them into killing zones. Newspapers and television stations covered the slaughter, and hunters were just as outraged as animal welfarists.[14]

In response to the negative press coverage and pressure from western congressmen, Secretary of the Interior Stewart Udall announced that the Special Advisory Board on Wildlife Management would review the Department of the Interior's wildlife management policies and practices. Notably, the board included botanist Stanley Cain (later appointed to the post of Assistant Secretary of the Interior for Fish, Wildlife, and Parks) and A. Starker Leopold, the eldest son of Aldo Leopold and a noted ecologist at the University of California, Berkeley. Both brought an ecosystem science perspective to the board's deliberations. They began with the National Park Service. The question at hand—posed by the elk culling controversy but with far-reaching implications—was "How far should the National Park Service go in utilizing the tools of management to maintain wildlife populations?"[15] This question, in a generalized form—To what degree should managers intervene in ecosystems?—would continue to demarcate the borders between preservation, conservation, and restoration into the 1980s.

The advisory board's 1963 report on National Park Service wildlife policy (today known as the "Leopold Report") made ecological restoration an official goal of the agency (detailed in Chapter 7). The advisory board's subsequent report, a 1964 review of the FWS's species management practices, would be just as consequential to the history of ecological restoration, and far more controversial. Most often, the goal of the FWS's ecological interventions had been to increase the populations of particular fish and game species of clear economic and recreational value. Wildlife managers routinely analogized wild species to crops, and saw it as their job to protect and nurture these crops. The Advisory Board on Wildlife Management's 1964 report on the FWS (confusingly, also called the "Leopold Report") recommended that the agency transition away from this agricultural analogy and toward an ecosystem-based approach.

This transition to policy grounded in ecosystem science would have long-term significance for the practice of ecological restoration nationwide and even internationally. More visible in the short term, however, was the report's concern with FWS's predator control practices and changing public attitudes toward wildlife. The board's research revealed that the budget of the FWS Branch of Predator and Rodent Control had nearly doubled from $3 million annually in the early 1950s to almost $6 million in 1963. That year, the advisory board estimated, federal and state wildlife managers had killed nearly two hundred thousand predators, including 20,780 lynx and bobcat, 2,779 wolves, and 842 bears. The authors argued that, on many federal lands, the cost of predator control exceeded the value of livestock that would otherwise be lost to predators. Further, they cautioned that the Branch of Predator and Rodent Control needed to recognize that "times and social values change," and that "for every person whose sheep may be molested by a coyote" there were "perhaps a thousand others who would thrill to hear a coyote chorus in the night." The Leopold Report recommended that the FWS shift its research program away from "methods of killing animals" and toward ways of "repelling, excluding, or frightening animals." They warned that if the FWS did not revise their management practices, "an even more drastic revision will sooner or later be forced by the public."[16]

The 1964 Leopold Report included a detailed discussion of "1080 bait stations," then a widespread and increasingly contested method of poisoning predators. During World War II, the U.S. Office of Scientific Research and Development sent Compound 1080, also known as sodium fluoroacetate, along with hundreds of other poisons, to the Patuxent Research Refuge in Laurel, Maryland, to be tested as rodenticides.[17] At Patuxent, and at the Eradication Methods Laboratory in Denver, Colorado, FWS scientists fed these poisons to mice, dogs, and other mammals, keeping detailed notes on how the animals died.[18] The FWS authorized Compound 1080 for widespread use in 1946, and by the mid-1950s it had become the method of choice for killing predatory species in the United States, Canada, South Africa, Australia, and New Zealand, replacing trapping, den hunting, and strychnine poisoning.[19]

To create a bait station, wildlife managers injected a dead animal—often a sheep—with Compound 1080. Predators and scavengers would then find and consume the poisoned carcass. By 1950, the FWS had set up more than 16,000 Compound 1080 bait stations in the western United States; they covered 91 percent of rangeland in Idaho. And while the FWS had negotiated with Monsanto Chemical Corporation, the manufacturer of Compound 1080, to sell the compound only to the FWS, county boards subverted the arrangement. In 1953, for example, officials from Campbell County, Wyoming, purchased 2.5 pounds of the chemical for local predator control—enough to bait the entire state.[20] In other words, Compound 1080 was being applied in enormous quantities that were not fully accounted for by the FWS.

6.1 Pelts of predators killed by the U.S. Bureau of Biological Survey in the state of Arizona in summer 1919.
Denver Public Library Special Collections

Through such widespread predator control, the FWS was wildly successful in promoting livestock at the expense of wild species. American ranchers depended on low-cost grazing on federal land, and poisons allowed them to reduce their spending on herders, fences, and lambing sheds. One FWS biologist marveled at a rancher who, in 1961, had raised "7,000 lambs that spring without a single known loss to predators! And this was accomplished in lambing on the open range, no herders, no fences!"[21]

As it turns out, Compound 1080 bait stations were also effective at killing pets and other "nontarget" species, including wild species of public interest. Early field trials recorded the deaths of eagles, crows, hawks, red foxes, and badgers. In the fall of 1963, two dead California condors were picked up in an area where 1080-laced grain, intended to poison ground squirrels, had recently been spread. Compound 1080 also purportedly killed the last grizzlies in southwestern North America, a remnant population in Chihuahua, Mexico. The Leopold Report noted that the black-footed ferret in the northern Great Plains was nearing extinction, and the primary cause was almost certainly poisoning campaigns against prairie dogs, the main prey of the ferret.[22]

Dead pets created even more controversy than dead wildlife. When dogs died from eating meat poisoned with Compound 1080, the owners often blamed the FWS, who, in turn, blamed local ranchers. Animals dying from 1080 ingestion convulse, which was especially distressing for pet owners to watch. One grievance that reached the FWS national office was that of two friends of Estella Leopold, a renowned paleobotanist and daughter of Aldo Leopold. They complained that their two "beautiful" Belgian Tervuren sheepdogs, Nicholas Columbine and Monsieur Beau de Fauve Sharbonne, had died of poisoning on a walk in the San Juan National Forest in southern Colorado.[23]

Faced with folders-worth of such letters, FWS biologists defended the use of Compound 1080 by maintaining that it was less toxic than many other chemicals "readily available to the public." The FWS adamantly resisted speculation that Compound 1080 accumulated up food chains. In correspondence with concerned citizens, the FWS insisted that it had "never found evidence of a so-called 'chain reaction' where one bait kills one animal after another as each in turn feeds on the previous victim—not even in the laboratory."[24] And yet in internal memos as early as 1945, Division of Predatory Animal and Rodent Control scientists had warned that "the danger of secondary poisoning cannot be overemphasized," especially the danger of canids feeding on rodents poisoned by Compound 1080.[25]

The authors of the 1964 Leopold Report recommended that the FWS completely reassess its purpose in light of changing public attitudes toward wildlife, and they singled out the Branch of Predator and Rodent Control. They were specifically critical of Compound 1080, although they called for stricter legal regulation of its use, rather than an outright ban. In response, Udall appointed new leadership and reorganized the Branch of Predatory Animal and Rodent Control as the Division of Wildlife Services. In establishing official policy in line with the Leopold Report, the Division of Wildlife Services had the tricky job of appeasing both conservation groups and ranching groups. The American Farm Bureau Federation, for example, complained that the division's draft mission statement portrayed pest control as "archaic" and "inappropriate."[26] Meanwhile, the National Audubon Society explained that they could not endorse a policy that spoke so broadly about "control of the animals themselves" rather than "control of animal damage."[27] Ultimately, Division of Wildlife Services leadership compromised. They struck a line from the draft which had stated that predators, birds, and other so-called nuisance species must be "accepted as important and valuable members of our native fauna." They also changed the phrase "animal control" to "animal damage control," framing it as an "essential function" of the FWS, one that accomplished specific goals like preserving public health and safety and improving agricultural production.[28]

In response to the Leopold Report, the FWS also began researching the use of hormonal birth control on herring gulls, starlings, house sparrows, and coyotes, with the idea that this mode of "control" would be more readily accepted by the public than killing extant individuals. In the first field trial of stilbestrol administered to coyotes in beef tallow, conducted in New Mexico in 1963, FWS biologists found that rodents and ravens also took the bait.[29] Nevertheless, they considered the field trials a success. In a subsequent press release, the FWS explained that "coyotes, the little 'bad guys' of the western plains," having survived shooting, trapping, and poisoning, now faced "an adversary that may prove their most formidable and at the same time the most humane: birth control." Unlike poisons, which the press release explained might harm hunters, hikers, or pets, hormonal birth control was described as "safe" and "acceptable."[30] Along with birth control, the FWS began experimenting with "biological control," or intentionally introducing predators, to limit the population of a prey species, as an alternative to direct killing. In 1964, the FWS began testing the effects of introducing foxes and raccoons on Massachusetts Bay islands with colonies of herring gulls, "a serious offender in the bird-aircraft problem."[31] Chemical sterilization and biological control both, then, share an origin in the search for alternatives to predator poisons. By the 1980s, both methods would be widely established restoration techniques.

In April 1967, Secretary Udall approved a new predator control policy. It specified that when using poisons, Department of the Interior agencies would "minimize hazards to non-target species." The policy stated that wildlife must be managed "not only for the consumer, the sportsman," but also "for the ever-increasing proportion of people who simply enjoy seeing and hearing wild animals in their native habitat, or for that matter, simply enjoy the knowledge that these animals do exist."[32] The human audience for wild species, in other words, was newly acknowledged in policy.

6.2 A coyote being tagged for a Fish and Wildlife Service field trial of birth control through baits laced with synthetic stilbestrol, a synthetic hormone, 1962. Records of the U.S. Fish and Wildlife Service, National Archives and Records

Despite the policy, however, the FWS continued to use Compound 1080 liberally, and while the overall amount used by the FWS declined steadily after 1963, this was as much about coyote decisions as it was about changing human values: for years, FWS biologists across the country had noticed that coyotes' interest in poisoned meat declined precipitously after just a short time. In 1956, for example, coyotes had almost completely rejected the 1080 baits in eastern Colorado, whereas in 1955 they had eaten hundreds of them. In other states, too, coyote acceptance of 1080 bait meat declined after only a few years of use.[33] The coyotes that survived the first years of 1080 bait traps were the ones that learned to detect the poison. Because the 1080 bait stations became less effective, the FWS eventually put fewer of them out.

The subject of predator control, which had exploded into the headlines in 1962, became a matter of public outrage again in 1970, the year of Coca-Cola's "Age of Ecology" advertisement. In another public relations disaster for the Department of the Interior, a troop of Boy Scouts found several dead eagles that May in Casper, Wyoming; the birds had eaten sheep carcasses laced with thallium.[34] Then, in June, the *New Yorker* published an article by Faith McNulty on conservationists' efforts to end widespread poisoning of prairie dogs in South Dakota because of the attendant harm to the endangered black-footed ferret.[35] Also in June, the Defenders of Wildlife sued the FWS, claiming that the National Environmental Policy Act, enacted that January, required an environmental impact statement for the predator control program. Soon thereafter, seven environmental organizations petitioned the newly formed U.S. Environmental Protection Agency to end the use of Compound 1080 and other poisons nationwide, on the grounds that they killed nontarget species. In response, the Department of the Interior announced yet another review of its predator control policy. The resulting report called for an end to aerial hunting, aerial broadcasting of poison baits, and rodent control in areas where secondary poisoning could occur. Along with the Leopold Report, it formed the basis for President Nixon announcing an immediate end to federal use of Compound 1080, strychnine, and cyanide in February 1972.[36] The power and influence of the Branch of Predator and Rodent Control—and its successor, the Division of Wildlife Services—were diminished within the FWS; meanwhile, another office, the Committee on Rare and Endangered Wildlife Species, prospered.

Managing for Endangered Species

The Department of the Interior in the 1960s faced pressure not only from the American public, but also from a growing international movement to protect rare and endangered species. For decades, groups like the American Committee for International Wild Life Protection (founded in 1930) and the International Union for Conservation of Nature (founded in 1948) had sought to craft international treaties to protect charismatic game species like bison, mountain gorillas, and northern fur seals.[37] The emergence of ecosystem theory in the 1960s, with its language of irreversible human-caused ecological change, inspired activists to justify their efforts with new, powerful rhetoric. As signaled by the bumper sticker "Extinct is Forever," they embraced a narrative of irreversibility, the idea that once a species is gone, it can never be brought back.[38] This growing emphasis on species extinctions contributed to the FWS's transformation from a predator-killing service to a service for restoring native wildlife, including species like wolves and black-footed ferrets that the FWS had been largely responsible for extirpating in the first place.

Whereas bison had been the focus and the prominent symbol of early game restoration in the United States, the whooping crane became the dominant symbol of species endangerment in the 1960s. And as bison restoration practices had far-reaching consequences, ultimately inaugurating the U.S. National Wildlife Refuge System, so whooping crane restoration practices mattered immensely: they directly informed the endangered species legislation that culminated in the Endangered Species Act of 1973, one of the most expansive and influential environmental laws to date, and one that contributed to the professionalization of ecological restoration.

Whooping crane restoration was the main purpose of the Committee on Rare and Endangered Wildlife Species (CREWS), established in 1964, the first office in the Fish and Wildlife Service dedicated to studying endangered species management. A memo attached to CREWS's charter stated that it was to assume the duties previously assigned to the FWS's informal whooping crane committee, and many CREWS members had attended the inaugural 1956 Whooping Crane Conference.[39]

Whooping cranes, a migratory species once found from the Canadian arctic tundra to Mexican grasslands, had been in decline since the early 1900s, owing to a combination of hunting and wetlands draining. The U.S. federal government first became involved in whooping crane protection with the 1916 Migratory Bird Treaty between the United States and Great Britain (on behalf of Canada, then part of the British Empire) that sought to "insur[e] the preservation of such migratory birds as are either useful to man or are harmless."[40] A clause in the treaty forbade the hunting of whooping cranes for ten years. Then, in 1937, President Franklin Roosevelt established by executive order the Aransas Migratory Waterfowl Refuge, protecting the wintering grounds of the whooping crane on the southern Gulf Coast of Texas. Nevertheless, whooping crane numbers remained perilously low, and in the early 1940s, the number of known whooping cranes in North America hit its nadir at fifteen, all of whom wintered at the Aransas Refuge.[41]

The seeming failure of hunting restrictions and habitat preservation to restore whooping crane populations spurred some biologists to advocate for more intensive interventions, including captive breeding. After years of discussing whooping crane numbers informally with concerned members of the National Audubon Society, the FWS held its first Whooping Crane Conference in October 1956, inviting zoo directors and members of environmental organizations that included the Audubon Society, The Nature Conservancy, and the International Union for Conservation of Nature (IUCN). Conference participants heatedly debated the idea of captive breeding. On one side, Audubon member Robert Porter Allen argued, "It is not our wish to see this noble species preserved behind wire, a faded, flightless, unhappy imitation of his wild, free-flying brethren. We are dedicated to preserving the whooping crane in a wild state."[42] Such participants believed that habitat preservation was the best bet for whooping crane survival, and that by taking eggs or chicks out of the wild and into captivity, managers might accidentally extinguish the species. On the other side, FWS biologist John J. Lynch argued in favor of taking eggs and birds from the wild to develop a captive flock, and Harold Coolidge read resolutions recently adopted by the IUCN in Edinburgh, Scotland, that urged the artificial propagation of endangered animals.[43] Proponents of captive breeding argued that the removal of rare animals from the wild should be standardized and carefully monitored by a coordinated network of professionals.

With conservationists weighing in on both sides, the FWS at first maintained a noninterventionist stance.[44] Somewhat ironically, given the FWS's emphasis on species of commercial value and its large role in protecting livestock, the chief of the Branch of Wildlife Refuges had written to a colleague a few months prior to the 1956 conference that the FWS's primary responsibility was "for the preservation of the species in the wild rather than under wire." The risk to the whooping cranes of disease, malnutrition, and predation in captivity seemed too great.[45] For the next few years, captive breeding attempts involved only three whooping cranes that were already in captivity: Crip and Josephine at the Audubon Park Zoo in New Orleans, and Rosie at the San Antonio Zoo.

Although the FWS had overseen fish hatchery and stocking programs for decades, it had not been directly responsible for breeding and releasing other species. Instead, the FWS had supported the translocation and reintroduction efforts of groups like the American Bison Society and Ducks Unlimited. In the 1950s, the FWS had also begun to encourage states to use funding from the Federal Aid in Wildlife Restoration Act of 1937 (The Pittman-Robertson Act) for the captive breeding of game species. These projects, however, were typically small-scale and focused on charismatic species like Canada geese, white-tailed deer, and wild turkeys. In one such project, the Hawaii Board of Commissioners of Agriculture and Forestry began a venture to breed the highly endangered nēnē, a goose endemic to Hawai'i, to maintain "something of 'Old Hawaii' for the education of tourists and local residents alike."[46] The nēnē breeding and reintroduction project reveals the casual and decentralized nature of early captive breeding programs, with the responsibility shared among federal scientists, NGO members, and volunteers. In 1962, a team released nēnē around the crater of Haleakalā, on Maui, that had been captively bred in three places: a ranch in Pōhakuloa, Hawai'i; the Severn Wildfowl Trust in England; and the Connecticut house of the secretary of the Smithsonian Institution, S. Dillon Ripley.[47] The nēnē program's apparent success fueled optimism that such decentralized projects would lead to quick population recovery.

The FWS first became directly involved with captive breeding in 1961, when Ray Erickson, a biologist in the FWS Branch of Wildlife Research, responded to an all-agency memo from President Kennedy that called for innovative ideas by drafting a proposal titled "Production and Survival of the Whooping Crane."[48] In it he proposed captive breeding experiments with the sandhill crane, a species closely related to the endangered whooping crane. The experiments, Erickson hoped, would inform the eventual establishment of a captive whooping crane population.[49] This was a marked departure from Erickson's policy as head of habitat improvement in the Branch of Wildlife Refuges, just a few years prior, that the present population of whooping cranes was "neither large enough to justify such experimentation nor small enough to require experimenting as a last resort."[50]

Encouraged by the initial captive breeding experiments with sandhill cranes, Erickson pitched the idea of a federal center for the management of species threatened with extinction.[51] In 1965, after the publication of the Leopold Report, he helped secure a $350,000 appropriation from Congress to establish an endangered species propagation center at Patuxent, Maryland—the same site at which Compound 1080 and other biocides had been developed. By May 1966, Erickson and his colleagues had built temporary pens for Canus (the young whooping crane described in the Introduction), who had been found injured two years prior in the Northwest Territories of Canada—along with cages for forty-five sandhill cranes (proxies for the whooping crane); eight Aleutian Canada geese; three tule white-fronted geese; eight masked bobwhite quail; four dusky grouse; several dozen silky bantam chickens ("used for incubation studies"); and eight South American snail kites (proxies for endangered Florida Everglades kites). The plan was that within ten years Patuxent would hold at least one hundred endangered species in captivity and produce more than five thousand birds for release "to the wild" annually.[52] A federal agency would take control of an activity that until this point had been pursued by a highly decentralized network of individuals and NGOs. Ray Erickson's Endangered Species Program at Patuxent represented a consolidation of power by federal wildlife biologists and academic ecologists, who argued that nonexperts should no longer be allowed to breed wild animals in their backyards in the name of restoration.

With the 1964 establishment of the Committee on Rare and Endangered Wildlife Species, endangered species research became an official function of the FWS, and one of CREWS's first tasks was to compile a national list of endangered species. Secretary Udall, having recently attended a meeting at which he learned about the IUCN's evolving list of globally endangered birds and mammals—known today as the Red Book or Red List—charged

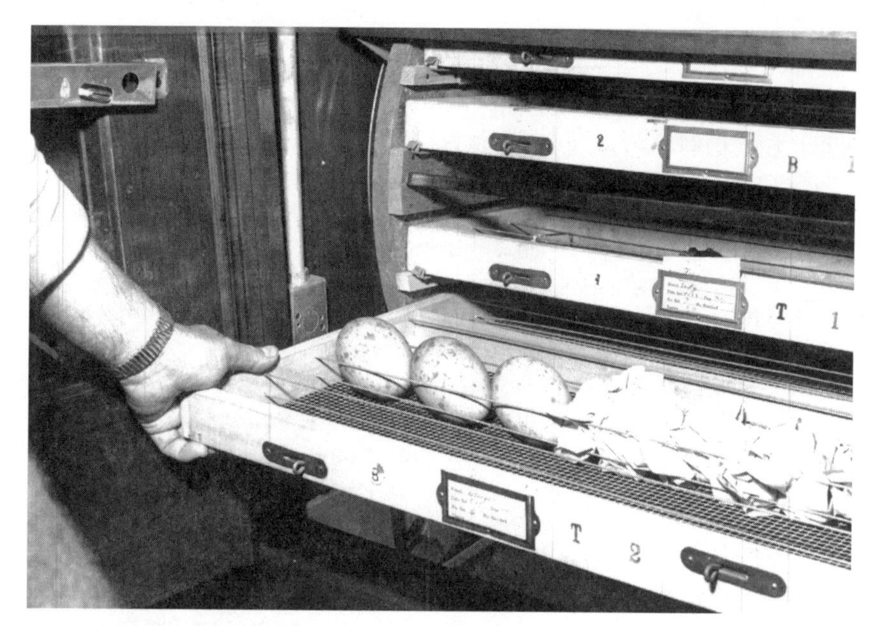

6.3 Whooping crane eggs collected in the Northwest Territories of Canada at the Patuxent Wildlife Research Center in 1971. The eggs were artificially incubated and the young reared in captivity. Records of the U.S. Fish and Wildlife Service, National Archives and Records

6.4 A newly hatched whooping crane chick accepts food from a wildlife technician at Patuxent Wildlife Research Center, June 1971. The young birds have just hatched from eggs taken from nests in Wood Buffalo National Park, Northwest Territories, Canada. Records of the U.S. Fish and Wildlife Service, National Archives and Records

CREWS with creating its own official list of native vertebrate species threatened with extinction.[53] The committee quickly and rather haphazardly compiled a draft list of 317 rare species, including 78 believed to be in immediate danger of extinction, such as the whooping crane, the nēnē, and species that the FWS bore responsibility for endangering, like the grizzly bear, timber wolf, and Florida panther.[54]

The FWS did not publicly admit its role in endangering any of the listed species. Rather, FWS rhetoric around endangered species care echoed the language of much earlier naturalists who saw the environmental results of white settlement of the United States as "inevitable." A *Washington Post* article covering the FWS's "campaign to rebuild native stocks of birds and mammals" at Patuxent, asked, "What is the cause of the vast depletion in the ranks of our native animals?" It continued: "The answer is a simple one: people, or as the officials of the U.S. Fish and Wildlife Service put it, 'the encroachment of man.'"[55] In other words, the article—and many others like it—universalized: rather than naming responsible individuals and institutions, it diffused responsibility for the loss of nearly forty species of birds and mammals.

In June 1965, in an effort to expand and formalize CREWS's work, Secretary Udall submitted identical bills to the House and Senate that would authorize the Department of the Interior "to initiate and carry out a comprehensive program to conserve, protect, restore, and, where necessary to establish wild populations, propagate selected species of native fish and wildlife, including game and nongame migratory birds, that are found to be threatened with extinction." The FWS was well prepared to undertake this restoration work, Udall argued, by its previous collaborations with conservation organizations and several states to save the American bison and whooping crane.[56] Perhaps surprisingly, in retrospect, members of both the House and Senate generally considered the bills uncontroversial "refuge bills." The bills prohibited the taking of endangered species, but only in federal wildlife refuges, and the Department of the Interior testified that only seventy-eight species were considered to be endangered. In Senate hearings, the only debate took place over a constitutional issue—federal appropriation of the states' historic rights to manage resident wildlife species. The bill was passed easily, as the Endangered Species Preservation Act of 1966, and was the crucial precursor to the Endangered Species Act of 1973.

The Endangered Species Preservation Act consolidated FWS refuges into one National Wildlife Refuge System, and it directed the interior secretary to compile a list of endangered species and authorized the FWS to spend up to $15 million per year to buy lands for a novel purpose: protecting native wildlife threatened with extinction. Nevertheless, it elided federal responsibility for species declines. The act described extinction as "one of the unfortunate consequences of growth and development in the United States" and did not list predator control among its four recognized causes of extinction: habitat destruction, overexploitation, disease, and predation.[57] It did, however, establish a federal mandate for the FWS's shift toward restoration work. In addition to managing desired fish and game species, the FWS was now also responsible for managing a variety of endangered species.

Despite the passage of the Endangered Species Preservation Act of 1966, and despite the revision of wildlife policies within the National Park Service (NPS) and the FWS, there was still no single wildlife policy for all Department of the Interior lands. While CREWS researched endangered species in the lab and in the field, the Branch of Predator and Rodent Control (called the Division of Wildlife Services after the Leopold Report) continued to kill bears, wolves, panthers, and other species on CREWS's list of rare and endangered species. But the idea of a unified policy, a holistic one, was made more appealing by the growing popularity of ecosystem theory. In remarks at a 1966 conference, Assistant Secretary of the Interior Stanley Cain—who had, one year prior, praised the promise of Lauren Donaldson's hybrid fish—noted that while "wildlife biologists are ecologists, and many of them are very good ones," they had problematically confined their attention to single species rather than working to understand and manage "the ecosystem as a whole." At a planning meeting that followed the conference, Cain noted that critics of the department "see schizophrenia and lack of policy in a Department which prohibits potshots at whooping cranes and other rare and endangered species, bans hunting in national parks, yet promotes the hunting of game on refuges, ranges, Indian lands, and public domain."[58]

In 1971, Congressman John Dingell of Michigan proposed amendments to the Endangered Species Preservation Act that would expand the Patuxent captive breeding program and fund further habitat acquisition.[59] The bill was referred to the Department of the Interior for comment, and this is likely how language on ecosystem restoration first entered federal legislation. Earl Baysinger, then head of the Office of Endangered Species (which incrementally replaced CREWS), later recalled that he, Ray Erickson, and other CREWS members drafted comments on the Dingell bill that extended its applications to all animals and plants, effectively endorsing ecosystem preservation and restoration.[60] A subsequent bill aimed both "to provide a program for the conservation, protection, restoration, or propagation of such endangered species and threatened species" and "to provide a means whereby the ecosystems upon which endangered species and threatened species depend may be conserved, protected, or restored."[61]

Such language was not, on its face, especially powerful. In his Environmental Message of February 8, 1972, President Nixon pointed out that the 1969 act "simply does not provide the kind of management tools needed to act early enough to save a vanishing species." In this message, Nixon also announced Executive Order 11643, which barred the use of poisons for predator control on all public lands. Noting that "even the animals and birds which sometimes prey on domesticated animals have their own value in maintaining the balance of nature," Nixon adopted the language of ecosystem theory.[62] Then, in spring 1973, the Nixon administration, Congressman Dingell, and Senator Harrison Williams of New Jersey submitted nearly identical bills that would become the Endangered Species Act.[63] In congressional hearings, the FWS testified that the extinction of one species could "affect or destroy other species in the long run ultimately damaging or eliminating an entire ecosystem."[64]

The Endangered Species Act of 1973, which repealed both the 1966 Endangered Species Preservation Act and its 1969 amendments, is still the law today. Two of its provisions, Section 4 and Section 9, represented a major erosion of the states' control over resident wildlife. Section 4 of the act required the interior secretary to establish a list of species, subspecies, and isolated populations that are considered to be in danger of extinction (endangered) or "likely to become an endangered species within the foreseeable future" (threatened). Any animal other than insect pests and any plant was now eligible for listing, whether resident or migratory, and a species could be considered endangered or threatened for any reason, whether "natural" or "manmade." Once listed as endangered or threatened under Section 4, a species came under Section 9 protection. Section 9 applied broad "take" prohibitions to listed endangered species, making it illegal to "harass, harm, pursue, hunt, shoot, wound, kill, trap, capture, or collect, or attempt to engage in any such conduct." Yet Congress barely debated the act. The final version passed in the Senate unanimously and in the House by a vote of 345 to 4.[65] Remarkably, Congress failed to anticipate the act's enormous implications for public and private land use, which became apparent soon after its passage. As Historian Peter Alagona has convincingly demonstrated, Sections 4 and 9 played key roles in the expansion of habitat preservation in the United States.[66]

The 1973 act also implemented new regulations across the federal government. Section 7 requires all federal agencies to consult with the interior secretary to ensure that any actions they authorize, fund, or carry out do not "jeopardize the continued existence of such endangered species and threatened species" or "result in the destruction or modification of [critical] habitat of such species." Section 7 has received less attention from scholars than Sections 4 and 9, but it played a key role in the national expansion of ecological restoration. Significantly, it threw an enormous responsibility on the shoulders of an agency whose internal policies about wildlife management were still being sorted out. The FWS would become the arbiter of how endangered species and ecosystems were managed across the nation, and it had to build up its resources of staff and policy in order to manage its new role. When a 1978 amendment to the act required official recovery plans for listed species, the FWS's workload jumped, and the Endangered Species Act solidified the role of ecological restoration in federal environmental practice.

Restoration and Section 7 of the Endangered Species Act

The implementation of the Endangered Species Act of 1973 was thus shaped by the mandates of the different agencies responsible for overseeing it, and ecologists had quite various roles in those agencies. The act vested the National Oceanic and Atmospheric Administration, within the Department of Commerce, with responsibility for marine and anadromous species. The Smithsonian Institution was tasked with preparing a list of endangered and threatened plant species. And the FWS, in the Department of the Interior, was placed in charge of creating a list of all other threatened and endangered species and for reviewing all proposed federal projects that might affect those species. For example, if the Army Corps of Engineers planned to build a dam on a river that contained a listed species, it would need to enter the Section 7 consultation process. The Office of Endangered Species would then deliver a "biological opinion," either clearing the project to go forward or suggesting modifications if it found that that the proposed project would adversely affect listed species.

The small Office of Endangered Species was responsible for this oversight. But its initial assignment had been to create an official list of endangered species and to set guidelines for the hunting and export of species like bobcat, American alligator, and river otter.[67] Although the FWS increased the size of the office from two to eight biologists in 1973, the staff was quickly overwhelmed with the responsibilities of listing species, and it spent little time on the interagency consultation required by Section 7.[68] Keith Schreiner, the endangered species program manager, remarked in 1975, "It takes us a minimum of 36 professional man days to list a single plant or animal species and I've only got six full time professionals who work at this—among other things—for the whole lot of them. It will take us, at this rate, the next 6,000 years just to list all the endangered plants and animals that need protection by the Endangered Species Act, not to mention developing programs for them."[69]

In 1978 Congress took up amendments to the Endangered Species Act, and in hearings the FWS estimated that it had been involved with more than 4,500 Section 7 consultations in the preceding five years, reviewing projects that included highways, dams, airports, coal mines, nuclear power plants, and bombing ranges.[70] The majority of consultations were resolved by determining that the project would not adversely affect listed species. In the remaining cases, the FWS recommended project alterations such as choosing an alternative site, lengthening a discharge pipe to avoid critical habitat, or completing construction prior to a nesting season.

From the beginning, parties to such consultations were under tremendous pressure to compromise. The agency in which a new project originated did not want to be taken to court over noncompliance with the act; the FWS, in turn, did not want to invite congressional backlash against its endangered species program.[71] This pressure to compromise in interagency negotiations under Section 7 gave rise to what is now known as "compensatory mitigation."[72] Development agencies, anticipating the negotiation process, began to describe plans to "offset" adverse environmental impacts by protecting endangered species in other ways. For example, a proposal for a federal aid project on Mona Island, Puerto Rico, anticipated that road and building construction might lead endangered Mona boas to be killed by motorists or by poachers, but suggested that the "negative aspects of road improvement" would be offset by new efforts to control "feral animals" that preyed on the species.[73] Prior to 1974, few development proposals involved mitigation measures, and the FWS rarely suggested them. This changed after a 1974 Government Accountability Office report encouraged the FWS to require development proposals to minimize losses to wildlife, or to "replace" wildlife through artificial propagation or habitat acquisition.[74]

The move toward compensatory mitigation was swift. By 1976, ecologists were organizing a meeting to discuss emerging mitigation practices spurred by the Endangered Species Act, the Fish and Wildlife Coordination Act, and the National Environmental Protection Act. The resulting 1979 Mitigation Symposium at Colorado State University in Fort Collins, Colorado, was sponsored by a suite of organizations including the Ecological Society of America, the FWS, the American Fisheries Society, and the Army Corps of Engineers. Presenters focused on defining what should count as compensation. Could a destroyed habitat be compensated for with the purchase and protection of habitat elsewhere? Could plans to restore one ecosystem offset plans to devastate another? One FWS field biologist described a plan to restore prairie grassland ecosystems as mitigation for a Bureau of Reclamation project in eastern North Dakota, noting that while "restoration itself is not a new concept to the biologist," it was, however, "a new approach when applied on the large scale necessary to offset extensive losses from a major irrigation project."[75] Another participant reasoned that, in the future, estuarine ecosystem loss due to housing development and mineral exploration could be compensated for through ecosystem restoration, or even by building marshes de novo, citing a list of 105 such projects planned or under way.[76] Restoration was appropriate compensation for development, one conference organizer explained, in cases when humans had interfered with an ecosystem to the point where it could not "restore its functioning systems to original conditions through natural processes alone"—an argument grounded in ecosystem theory.[77]

The first high-profile endangered species case involving compensatory mitigation concerned the whooping crane. The proposed Grayrocks Dam near Wheatland, Wyoming, would be part of a $1.6 billion Missouri Basin Power Project that would serve eight states. The State of Nebraska opposed the project, arguing that Wyoming would consume more than its share of North Platte River water. Meanwhile, the National Audubon Society, the National Wildlife Federation, and other various nongovernmental environmental organizations at the local level argued that the reduction in water would adversely impact habitat some 270 miles downstream from the dam that was critical for the endangered whooping crane. The NWF and State of Nebraska filed suit to stop the project and won a temporary injunction in October 1978.[78] The injunction led to an elaborate out-of-court settlement between the NWF, the Missouri Basin Power Project, the Army Corps of Engineers, and the Rural Electrification Administration that guaranteed a minimum water flow and established a $7.5 million trust fund for the "maintenance and enhancement" of whooping crane habitat.[79] The FWS Office of Endangered Species issued a biological opinion in December 1978, stating that if the project followed the terms of agreement, there would be no jeopardy to the whooping crane population.[80] It held, in other words, that the promise of continued intervention ("maintenance and enhancement") would offset any immediate habitat damage.

The resulting Platte River Whooping Crane Maintenance Trust would go on to fund The Nature Conservancy's acquisition of land along the Platte River, where it experimented with controlled burns, herbicides, and hand-pulling to try to eliminate willows, cottonwoods, dogwood, and other woody species in an effort to restore open wet meadow. The restoration project was described at the Mitigation Symposium and, two years later, in the inaugural issue of *Restoration & Management Notes,* the flagship journal of the Society for Restoration Ecology.[81] Restoration thus became compensation for ecological harm.

Restoration and Recovery Plans

The Grayrocks Dam controversy, along with the better-known Tellico Dam controversy, in which the snail darter, a small endangered fish, halted construction of a dam on the Little Tennessee River, led members of Congress to push to weaken the Endangered Species Act. A 1977 *Washington Post* article reported that "misty-eyed environmentalists" and "dam-fighters" were using the Endangered Species Act as a "weapon of last resort" against projects they opposed on broader grounds. Already some members of Congress were "grumbling that when they approved the act" they meant to save charismatic species like bald eagles, not "a whole assortment of undistinguished flora and fauna with precarious existences and funny names."[82] Some members of Congress, in other words, had voted for the Endangered Species Act without an accurate sense of how long the list of endangered species might grow or how ecosystem theory would affect the impact assessments of the FWS's Office of Endangered Species.

In 1978 Congress passed amendments that created a cabinet-level committee that could exempt projects from Section 7 provisions.[83] The 1978 amendments also mandated that the interior secretary develop and implement "recovery plans" for endangered and threatened species. Congress added this provision at the urging of environmental groups and with the support of the FWS, which wanted Congress to authorize funding for plans that were, in some cases, already under way (for example at the Patuxent Wildlife Research Center). The amendments authorized the FWS and the National Marine Fisheries Service (under NOAA) to "procure the services" of experts outside the agency in preparing recovery plans.[84] This provision created a formal demand for experts in ecological restoration.

In 1979, the FWS reported to Congress that it had developed recovery plans for twenty-six species and was in the process of developing plans for sixty-three more.[85] Each "recovery team" was tasked with drafting a plan that identified and scheduled "actions required for securing or restoring an endangered or threatened species as a viable self-sustaining member of its ecosystem."[86] Recovery teams brought together agency biologists with academic and NGO ecologists, and this helped professionalize restoration ecology. During the course of that professionalization, however, restoration practices were both various and contested. Many of the early species recovery plans reflected the view that had prevailed prior to the 1960s, that wild species would recover if they were provided "unmolested" habitat. But some recovery plans reflected the mounting sense that ecosystems, too, were threatened, and that the species of concern would not recover without interventionist management. The draft plan for the Aleutian Canada goose, for example, called for eradicating introduced arctic foxes on three nesting islands before reintroducing the geese. The draft plan for the palila, an endangered Hawaiian honeycreeper (a songbird), called for the removal of feral goats, cattle, and sheep to promote the growth of māmane, an important food source for the palila.[87] A few early draft recovery plans also mentioned the possibility of captive breeding and reintroduction. The set of early recovery plans indicates a period of overlap between single-species restoration and ecosystem restoration.

The goal of recovery was not legislatively defined in ecological terms. Rather, through subsequent regulation, the goal of recovery became official delisting.[88] The 1978 amendments specified that the interior secretary, at least once every five years, review all listed species and determine which species, if any, should be removed from the list (delisted), changed from endangered to threatened (downlisted), or changed from threatened to endangered (uplisted). Regulations made delisting permissible if the FWS determined that a species was already extinct, that it had sufficiently recovered, or that the original data for classification was in error.[89] Two of the earliest species to be delisted because of recovery were the eastern subspecies of brown pelican, in a portion of its range, in February 1985, and the American alligator across all of its range in June 1987.[90] The FWS attributed brown pelican recovery to the banning of DDT in 1972, and they attributed alligator recovery to hunting restrictions. Thus, despite the inroads that ecological restoration was making within the FWS, the service's public statements continued to reflect the view that nature, if left alone, would do the work of recovery itself. "All we had to do was stop the poachers, and the gators did the rest," a FWS press officer told *Time* magazine in 1987.[91] Not until 1999 would a species, the peregrine falcon, be delisted because of a restoration effort in the form of captive breeding and release. The Aleutian Canada goose and Robbins's cinquefoil (a member of the rose family) followed as restoration success stories in 2001 and 2002.[92]

Reintroducing Predators to "the Wild"

The peregrine falcon's reintroduction to the wild was one of the first reintroductions under the Endangered Species Act of 1973, and it was highly publicized. The reintroduction was especially remarkable because the peregrine falcon is a predator. It was also an early instance of the regulatory and philosophical pitfalls that could trouble captive breeding programs.

Peregrine falcons were historically found throughout North America, and unlike whooping cranes, they were also found on other continents. Taxonomists recognized some twenty or so subspecies of peregrine falcon in the early 1970s. In the eastern United States, peregrines had nested from the Great Lakes to eastern Maine and south to Georgia. This eastern peregrine population was extinct by 1964, however, and in the late 1960s scientists agreed that the cause was the persistent insecticide DDT. The poison's fat-soluble metabolites had caused eggshell thinning, leading to a precipitous decline in the survival of young peregrines and other raptors. Peregrines still nesting in the western United States were determined to be an endangered species in 1970, two years before the federal government banned the use of DDT.

In 1974, shortly after the passage of the Endangered Species Act, the National Audubon Society sponsored a meeting of experts in peregrine biology, including representatives from the FWS. There it was decided that to improve the odds of the reintroduced peregrines surviving in the wild, they should be bred from as diverse a genetic pool as possible. Beginning

6.5 Jim Weaver and Tom Cade handling a falcon for captive breeding at the Peregrine Fund at Cornell University, 1976. The Peregrine Fund

that year, Tom Cade, an ornithologist at Cornell University, began experimenting with breeding peregrine falcons (*Falco peregrinus*) in captivity, assembling his breeding stock from three western North American subspecies (*tundrius, pealei,* and *anatum*) as well as two European subspecies (*peregrinus* from Scotland and *brookei* from Spain). By winter 1977, Cade and his collaborators, the Peregrine Fund, had raised 229 hybrid peregrines and released 133 in eight eastern states.[93] The release sites included four national wildlife refuges: Brigantine and Barnegat in New Jersey and Chincoteague and Fisherman Island in Virginia. A *New York Times* reporter wrote that Cade "looked like a proud mother as he sat aboard a commercial jetliner cradling a pasteboard box with three fuzzy peregrine falcon chicks nestled inside."[94]

The program hit a serious snag, however, amid frequently changing wildlife policies and regulations. On June 30, 1977, Cade received an urgent cable from the FWS: "Effective immediately, you are not to release peregrine falcon subspecies *brookei* or *peregrinus.* Executive Order 11987 signed by President Carter on May 24 prohibits the release of exotic species." Under the executive order, the captive-bred hybrid peregrines, intended to restore a wildlife population to its historic range, were deemed exotic rather than native, and the careful efforts to aid falcon restoration came to a halt. Cade was incensed; he argued that native species were better adapted to their environments than introduced nonnative species were, and that there was "very little concrete evidence to support the widespread notion among protectionists that exotic bird species have competitively excluded native North American species."[95] Backing Cade, the World Wildlife Fund, the Smithsonian Institution, the National Wildlife Federation, and the National Audubon Society wrote to the FWS in support of the Peregrine Fund's breeding and release of hybrid peregrines.[96]

The FWS found itself in a unique predicament. The peregrine reintroduction program enjoyed wide public support, but President Carter's executive order restricted "the introduction of exotic species into natural ecosystems" on lands owned or administered by federal agencies. In a message to Congress accompanying the order, President Jimmy Carter echoed the language of ecosystem ecology, writing, "Americans long thought that nature could take care of itself. [. . .] As we know now, that assumption was wrong."[97] The order was part of the Carter administration's broader work to streamline environmental administration and to direct attention to nongame species at a time when ninety-seven of every hundred federal wildlife dollars still went to species that were hunted or fished for commerce or sport.[98] The Council on Environmental Quality, which drafted the executive order, contended that certain plants and animals that had been

6.6 Peregrine Fund employees releasing three female peregrine falcons and one male, all about five weeks old, atop the Manhattan Life Insurance Building on 57th Street, in July 1980. The Peregrine Fund

"introduced into the natural ecosystems of the United States," such as the gypsy moth, the European starling, and the water hyacinth, were detrimental to agriculture and native wildlife, and that it was in the public interest to restrict the introduction of further exotic species.[99]

Despite legal uncertainty over the status of Cade's hybrid peregrines, the FWS reversed its position toward Cade's project in 1978 and confirmed their support for the release of hybrid peregrines to establish an eastern population. In June 1979, the Department of the Interior, in conjunction with Cade's Peregrine Fund, released four young peregrines from the roof of their main building to much fanfare. "The prospects of seeing this magnificent bird once again soaring above the Nation's Capital testifies to the fact that all the news about endangered species is not gloom and doom," said Cecil Andrus, the secretary of the interior appointed by President Carter.[100] One of the four peregrines was named Rachel, after Rachel Carson, the former Department of the Interior employee whose 1962 book, *Silent Spring,* had brought global attention to the issue of DDT. The following year, recognizing that skyscrapers made good alternatives to cliffs, the Peregrine Fund released four young falcons on top of the Manhattan Life Insurance Building in New York City. One of the females fell down an air shaft on the building in August, but biologists found her unharmed. Her release to "the wild" meant the city: wildness referred not to an unpeopled place, but to the condition of living unmanaged by humans. That spring peregrine falcons successfully bred in eastern North America for the first time in more than twenty years, when chicks hatched to two pairs previously released at two national wildlife refuges in New Jersey. By the spring of 1983, peregrine falcons—presumedly those released at the Manhattan Life Building—were discovered nesting at two New York City bridges.[101] Approximately 2,500 captive-bred hybrid peregrines were captively bred and released in the eastern United States between 1974 and August 1999, when the Department of the Interior ultimately judged the program a success and removed all peregrine falcons in the continental United States from the Endangered Species List.[102] Today peregrine falcon reintroduction is considered a major restoration success story.

New Meanings of Wildness

The Fish and Wildlife Service's shift from predator control to species reintroduction brought many kinds of conflict to the surface. In addition to legal questions of whether hybrid animals could be reintroduced (and whether such restored populations were subsequently protected under the Endangered Species Act), the FWS faced political resistance to reintroductions, even of nonpredatory species. Private landowners, states, and even other federal agencies worried that introducing an endangered species to habitat that it did not already occupy would restrict land uses and infringe on private property rights. The FWS endorsed amendments to the Endangered Species Act that gave the service some flexibility in its negotiations. Indeed, the Department of the Interior very much hoped to encourage states and NGOs to reintroduce species voluntarily, as the federal budget for wildlife restoration was quite limited, and at least 70 percent of listed species depended on nonfederal land for the majority of their habitat.[103] A provision on these matters, enacted by Congress in 1982, confirmed the department's authority to move species to areas beyond their current range; it also allowed the reintroduction of "experimental populations" of threatened or endangered species into their historic ranges without requiring strict compliance with many of the act's restrictions.[104]

However, even with the leeway provided by the 1982 amendment, the FWS preferred not to alarm landowners if it could help it; it focused its growing reintroduction program on charismatic species that enjoyed broad support, and it went to great lengths to ensure that reintroduced animals would stay on federal lands—no easy task for wide-ranging animals. The FWS even fit red wolves with "capture collars" that could deliver a sedative dose in response to a radio signal before reintroducing the wolves in North Carolina. And the California sea otter relocation program included, at Congress's direction, a commitment to maintain an "otter free" zone between the reintroduced otters and an existing population.[105] Plant reintroductions generated much less controversy than animal reintroductions, because individual plants stay where they are placed and also because the Endangered Species Act does not forbid the taking of listed plants on private property. Nevertheless, by the early 1990s, nearly two-thirds of species recovery plans called for captive breeding and reintroduction of animals, and an even greater proportion called for some kind of translocation of individuals or populations.[106]

The controversies over captive breeding and reintroduction indicated a change in how wildness itself was understood. In early game restoration, individual species were the objects of concern. American Bison Society members, in their appeals for bison restoration, promised future domestication, and the Committee on Breeding Wild Mammals of the American Breeder's Association hoped to discover wild animals that could be bred for food, clothing, labor, and companionship. Although some game restorationists did worry that captivity would cause species to lose their "wild character," they imagined this character to be conferred by large spaces. Wildness, for them, was more a behavior conferred by environment than an inherent trait. In the 1960s, however, *wildness* came to be seen as endangered. When ecologists initially objected to breeding whooping cranes "under wire," for example, they feared for the wildness of the cranes. In the same vein, Canadian ecologist Ian Cowan wrote in *Nature* in 1965 that the "recent approach to the restoration of endangered species"—captive breeding for later release into the wild—"cannot be regarded as the perfect answer" because some species had been shown to possess "a heritable factor for wildness which may be selected against in captive breeding."[107] When the FWS endangered species program set the goal of restoring "self-sustaining wild populations" in 1971, it specified that captive propagation for reintroduction must retain "the inherent wild qualities of appearance and behavior" and that captive propagation was only appropriate when other restoration measures were not sufficient to assure the species' survival.[108] Such worry over heritable wildness led to a hardening of the distinction between natural areas and zoological parks, and between restoration ecologists and zoologists, and the FWS uncomfortably straddled this divide. In a 1967 presentation, Assistant Secretary of the Interior Stanley Cain described the Wichita Mountains Wildlife Refuge—with its origins in the American Bison Society—as "a combination of refuge, wilderness, farm and zoo."[109]

The Endangered Species Act ultimately consolidated the power to manipulate endangered or threatened species, especially animal species, in the U.S. Fish and Wildlife Service. Whereas before the 1970s individuals and organizations like the American Bison Society, the National Audubon Society, and the World Wildlife Fund—and smaller organizations like Tom Cade's Peregrine Fund—had partnered with the federal government to provide funding and organisms to be reintroduced, by 1980, FWS scientists at the Patuxent Wildlife Research Center in Maryland were researching and propagating more than sixty endangered and threatened species, including bald eagles, Everglades kites, gray wolves, black-footed ferrets, and whooping cranes.[110] With the establishment of species reintroduction as a core ecological restoration practice, the groundwork was laid for the reintroduction of a large mammalian predator—wolves—to Yellowstone National Park in 1995. On January 12, eight gray wolves from Jasper

6.7 A Fish and Wildlife Service biologist prepares a gray wolf captured in Alberta, Canada, for transport and introduction to Yellowstone National Park, January 1995. Records of the U.S. Fish and Wildlife Service, National Archives and Records

National Park in Alberta, Canada, became the first wolves to set paw in Yellowstone since the federal government had eradicated the last pack in 1923.[111]

But even as species reintroduction practices under the Endangered Species Act solidified in the first decade after the act's passage, concern about nonnative species was spreading among federal agencies involved in the management of plants and animals. President Carter's Executive Order 11987 had very much complicated the FWS's single-species reintroduction efforts when captive-bred hybrid peregrines, intended to restore a wildlife population to its historic range, were deemed exotic, and the careful efforts of the FWS to aid falcon restoration came to an unanticipated (though temporary) halt. But the National Park Service's management policies and practices were better aligned with the order. The NPS had worked to reduce populations of nonnative species in the parks following the adoption of the NPS-focused Leopold Report as official policy in 1963. A 1967 report listed thirty parks with active programs to control nonnative plant species, and nine with nonnative mammal control programs.[112] By the 1980s, federal agencies and land trusts like The Nature Conservancy had begun aggressive campaigns to kill nonnative species, which were newly framed as the main threat to wild species. Through these practices, nativity would become a precondition to wildness—of plants and animals both.

7

The Mood of Wild America

On a warm April day in 2004, President George W. Bush picked up a pair of shears and attacked an earleaf acacia, a species originally from Australia that was now found in Florida. It was an election year, and the Democratic challenger, John Kerry, had bashed the president for not prioritizing the environment.[1] Taking on a nonnative species made political sense. So-called invasive species were highly recognizable environmental threats; recent *New York Times* headlines included "Green Invaders Spread Their Tentacles" (on kudzu) and "An Invasion of Hungrier, Bigger Worms" (on Asian earthworms). Ecologist David Lodge warned in a *Times* op-ed that each day, international trade brought "thousands more hitchhiking species into our country," species that threatened to destroy crops, wildlife, and entire ecosystems. Unlike air and water pollution, which were "often correctable," Lodge argued, biological invasions were "usually irreversible, for the simple fact that alien organisms reproduce."[2] Capitalizing on this media trend, Bush's campaign set up the photo-op with an ecological restoration project at the Rookery Bay National Estuarine Research Reserve near Naples, Florida.

The idea that nonnative species were ecologically threatening was a relatively new one. Indeed, from the beginning of European colonization of North America until the mid-twentieth century, many settlers believed nonnative species improved the landscape; indeed, they considered species introduction to be crucial to nation-building. The 1862 legislation establishing

7.1 President George W. Bush holding a pair of pruning shears as he helps clear nonnative invasive plants during a visit to the Rookery Bay National Estuarine Research Reserve in Naples, Florida, April 23, 2004. REUTERS the Department of Agriculture described a direct federal role in the introduction and distribution of nonnative plants. At the local level, species introduction was pursued by horticultural groups and by organizations such as the Cincinnati Acclimatization Society, founded in 1873 by a German immigrant with the goal to "introduce to this country all useful, insect eating European birds, as well as the best singers."[3] Among the many species purposefully introduced to the United States are kudzu, European starlings, and gypsy moths, all of which today are considered ecologically destructive "invasives."

How did nonnative species become such a recognized threat that President Bush could enhance his environmental bona fides with a pair of shears? Invasive nonnative species can lodge themselves in ecosystems, competing with or preying on native species and changing the ecological and aesthetic qualities of places people know and care deeply about.[4] Fears in the 1980s and 1990s about deregulation, the dismantling of national borders, and the rise of global markets also certainly shaped American perceptions of invasive species. Those concerns alone, however, do not explain the proliferation of invasive species control programs in the United States and, eventually, worldwide.[5] To understand the ubiquity of invasive species management, we must also consider the entwinement of restoration practice with a particular form of American nativism.

Like other organizations that managed wild species, The Nature Conservancy (TNC) was driven toward nonnative species control by the Endangered Species Act of 1973, which consolidated the power to handle, reintroduce, and otherwise manipulate endangered and threatened native species in the U.S. Fish and Wildlife Service and the National Oceanic and Atmospheric Administration. One result of this consolidation was that the National Park Service, state parks, and natural areas organizations came to focus their restoration efforts on plant species, for two reasons. First, whereas in the United States wild animals are considered to be owned by the public and entrusted in the care of government, wild plants are generally considered to be the property of the landowner.[6] If a land trust bought a parcel of land, it could then manage nonendangered plant species as it wished. Second, the federal listing of endangered plants lagged significantly behind that of animals. Land trusts and state environmental agencies thus had greater latitude to experiment with restoring rare and endangered plants, for which the legal protocol was murkier and the possibility of public criticism was considerably less. A 1981 survey of state fish and wildlife agencies and TNC natural areas managers found that nearly 30 percent had experimented with propagating and reintroducing endangered plants, whereas none had experimented with endangered animals.[7]

Ecological restoration professionalized in the 1980s as natural areas protection took off in the United States, driven largely by growth in the number and size of land trusts, nonprofit organizations that own and manage land. More than sixty years after Emma Lucy Braun, Victor Shelford, and other Ecological Society of America members founded the Committee on the Preservation of Natural Conditions for Ecological Study, an expanding network of land trusts began acquiring and managing land for the express purposes of ecological study and biodiversity protection. By the early 2000s, most natural areas managers were working to restore native plant species. This work largely involved killing nonnative ones. While it was difficult to get federal permission to handle endangered and threatened native species, federal and state environmental agencies permitted, and even encouraged, the killing of nonnative species. Organizations were motivated, therefore, to theorize the killing of nonthreatened species as a form of ecological care.

As private and state natural areas programs expanded, managers exchanged stories about their experiences propagating native plants and eradicating nonnative ones. In 1988, a group of managers established the Society for Ecological Restoration and Management with the goal of "promoting the scientific investigation and execution of restoration."[8] The organization grew rapidly; it is now called the Society for Ecological

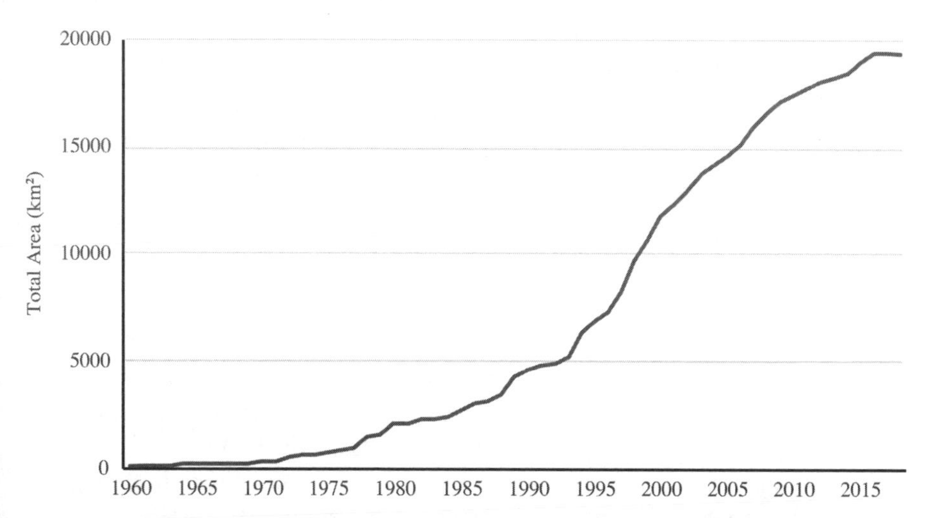

7.2 Total protected area (km²) managed by nongovernmental organizations in the United States by year. This growth in protected natural areas contributed to the professionalization of restoration ecology as a management practice and academic discipline. Data source: World Database on Protected Areas

Restoration, with more than three thousand members in seventy-six countries, and restoration ecology is a thriving scientific discipline with its own academic journals and concerns.[9] Not only did ecologists consider nonnative species a threat to native species—they also saw them as threatening the ideal of ecological restoration without continuous human intervention. Once a nonnative species was established in an ecosystem, it could prove difficult or impossible to remove it. Additionally, the possibility of the arrival of new species every day would require ongoing monitoring and vigilance.

Efforts to eradicate nonnative species entrenched the use of historical baselines in ecological restoration. The word "restoration" connotes a return to a former condition, and today many people think of ecological restoration as an attempt to reestablish a historical species assemblage. Surprisingly, however, historical fidelity did not become a widespread restoration goal among ecologists and environmental organizations until the 1980s. In the United States, historical fidelity often meant the pursuit of precolonial ecologies. Species that arrived after 1492 were deemed nonnative, unwanted reminders of human (colonist) presence and activity. A mode of ecological restoration emerged that attempted to minimize intervention by dividing species into those it was permissible to interfere with (nonnative species) and those that were protected: native species—especially threatened and endangered ones. Killing nonnative species was a form of distanced design, a less direct mode of care for wild species than, say, captive breeding. Rather than manipulating the native species of concern, the manager controlled the nonnative species around it, framing those species as un-wild, as "biological pollution" in the environment of the species of concern. Ecological managers, in turn, naturalized the precolonial baseline, obfuscating their role in designing native nature.

The End of "Do Nothing" Management

The Nature Conservancy evolved from the Ecological Society of America's Committee on the Preservation of Natural Conditions, which had played a central role in framing succession theory as a tool for ecological restoration. The role of the preservation committee within the ESA shifted during World War II, when changes to the U.S. tax code strictly limited lobbying by nonprofit organizations.[10] After a 1945 referendum, the society's executive committee decided that, to protect its nonprofit status, the ESA should "maintain a position as an unbiased scientific expert"; it had "neither the numbers nor the resources to engage effectively in political action."[11] With the preservation committee effectively shut down, eighty-three scientists, including Victor Shelford and G. Evelyn Hutchinson, formed The Ecologists Union, an advocacy organization independent of the ESA devoted to "the preservation of natural biotic communities for scientific use." The Ecologists Union planned to distinguish itself from the growing wilderness preservation movement; instead of focusing on "large, spectacular, and scenic" places, their organization would save examples of "the typical"—small wetlands and prairie remnants, for example, that contained species of interest to ecologists. These were nature reservations.[12]

In 1950, George Fell, a former student of Victor Shelford, convinced the Ecologist Union's leaders to open membership to nonscientists and to establish a fund to purchase natural areas.[13] That year the Ecologists Union opened an office in Washington, DC, and adopted a new name: The Nature Conservancy. The Nature Conservancy's first promotional pamphlets described a need to protect areas as "living museums" of the ecological past.[14] Echoing the earlier justifications of the preservation committee, TNC argued that nature reserves would serve as experimental controls or "check-areas" for federal forestry and wildlife management. Additionally, they imagined that visitors would be "intrigued and thrilled by the possibility of seeing the American Landscape as it was before the whiteman arrived."[15] Thus TNC framed its work as preserving museum specimens of a precolonial ecological past. The natural areas it protected would be "islands of primeval conditions" and "fragments of wild America."[16] By "wild," then, TNC founders meant *appearing* uncolonized. The organization acquired its first parcel of land in June 1955, a sixty-acre tract along the Mianus River, near the border of New York and Connecticut.[17] By 1958, TNC owned sixteen natural areas in New York, Pennsylvania, Maryland, Connecticut, Minnesota, and Illinois.[18]

Today TNC has the greatest revenue of any U.S. environmental organization and is arguably the most influential environmental NGO in the world. It has more than one million members and ownership or oversight of thousands of protected areas worldwide that, combined, cover an area larger than Sweden.[19] But by 1960, TNC had preserved only about 4,000 acres and had an operating deficit. George Fell, by then the director of TNC, quipped, "You couldn't make a go of a modern family farm if you were trying to operate on the scale Nature Conservancy is trying to."[20] The organization's trajectory shifted significantly in 1960, when a new director reoriented TNC under the motto "land preservation through private action" and worked on building alliances with major businesses. For the next decade or so, TNC focused on land acquisition and fundraising. Employing the language of survival that defined American environmentalism at the time, a 1969 TNC poster described money as a "super-weapon" against the indifference that was "tipping the balance away from survival for much of our remaining wilderness as well as man himself."[21] From 1960 to 1969, TNC increased its assets from $750,000 to $20 million.[22]

The Nature Conservancy had been founded by ecologists, but by 1969, ecologists constituted only one-fifth of the board of governors. Upset by their lack of institutional influence, and wanting to take advantage of funding opportunities under the rapidly expanding National Science Foundation, which would not fund land acquisition, the remaining ecologists on the board demanded that a scientist be added to TNC's permanent staff.[23] Their hiring of Robert Jenkins as full-time "ecology advisor" in 1971 had the consequential, if gradual, effect of shifting TNC from passive management of its lands to interventionist ecological restoration. This interventionist work eventually included nonnative species removal at remarkable scales.

In 1971, Jenkins had just defended his doctoral thesis in ecology at Harvard University under E. O. Wilson.[24] The notes archived in Jenkins's "ideas folder" suggest that he began his job at TNC eager to introduce the ideas of ecosystem theory and island biogeography to the organization. Like many ecologists at the time, Jenkins thought of protected areas as islands in a sea of degraded land. Island biogeography theory suggested that smaller islands supported fewer species than larger islands, so conservation biologists argued by analogy that small preserves would support fewer species than large preserves. In 1971, 30 percent of TNC's preserves were smaller than ten acres, and so Jenkins advocated within TNC for the acquisition of larger parcels and the expansion of existing ones.

Jenkins viewed ecological restoration as another method of expanding TNC's preserve system, and he conceptualized restoration in way that he, at least, believed to be novel. In his "ideas folder"—along with notes on plans "to create potholes in the mid-western prairies by carpet bombing" and to "start day school for children of working mothers"—he wrote to himself: "Restoration of natural areas where these have been destroyed: This would involve doing vegetation rehabilitation, erosion control, accelaration [*sic*] of plant succession, reconstitution of original streamflow conditions, etc. Other people are involved in restoring blasted landscapes but none that I know of attempt to recreate the supposed original conditions on the land."[25] In a proposal to TNC's leadership in his first year on the job, Jenkins described the immediate need for "restoration and maintenance of diversity and the resulting stability of the ecosystem."[26] It is notable that Jenkins proposed to restore species diversity and ecosystem stability, rather than individual species like the peregrine falcon. As we will see, the idea of restoring an ecosystem to its "supposed original conditions" would also matter enormously.

An opportunity to attempt ecosystem restoration arose soon after Jenkins arrived at TNC, when Edgar Garbisch requested TNC's help in acquiring Hambleton Island, a rapidly eroding, formerly farmed island on Maryland's eastern shore. Garbisch, a professor of chemistry at the University of Minnesota, had spent a recent sabbatical at his wife's family's eastern shore house. While there, he read Mildred and John Teal's influential 1969 book, *Life and Death of the Salt Marsh,* which argued for the economic and aesthetic value of the "green ribbon of soft, salty, wet low-lying land" on the eastern shore of the United States. Ready for a career change, Garbisch decided to try to rebuild wetlands. "The book and some other writings I came across suggested that wetlands were a renewable resource, unlike coal or oil," Garbisch later explained to a journalist.[27] Through the act of restoration, he would attempt to renew them.

The Nature Conservancy helped Garbisch purchase Hambleton Island, and in October 1971, TNC created the Center for Applied Research in Environmental Sciences (CARES) under Garbisch's and Jenkins's direction. That spring, six staff members collected seeds of *Spartina, Juncus, Typha, Scirpus,* and other native marsh plants from around the Chesapeake Bay and planted them in indoor growth chambers. They next transplanted the seedlings into more than five hundred barge loads of sand and mud, which were transported to Hambleton Island.[28] To avoid sinking into the tidal mud flats, they wore snowshoes. Within a few months the CARES team had planted sixty thousand seedlings on Hambleton Island, covering around 1.5 acres. An article in *Sports Illustrated* asked of Garbisch, "Can a former professor of chemistry at the University of Minnesota create in half a year what it takes nature 1,000 years to accomplish?"[29]

For his part, Jenkins believed that salt marshes were good candidate ecosystems for restoration. Because salt marshes are simple "in terms of structure and diversity," Jenkins wrote in a 1971 report, it should be possible "to reconstruct a salt marsh ecosystem with every hope of rapid reversion to natural conditions."[30] Jenkins imagined that after working on salt marshes, CARES would go on to research the restoration of prairie and forest plant communities that contained more species. At the Midwest Prairie Conference at Kansas State University in 1972, Jenkins described CARES's mission as essential for ecological science: to save representative ecosystems of every type as "benchmarks of naturalness"—language that echoed ecologists' defense of "check-areas" in the 1930s—TNC would have to begin restoring "disturbed areas," because it would be impossible to find, or prohibitively expensive to acquire, "pristine" preserves of many ecosystems.

7.3 **Aerial views of The Nature Conservancy's CARES Hambleton Island restoration project before the start of work in 1971 and after the work was completed in 1976.** Reproduced with permission from Elsevier from Edgar Garbisch, "Hambleton Island Restoration: Environmental Concern's First Wetland Creation Project," *Ecological Engineering* 24 (2005): 289–307, Figure 13.

"So far as I know," Jenkins concluded, CARES "is the only such center wholly devoted to ecosystem restoration in existence anywhere, though hopefully it will not be the last."[31] Indeed, it would not be.

Despite Jenkins's enthusiasm, the CARES program was short-lived at TNC. Garbisch and his collaborators constructed a large greenhouse and staff offices in St. Michaels, Maryland, and in 1972 cultivated more than one hundred thousand seedlings. But for reasons undocumented, Garbisch chose to disaffiliate from TNC, and CARES became the public, not-for-profit corporation Environmental Concern, Inc.[32] The Clean Water Act of 1972 (whose importance to the emergence of off-site ecological mitigation is detailed in Chapter 8) would create a constant source of business for Environmental Concern, as developers were required by law to mitigate damage to wetlands. Over the next decades, Environmental Concern was hired by businesses, the U.S. Army Corps of Engineers, and private landowners to restore and create hundreds of wetlands along the Atlantic Coast.[33]

7.4 Edgar Garbisch and members of Environmental Concern planting native grasses in a tidal flat near Huntington, Long Island, New York, c. 1975. Courtesy of Environmental Concern Inc.

Even though CARES was under the auspices of TNC for only a couple of years, it instigated TNC's slow shift away from hands-off management and toward interventionist management. Between 1954, when TNC acquired its first preserve on the Mianus River, and 1971, when Jenkins joined TNC as the first staff scientist, TNC's natural areas had been managed by a loose network of volunteers. These volunteers visited preserves periodically to check for incidents of vandalism, timber theft, and littering. Adding or removing species from the landscape was against TNC's policy; any such intervention was seen as decreasing the "wildness" of a site. The leadership of TNC even debated whether hiking should be allowed, since it might threaten that wildness or naturalness. "The idea of even limited management of natural areas is a hard mental hurdle for many of our members," one member of the board of governors wrote in 1958, "We heartily subscribe to the principle, 'If in doubt, do nothing.'"[34]

In contrast, the 1972 *TNC Preserve Management Manual* stated, "Simply allowing nature to take its course on our many preserves is not always the best guarantee of achieving our conservation goals." Unlike other land preservation organizations, which "try to avoid manipulating natural processes," TNC would begin "a program of active restoration," manipulating sites in which past disturbances "have left scars or deteriorating conditions which are unlikely to be self-healing."[35] These manipulations could include mowing or burning grasslands to "control succession" and eradicating nonnative species or overly abundant herbivores. No longer trusting in the homeostatic capacity for ecosystems to maintain themselves, managers would take on the responsibility of maintenance.

The Nature Conservancy's ecologist founders had likened natural areas to museums; the purpose of TNC's preserves was "maximum protection from human interference," a 1958 white paper on TNC's objectives explained. Like museum pieces, natural areas would be "outdoor exhibits" available only for "limited observation."[36] Jenkins, fresh from his ecological training in 1971, analogized TNC's natural areas holdings not to a museum, but to a database, describing it as "the biggest information storage and retrieval system we are ever likely to encounter," one that "must be put to use in environmental problem solving."[37] By the museum metaphor, TNC's natural areas had been objects that TNC protected so that they could be observed by scientists. By the database metaphor, natural areas were systems that could be queried experimentally and that required experts to organize and steward them. TNC hired a full-time director of stewardship in 1973, and the following year it renamed the Office of Preserve Management the Stewardship Office.[38]

Robert Jenkins and others at TNC envisioned ecosystem restoration as a process in which humans facilitated natural recovery rather than dictating its final form: "For the most part, ecosystems must restore themselves and our role should be to subsidize more than to guide," Jenkins explained in 1972. First, natural areas managers would remove "the disturbance," the "obvious barriers to natural ecological recovery," such as livestock or pavement, to "hasten natural succession processes." Next, they would reassemble the ecological community. If populations of native species were too far away from the site to recolonize it themselves, then managers would transport species from locations where they still existed. If nonnative species had colonized the site, managers would eliminate them.[39] This was "a science which has not even been born yet," Jenkins declared, and TNC was "the obvious agent for midwifing the birth before it is too late."[40]

The new emphasis on stewardship at TNC did not mean giving up the goal of maintaining "wild" areas. At the aforementioned Midwest Prairie Conference and elsewhere, Jenkins publicly distinguished the work of CARES from the management practices of government environmental agencies, which, in his view, demonstrated "little concern with environmental naturalness." Jenkins contended that, in many cases, government reforestation and soil conservation projects caused "additional artificialization of the environments they treat." In contradistinction, TNC's ecosystem restoration projects would attempt to achieve "ecological stability and function." If a historical ecological community no longer existed, Jenkins explained, as was the case in the eastern United States, where "what remains hardly measures up to the descriptions of the early explorers," TNC would "attempt the recreation of such communities through the reassembly of their scattered components."[41]

In these early-career talks, Jenkins articulated two key ideas that would structure the science and practice of restoration ecology from the mid-1980s until the mid-2000s. First, ecological restoration was to be a restrained activity, distinct from species management practices that had come before. Ecological restorationists would not "do nothing," but they should intervene only as much as was required to restore an ecosystem's original ability to maintain itself. Restoration represented a third way, a mode of management that was more active than preservation but more restrained than conservation. As a 1990 TNC training handbook put it, "the ideal goal of any restoration project is a naturally functioning landscape which once planting and establishment has taken place, then requires no human input."[42] The word "ideal" is telling, of course; ecologists had been arguing for decades already that some ecosystem damage was irreversible. Second, the goal of restoration should be the re-creation of a precolonial "baseline" community such as "early explorers" would have encountered. How to pursue these two ideas simultaneously—and without running afoul of the developing federal regulatory regime—became the central puzzle of the discipline of restoration ecology.

Re-Creating the Ecological Scene

The idea of restoring ecological communities to a precolonial baseline emerged in parallel in multiple environmental organizations. The Nature Conservancy had imagined as early as 1953 that visitors to their preserves would see "the American Landscape as it was before the whiteman arrived."[43] Similarly, the 1963 Leopold Report on National Park Service policy enshrined the management goal of restoring ecological communities to "the condition that prevailed when the area was first visited by the white man," so that each national park would "create the mood of wild America."[44] For decades before that, restorationists had pursued goals other than recreating historical assemblages of plants and animals. Game restorationists hoped to increase populations of individual species, whether through captive breeding, or planting food and cover crops, or, in Lauren Donaldson's case, accelerating evolution. Naturalistic gardeners cultivated scientifically interesting plant species, with nativity as both ethos and aesthetic. Before the 1980s, most ecological restoration was about saving species in the present and for the future, with little notion of turning back the clock.

The Special Advisory Board assembled by Secretary Stewart Udall in response to the Yellowstone elk culling controversy gave itself wide latitude to rethink species management under the auspices of the National Park Service. The congressional act that created the NPS in 1916 stated that the purpose of national parks was "to conserve the scenery and the natural and historic objects and the wild life therein and to provide for the enjoyment of the same in such manner and by such means as will leave them unimpaired for the enjoyment of future generations." In their 1963 report, "Wildlife Management in the National Parks" (the Leopold Report), Starker Leopold and his colleagues argued that, over the past forty years, the NPS had developed a philosophy of wildlife preservation in which parks were refuges and managers protected certain animals from fire and predators. But recent ecological research suggested that few parks were large enough to be "self-regulatory ecological units." Thus, the authors recommended that the NPS embrace management, including the "active manipulation of the plant and animal communities."[45]

Active ecological management, the Leopold Report continued, required a goal. As a primary goal, the authors recommended that the NPS manage its lands "to preserve, or where necessary to re-create, the ecologic scene as viewed by the first European visitors." They argued that the ecological communities then found in the national parks were "artifacts, pure and simple" of logging, fire suppression, livestock grazing, wetland draining, and predator control. Protection alone, therefore, would not guarantee the authenticity of this "primitive scene" or the presence of a diversity of native animals. Instead, the National Park Service would need to pursue "active management aimed at restoration of natural communities of plants and animals," a task demanding "skills and knowledge not now in existence."

The Department of the Interior, in other words, would need to invest in restoration research.

The "primitive scene" imagined by the Leopold Report authors was one in which colonization was perpetually beginning. Tourists would view ecosystems as they appeared when they were "first visited by the white man," a scene like that "viewed by the first European visitors." Notably, the report employed the language of "visitor"—not "colonist" or "settler." Noting that there were no longer antelope in Grand Teton National Park, the authors wrote, "If the mountain men who gathered here in rendezvous fed their squaws an antelope, a 20th-century tourist at least should be able to see a band of these animals." This racial slur was the report's only mention of Native Americans, as the possession of white mountain men. The purpose of management, then, would be to create a world that appeared never colonized but imminently colonizable. Preservation was not enough to create this colonial fantasy; national park management would require active intervention to reverse (or at least to hide) the ecological consequences of colonization. Through the implementation of the board's recommendations, ecological management in the National Park System, and ultimately across the United States, became enmeshed with an effort to undo the ecological effects—but not the social and political effects—of settler colonialism.

The first step in restoring ecosystems to precolonial conditions, the Leopold Report argued, was to use paleoecological methods to ascertain which plants, animals, and biotic associations "existed originally in each locality." This would be a novel application of paleoecological data, which scientists had previously used to reconstruct past climates, not past ecological communities. Using this information, ecologists could experiment with reintroducing plant and animal species. "A reasonable illusion of primitive America could be re-created," the report's authors contended, "using the utmost in skill, judgment, and ecologic sensitivity," language reminiscent of Edith Roberts and Elsa Rehmann's *American Plants for American Gardens.*

The influence of Aldo Leopold's work on the (Starker) Leopold Report is clear. The report advocated manipulating plant species, not reducing predators, as the method for restoring desired species. It suggested restoring antelope in Jackson Hole, Wyoming, for example, by planting native forage plants. Unlike wilderness preservation, such management would certainly require active intervention, but it was an intervention far less controversial than killing predators. Indeed, Starker Leopold and his coauthors recommended that while ecological interventions in national parks "may at times call for the use of the tractor, chain-saw, rifle, or flame-thrower," such interventions should be hidden from visitors if at all possible. This recommendation would guide the future of federal restoration practice. Subsequent restoration work remained, and remains, outside of public view, preserving in the American public's imagination the national park as yet-to-be-colonized landscape.

In May 1963, just two months after the Leopold Report's completion, Secretary Udall declared it official NPS policy. This promised a significant departure from the service's previous practices: up until this point, the NPS had relied on myriad and conflicting "handbooks" for guidance on wildlife management, road and trail management, and master planning. Park superintendents banned poaching and grazing, but they also fed elk, corralled bison, and poisoned coyotes. Slowly but surely, the Leopold Report led to the implementation of ecological restoration, deeply reshaping the service's activities.[46]

That is not to say that the Leopold Report was uncontroversial. The report was vocally resisted by those managers and wilderness advocates who believed that nature should be "let alone" in national parks. Howard Zahniser, executive secretary of the Wilderness Society and principal author of the Wilderness Act, wrote that the Leopold Report "poses a serious threat to the wilderness within the national park system and indeed to the wilderness concept itself." An essential feature of wilderness, Zahniser went on to argue, was the ability of "natural forces" to operate without human influence. "With regard to areas of wilderness, we should be guardians not gardeners," he concluded.[47] "This is the most extreme anti-park policy statement I have yet encountered," wrote NPS biologist Adolph Murie of Starker Leopold's call for scientists to manipulate vegetation in order to simulate "primeval America." Murie continued, "Is a scene natural when you chop trees down or plant trees! Is this an honest presentation! Do we want to make Disney Lands out of our roadsides!"[48]

Nevertheless, the Leopold Report helped reorient the National Park Service from a preservation approach to a restoration approach. This reorientation is clearest in the domains of fire and nonnative species management. As historian Stephen Pyne details, the Leopold Report was central to a shift in federal fire management.[49] The report described the "controlled use of fire" as the "most natural and much the cheapest and easiest" method for manipulating plant species. Starker Leopold would later reflect that he began to think about the role of fire in semiarid ecosystems on a 1937 trip in the Sierra Madre with his father, Aldo Leopold, in which they watched their Mormon guides toss lighted matches into the grass as they traveled. At a Sierra Club conference in 1955, Starker, then an assistant professor of zoology at the University of California Berkeley, argued that the persistence of American grasslands, chapparal, and certain conifer forests was dependent on frequent fires, and that those tasked with preserving natural areas were "going to have to accept responsibility for some experimentation and management."[50] Biologists at Kings Canyon National Park began experimenting with controlled burns of giant sequoia groves, which had stopped reproducing after eighty years of fire suppression, in 1964. The first controlled burn experiments at Yellowstone National Park began the following year. By 1968, the NPS acknowledged that fires resulting from natural causes contributed "to the perpetuation of plants and animals native to [fire-prone] habitat," and it began to allow "natural" fires to "run their course," and to approve prescribed burning as a valid substitute for natural fire.[51] Controlled burns would become an important plant restoration technique and a central area of inquiry in the discipline of restoration ecology.

The Leopold Report also led the NPS to actively kill nonnative species. A 1967 report listed thirty parks with active programs to control nonnative plant species and nine with nonnative mammal control programs.[52] Both methodologically and ideologically, this work was connected to the earlier work of the American Bison Society and similar organizations. While those organizations had reintroduced single species for their connections to a mythic white settler past, the NPS now restored groups of species with the goal of creating a settler scene. The audience for restoration was park visitors who would be enabled to imagine themselves as white settlers. It was a form of reenactment.

In its early form, the goal of precolonial restoration was explicitly cultural and aesthetic. In the 1980s, however, ecologists offered a second justification for the precolonial baseline: its scientific value. David Graber, a research scientist at Sequoia–Kings Canyon National Park in California and former student of Starker Leopold, argued that the value of "wild ecosystems" was not that they reminded nostalgic visitors of the past, but rather that they could serve as controls against which to measure recent human-induced changes like acid rain and climate change.[53] Likewise, in an influential 1985 article, ecologist Reed Noss argued that "presettlement-type" ecosystems were relatively stable and could provide "a baseline against which to measure the vicissitudes of humanized landscapes."[54] Increasingly, ecologists argued that the restoration of precolonial ecosystems would allow timeless, natural processes to prevail once again. The scientist was now the audience. The cultural goal of settler fantasy was sublimated into the technical goal of returning ecosystems to a "wild" or "natural" state.

It was not foreordained that ecologists and wildlife managers concerned about recent environmental changes in the 1980s would choose the beginning of colonization as the baseline for ecological restoration. Other options, for example, could have included the rise of American industrialization or the postwar sprawl of the American suburbs. Restoration work is an implied critique: an attempt to undo environmental damage for which something—or someone—was to blame. The precolonial baseline for ecological "originality" suggested that culpability for environmental destruction lay with settler colonists. Projecting the moment of environmental destruction into the past was politically safer than critiquing more recent projects or policies. It also conveniently implied that settler colonialism was a completed event, rather than an ongoing project, and it dovetailed with the developing emphasis in the 1980s on nonnative invasive species: it was easy to blame the earliest white settlers for breaching the continent's supposed prior ecological isolation. Robert Jenkins, for instance, held that "native ecosystems were still basically intact" until the first European settlers had "stripped the forests, burned the prairies, and laid bare the soil in their destructive drive westward." But it should be stated clearly that the most important casualties were Indigenous people, whose presence Jenkins omits.

The omission is significant, because many scientists by the 1980s acknowledged the cultural, historical, archeological, and ecological evidence that Native Americans had managed environments intensively, and continued to do so.[55] In 1982, two forestry professors published a paper in *Ecology* challenging the idea that prescribed burning would restore giant sequoia ecosystems to their precolonial condition.[56] Instead, they argued, the NPS would be encouraging a new type of age-stratified forest that contained few white fir seedlings but many large white firs, which had come to dominate during decades of fire suppression. The only thing to do to create space for giant sequoias to germinate, they concluded, would be to take down white pines with chainsaws, a suggestion that was anathema to some NPS managers. David Graber mailed the article to Starker Leopold, writing, "The Park Service will find itself with the task of simulating Indian burning in perpetuity." He continued, "If Indian burning must be simulated, must also other ecological roles played by Indians, as predators for example, be likewise simulated?" Starker replied that "it makes little difference to me whether the fire is set by lightning, by an Indian, or by [NPS biologist] Dave Parsons, as long as the result approximates the goal of perpetuating a natural community."[57] A few days later, a volunteer from the Sierra Club visited Starker to record an oral history. Starker reflected,

> Some of the Park Service biologists, including Dave Graber—who was one of *my* products, one of my own kids, and a damn good boy—they're uneasy about arbitrary decisions. You decide to cut down a tree; who's to decide which tree to cut? Should you cut any tree at all? And they'll all go for the idea of letting natural fires run. If lightning starts a fire, then, this is something that has to do with God, and you didn't have to make a decision. But they were

really concerned in a genuine way with arbitrary decisions that have to be made. As soon as you move into management, you're going to have to manage for something, you're going to try to re-create it, try to maintain a given type of ecosystem. And I still defend it. OK, so you make some arbitrary decisions. So what? They may be arbitrary, but that doesn't necessarily mean they're capricious, as long as your objective is a goal of what you construe to be a natural ecosystem.[58]

Ecologists promoting the precolonial baseline thus depended either on naturalizing Native Americans—diminishing their agency by considering them "part of nature" like nonhuman animals—or on trivializing their established impact on the land. Ecologists and land managers routinely suggested that Native Americans, unlike white settlers, were part of "wild" or "primeval" nature. Jenkins wrote that "the Indian was a resident rather than manager or destroyer of the wilderness."[59] Reed Noss held open "the question of whether Indians should be considered a natural and beneficial component of their ecosystems, perhaps coadapted with fire and the vegetation since the Pleistocene."[60] A 2003 article in the journal *Ecological Restoration* acknowledged that many national parks "were shaped by the unremitting labor of generations of indigenous peoples." But rather than calling for restoration of Indigenous sovereignty, it suggested that ecologists should "simulate the actions of indigenous peoples" in areas "no longer used by tribal peoples" in order to restore biodiversity to a precolonial state. The article alternately described Native American actions in definitely agential terms ("management regimes") and in arguably naturalizing terms ("indigenous disturbances"), and suggested that ecologists could learn about, and then "mimic" or "simulate," Native American harvest and fire practices.[61]

Few ecologists in the 1980s, then, debated the politics of the goal of restoring ecosystems to a precolonial baseline.[62] But they did debate the feasibility of achieving that goal. When NPS scientists and their academic collaborators attempted to establish a precolonial baseline, they envisioned past plant (and, much more rarely, animal) communities using both historical documents (settler diaries, land surveys, photographs) and paleoecological studies of fossil pollen, tree rings, and charcoal deposits.[63] This created a role for paleoecologists in species management.[64] Colonial records of species were rare and notoriously unreliable, however, and paleoecological data was biased toward those species whose pollen was best preserved. Uncertainty about the composition of precolonial ecological communities proved an impediment to restoration. For instance, following the Leopold Report, the National Park Service set January 1778 as their restoration baseline in Hawai'i, permitting species introduced by ancestors of native Hawaiians but excluding "those that are post-Cook." But one University of Hawai'i ecologist argued that, even after removing nonnative plants and animals and reintroducing native ones, it would not be possible to restore ecosystems in national parks to late-eighteenth-century stages, "simply because we don't know what these were."[65] There were significant practical impediments, too, including conflicts over how to restore precolonial plant species while minimizing intervention. Between 1971 and 1975, the NPS killed 12,976 introduced goats in Volcanoes National Park in Hawai'i. But efforts to eradicate goats from Volcanoes National Park would pit scientists against one another. While many ecologists believed that goats preferentially ate native plants like the endemic koa tree, others believed that goats could be a useful tool in controlling the spread of nonnative plant species.[66] Threats like species introductions and climate change were also practical impediments to minimalist restoration, and remain so, because they are diffuse and ongoing. Today, many ecologists ask whether it is possible, or wise, to re-create historical states when the climate is changing, but in the 1980s and 1990s, the more pressing question for many land managers was how to limit their interventions when new nonnative species were constantly arriving at their sites.

For decades, ecologists had criticized federal environmental agencies, especially the Fish and Wildlife Service, for manipulating ecosystems so as to make them less natural. At a 1976 symposium on "ecological reserves," Carl Hubbs of the Scripps Institute of Oceanography criticized the use of airplanes to restock trout in wilderness areas, speculating that such introductions would lead to the widespread extinction of native species—"Vanishing Americans"—and even to the extinction of "whole ecosystems."[67] Such criticisms only intensified as ecologists took on leadership roles in environmental NGOs like TNC. To restore wild ecosystems, in David Graber's view, managers should use only "subtle tools necessary to reverse or mitigate anthropogenic influences without excessive interference in natural processes."[68]

During the 1970s, TNC largely adhered to the view that any human action could make a natural area unnatural. Preserve managers were not to intervene. When Jenkins was hired as "Ecology Advisor" at TNC in 1971, TNC was in many ways a decentralized institution, with management decisions made at the level of regional offices or individual preserves. For most of a decade, articles in the internal magazine *Stewardship* focused on established concerns like trail maintenance and patrolling for litterers, and the official Stewardship Guide prohibited "control of exotic species," along with enhancing the "neatness" of an area. Dead animals were not to be moved, unless they were in a "public use area."[69]

This would change in the 1980s with the rise of concern about nonnative species. In response to the passage of the Endangered Species Act, natural areas managers inventoried protected lands. As they did, many recorded the presence of multiple nonnative species in areas that were supposedly "undisturbed." Their observations challenged the long-held view that nonnative species could establish themselves only after a human-caused disturbance to an ecosystem, such as plowing or clear-cutting—or nuclear bombing.[70] Notably, ecologists came to consider invasive species *to be* ecosystem disturbances, rather than to *result from* ecosystem disturbances, and thus to be existential threats to native species.[71] This turn is exemplified by a 1990 article in which ecologist Peter Vitousek argued that invasions by nonnative species represented an "ecosystem disturbance" akin to those studied by Eugene Odum in the 1960s and at the Hubbard Brook Experimental Forest in the 1970s.[72] Threat discourse would thus come to dominate the field of invasion biology. Nearly half of the articles published in the academic journal *Biological Invasions* in the first five years after its founding in 1999 used the word "threat."[73] A 2000 report by the International Union for Conservation of Nature concluded that the impacts of nonnative invasive species were "immense, insidious, and usually irreversible."[74] Ecologists described nonnative species as "biological pollution," a persistent, unwanted reminder of human economy, one that necessitated intervention. As the interventionist ideas of Jenkins and other staff scientists spread through TNC, it and other environmental organizations would begin killing nonnative species on a massive scale.

Professionalizing Natural Areas Management

As natural areas managers pursued ecological interventions distinct from those of federal environmental agencies, invasion biology, once the pursuit of a handful of economic entomologists, became a robust subdiscipline of ecology and contributed to the professionalization of restoration ecology. Lack of scientific concern about nonnative species prior to the 1980s cannot be explained by the absence of those species. Some of today's most widely recognized invasive nonnative species were well established by the nineteenth century. Purple loosestrife proliferated in New England wetlands by the 1830s, having likely been introduced from Europe via ship ballast soil.[75] Leopold Trouvelot introduced the gypsy moth to Massachusetts in 1868 or 1869; he hoped to use the species to produce silk.[76] It is even unclear whether rates of species introductions were higher in the twentieth century than they had been in prior centuries, in large part because detection records are reflections of changing scientific interests and practices as much as they are reflections of biological trends. In other words, it may well be the case that there are more twentieth-century records of species introductions because more people thought to keep those records—not because more species were introduced.[77]

Although ecologists had suggested that nonnative species competed with native species prior to the 1980s, they generally held, as Tom Cade had argued in defense of his hybrid peregrine falcons, that native species were better adapted to their environments, and therefore able to persist. In a 1942 article, for instance, ecologist Frank Egler argued that, in the absence of the "anthropic disturbances" of fire and grazing, most of the species introduced to Hawai'i would be "destroyed by the indigenes."[78] Such views were consistent with those of British ecologist Charles Elton, who is often credited with writing the first treatise on invasion biology, his 1958 *The Ecology of Invasions by Plants and Animals.*[79] In it, Elton speculated that, upon arrival to a new location, an introduced species would be confronted with "a complex system, rather as an immigrant might try to find a job and a house and start a family in a new country or big city."[80] He quipped that in the subsequent competition for resources, species would "be bumped into—and often be bumped off." He did concede, however, that many introduced species were able to "co-exist" with their new neighbors and suggested that a "rich and interesting" ecological community could include "a careful selection of exotic forms."[81]

Interest in competition between native and introduced species was piqued in 1960 when Garrett Hardin, who is today best known for his 1968 essay, "The Tragedy of the Commons," published a paper in the journal *Science* on the "competitive exclusion principle," contending that if two species that occupied "precisely the same ecological niche" ended up in the same location, one of the species would become extinct.[82] Hardin's principle gained traction during the Cold War, as capitalist and communist blocs competed for exclusive influence over newly independent states. This coincided with ecological work on "disturbed" ecosystems indicating that these systems were less resilient than had previously been thought and were perhaps vulnerable to invasion by new species.

These views gradually made their way into natural areas management. A 1975 article in the journal *Bioscience* is indicative of the ambivalence toward nonnative species that characterized the 1970s and that would trouble peregrine falcon reintroduction. The authors argued that exotic animals could "serve man's interests well" if they "provide a new food or game resource," as in the case of trout introductions, while also noting that introduced species could become "pests, destroyers of habitat, or unwanted competitors of native species."[83] As concern about nonnative species grew in ecological circles, ecologists began studying nonnative species in natural areas as well as on agricultural land. Opportunities for such work grew enormously between 1970 and 1990, as the number of land trusts and state natural areas programs exploded. As of 1970 there were state natural areas programs in twelve states: Connecticut, Florida, Hawai'i, Illinois, Indiana, Iowa, New Jersey, New York, North Carolina, Ohio, Pennsylvania, and Wisconsin. By 1975, eleven more states had joined this list. During this same period, TNC's holdings increased from 140,000 acres to 1,586,706 acres.[84]

The rapid growth of protected areas under state and private management drove the professionalization of natural areas management, an applied field that, alongside academic ecology, played a central role in the rise of ecological restoration. In 1974, a group of natural areas managers met at Wyalusing State Park in Wisconsin to discuss creating a network for people working for natural areas programs to exchange ideas and compare management outcomes. Four years later, at the fifth Natural Areas Conference, fifty-three attendees decided to found the Natural Areas Association, a professional society for "people working in the natural area preservation and management field." In 1981, the association launched the *Natural Areas Journal,* proclaiming that "the field of natural areas preservation and management is relatively young, but growing fast."[85] Incidentally, that same year, botanists at the University of Wisconsin–Madison Arboretum founded another journal, *Restoration & Management Notes.*[86] In his opening editorial, William Jordan III explained, "We believe that a new discipline is taking shape in response to the growing challenge of developing better, more effective and more economical ways of restoring and managing ecosystems." The new journal, Jordan continued, would "deal only with the development and management of communities that are native or at least ecologically appropriate to their site."[87] Thus restoration ecologists began to distinguish themselves from other types of environmental managers through a commitment to historical fidelity.

In their first few years of publication, the majority of articles in *Natural Areas Journal* and *Restoration & Management Notes* focused on land acquisition and on strategies for inventorying endangered species. By 1984, however, they were publishing articles on how to exterminate widespread invasive species like Japanese honeysuckle and Canada thistle.[88] That same year, the international Scientific Committee on Problems of the Environment (SCOPE) convened an advisory committee on the impacts of biological invasions on ecosystems. Bringing together ecologists from the United States, South Africa, New Zealand, and Australia, SCOPE focused on three questions: (1) What factors determine whether a species will be an invader or not? (2) What are the characteristics of an environment that make it vulnerable or resistant to invasions? (3) How can insights from invasion ecology be used to develop effective management strategies?[89] This last question speaks to the increasing ties between academic ecology and species management in the 1980s.

This growing emphasis on invasive species was spurred by the fact that NGOs and states were eclipsing the federal government as the primary managers of natural areas, even as the government consolidated its authority over handling endangered species. Because the threat of invasive species was one that state and private land managers were authorized to address, professional ecological restorationists increasingly pursued nonnative species eradication. In 1984, the same year that academic ecologists initiated the SCOPE invasion program, natural areas managers convened a Native Plant Revegetation Symposium in San Diego, California. The symposium brought together managers and scientists from state, federal, and private environmental organizations, including the FWS, the NPS, the U.S. Army Corps of Engineers, the California Department of Parks and Recreation, and TNC. Each agency offered different rationales for native plant restoration. The representative from the U.S. Army Corps of Engineers was interested in revegetation "not to create pristine examples of pre-project flood plains," but instead to provide "high quality wildlife habitat" as required under the Endangered Species Act. In contrast, the NPS representative reported on a recent project in which more than 23,000 native plants had been planted at Golden Gate National Recreation Area to maintain "the integrity and character of the parks," in line with the recommendations of the Leopold Report of 1963. Tight budgets, in turn, spurred the California Department of Parks and Recreation's interest in native plant restoration; in his presentation, Wayne Tyson explained: "Why are we interested in native plant revegetation? Well, anybody can see that the wild vegetation on natural hillsides is permanent and can shift for itself. Nobody had to pay a dime to put it there, and its hasn't cost anybody a cent to irrigate and maintain it since The Beginning. We've looked at our water bills, our limited supply of water and the skyrocketing cost of major water projects.[. . .] There must be some way to take advantage of the operant processes that produce such self-sufficiency."[90]

The reasons for investing in native plants thus varied substantially by organizational mandate. But this shared investment was more than sufficient for nonnative species removal to become the organizations' major point of overlap, promoted by natural areas managers through conferences, professional journals, and newsletters. In 1987, at one of TNC's properties in Magnuson Butte, Washington, volunteers applied herbicides and participated in the "first annual weed pull," targeting nonnative Dalmatian toad flax, Canadian thistle, and knapweed.[91] That year, TNC stewards at Waikamoi Preserve in Hawai'i managed a $150,000 budget to remove feral goats and pigs.[92] While any manipulation of endangered species was strictly regulated under the Endangered Species Act, killing nonnative plants on privately owned lands required no permissions. And while killing nonnative animals generally required the approval of state fish and wildlife departments, these departments increasingly endorsed such measures: their scientists were attending the same conferences and reading the same journals as the scientists at NGOs. Nonnative species removal was easy to promote, good for fundraising, and fairly unregulated. Perhaps most importantly, it was a highly visible task of ecological care. Nonnative species removal began to take its place at the center of natural areas management.

With the increasing visibility and popularity of nonnative species removal, many state park and natural area systems adopted the goal of restoring entire ecological communities to their precolonial compositions, despite the epistemic and practical difficulties. In 1985, for example, the policy of the Florida Department of Natural Resources was to manage its properties "to appear as they did when the first Europeans arrived."[93] In another project, TNC planted hundreds of acres of "lost plant species" and reintroduced three hundred bison at the former Barnard Ranch (former Wichita and Osage homeland), in northern Oklahoma. The 1989 press release, titled "Extinct Ecosystem to Be Restored," described the effort as "one of the last opportunities to restore a tallgrass prairie ecosystem to presettlement conditions."[94]

The quest to restore precolonial conditions in private and federal natural areas involved (and continues to involve) a wide range of methods. Fire, hand-pulling, herbicides, and mowing were all used for plant control. By 1992, TNC preserve managers in California were also hunting hogs with dogs, poisoning honeybees, and shooting sheep in the name of restoring nature. In Hawai'i, natural areas managers led aerial hunts of sheep and goats.[95] Conservancy managers led volunteers in Connecticut's Chapman Pond and Bauer Preserves to help control the spread "of the evil phragmites and purple loosestrife."[96]

Sometimes invasive species management proved controversial, especially on federal lands, and especially if it involved the deaths of animals. Animal welfare activists found themselves increasingly at odds with restoration ecologists who wished to exterminate nonnative species.[97] Management of native species sometimes caused controversy as well, for example when the NPS refused, in 1994, to let wildlife veterinarians treat bighorn sheep that were falling from cliffs because the sheep's partial blindness was caused by a "native" disease.[98] Despite these occasional controversies, however, invasive species management became entrenched both in practice and in the emerging scientific discipline of restoration ecology.

Founding the Society for Ecological Restoration

In April 1987, during his keynote address to the second Native Plant Revegetation Symposium in San Diego, California, William Jordan III suggested the need for a national restoration organization. Later during the symposium, Jordan and other attendees sketched out a plan for the Society for Ecological Restoration and Management, an organization "with the goal of promoting the scientific investigation and execution of restoration."[99] The society conducted a membership drive at the January 1988 conference Restoring the Earth, held in Berkeley, California. Organized by John Berger, the conference attracted more than eight hundred attendees, scientists, corporate leaders, and government officials, representing hundreds of projects.[100] Edgar Garbisch's organization, Environmental Concern, reported that they had created more than two hundred marshes from Maine to Virginia. Ecologist Dan Janzen described his plans to restore 150 miles of tropical forest in ranching areas of Guanacaste, Costa Rica.[101] A *Newsweek* article on the conference reported, "The fix-it men of the environment are here. Not content to merely lobby for anti-pollution laws or sue to keep a developer from building on a bird sanctuary, they are repairing what man has already damaged. They are determined to do no less than turn back the calendar to the days when buffalo roamed the prairie and salmon ran as thick as molasses."[102]

The Society for Ecological Restoration and Management (SERM) held its first meeting in Oakland, California, in January 1989.[103] Among its stated goals were "to promote research into all areas (scientific, technical, social, political, economic and philosophical) related to the restoration, creation and subsequent management of biotic communities" and "to facilitate communication and the exchange of restoration technologies between restorationists." These goals reflected Jordan's assertion with the founding of *Restoration & Management Notes* that restoration ecology would be "a new discipline in its own right," one concerned with the development of ideas and techniques "peculiarly its own."[104] The members of SERM would distinguish themselves from those working on "reclamation in a more general sense" through their "commitment to a historical model—to putting something back the way it <u>was,</u> whether or not that is the way you happen to prefer it."[105] Thus, many early members of SERM differentiated their work from that of, say, the Fish and Wildlife Service, by a commitment to historical fidelity, or at least a certain imaginary of a precolonial native ecosystem.

Not surprisingly, professional natural areas managers and academic ecologists vied to define the goals of the new discipline: Was it to protect native species in perpetuity, or to better understand how ecological communities were assembled and how ecosystems functioned? Some restoration ecologists were more concerned about the threat of nonnative species than others, but members of SERM were in agreement that natural areas preservation would henceforth depend on restoration.[106] Ecosystem ecologists also strove to define the new discipline's parameters. John Cairns wrote that "damaged ecosystems" were "just as interesting as pristine ones," as evidenced by the work of ecologists like Daniel Simberloff, E. O. Wilson, and Gene Likens, who had "deliberately disturbed or manipulated ecosystems in order to study their response."[107] Ecologist Michael Gilpin defined restoration ecologists especially broadly, as "anyone manipulating or managing a community by adding, removing, or manipulating species."[108] British ecologist Anthony D. Bradshaw suggested that restoration would be a testing ground for ecological theory: whereas academic ecologists spent their time "taking ecosystems to bits and examining their component parts," restoration ecologists would "put the parts together again to see if they work."[109]

Notably, restoration ecology emerged as a discipline at the same time as conservation biology, and a number of people participated in both fields. The Society for Conservation Biology was founded in 1985 by scientists working on the genetics and population modeling of endangered species—among them Michael Soulé, Daniel Simberloff, Jared Diamond, and David Ehrenfeld—and the first issue of the journal *Conservation Biology* was published in 1987.[110] In subsequent years, conservation biology would retain a focus on zoology and genetics, whereas restoration ecology would emphasize botany and ecosystem ecology.[111] Thus the old split between zoologists and botanists, which the discipline of ecology had partially bridged in the 1910s, reemerged again, a product of the needs for specialization in training and of the different regulatory agencies charged with managing animals and plants.

"Active Biological Management"

Invasive species eradication was good for fundraising: a 2001 TNC business plan maintained that no other major environmental NGO had focused on invasive species as a priority issue, and TNC should take advantage of the "wide, bipartisan support" in the United States and foundation community to "elevate the political profile of the invasive alien species issue to establish new funding and policy support [. . .]."[112] The new focus on nonnative species also coincided with increasingly globalized trade and benefited from (and helped stoke) public concerns over national sovereignty and the ability of states to control flows of goods, capital, and people.[113] In 1999, with the European Union, the North American Free Trade Agreement, and the World Trade Organization having been established in the prior five years, President Clinton's Executive Order 13112 mandated that federal agencies prevent the introduction of invasive species, provide for their control, and minimize invasive species' deleterious effects, whether economic, ecological, or medical.[114] That same year, the scientific periodical *Biological Invasions* published its first issue.[115] A number of prominent articles in the early 2000s linked the threat of future biological invasion to continued integration of the global economy, and ecologists continued to call for expanded regulatory frameworks to prevent the introduction of more nonnative species.[116]

Peculiarly, concern over invasive species can be seen as bootstrapping its way into prominence: a 1998 article by David Wilcove and colleagues, "Quantifying Threats to Imperiled Species in the United States," argued that competition with alien species was the second greatest threat to biodiversity, behind only habitat destruction. Wilcove et al.'s conclusion has been cited thousands of times by ecologists, natural areas managers, and policy makers. And yet, as the authors emphasized, they reached their conclusions not through experimental evidence or quantitative data, but through compiling managers' perceptions of the threats facing "imperiled" native species.[117] The archive of annual management surveys reveals that the number of TNC preserves listing invasive species as a management challenge jumped from 6 percent in 1980 to 60 percent in 1992 to 94 percent in 2000.[118] Thus, circularly, managers came to justify nonnative species removal based on a paper that took their own perceptions as its evidence.

Instead of shielding natural sites from human action, managers saw themselves as protecting native nature from nonnative plants and animals. This represented a significant shift in consensus about how wildness could be achieved. In 1937, Richard Pough, an ornithologist who would go on to become TNC's first president, had written, "The restoration of an abundance of a species is simply a matter of making available to it suitable environment and freeing it from human molestation or other unnatural disturbing factors."[119] In earlier decades, federal agencies and the TNC had based their management actions on such beliefs. Indeed, in the 1970s it was explicit TNC policy not to remove a dead bird from the place it had died.

7.5 A sign marking work by the Massachusetts Department of Conservation and Recreation to kill nonnative bittersweet, honeysuckle, and garlic mustard, April 2015. Photo by Ben Garver republished with permission of The Berkshire Eagle

In contrast, the 1995 TNC *Steward's Handbook* stated that "most preserves, maybe all, need *active biological management* to maintain their native species and natural communities."[120] By the 1990s, the majority of managers and ecologists believed that physical and legal protection were not enough to maintain natural areas.

Although the role of invasive species in native species extinction has since been challenged by some ecologists, the influence of this fear on species management has been enormous.[121] In the United States and in countries around the world, invasive species management is today a key mode of environmental management. The U.S. federal budget for invasive species management increased by $400 million between 2002 and 2006, for example. To treat one nonnative species, a subspecies of the common reed (*Phragmites australis*), natural areas managers in the mid-2000s were spending millions of dollars per year and spraying herbicide on more than 200,000 acres of wetlands.[122] Invasive nonnative species remain at the forefront of international environmental organizations' concerns, and the frequency and extent of species invasions have recently been construed as a "global change," the same language used to describe climate change, population growth, deforestation, nutrient pollution, and stratospheric ozone depletion.[123] Tellingly, the language around controlling invasive species has tracked public discourse on national security. After the terrorist attacks of September 11, 2001, invasive species management adopted the language of counterterrorism: where once invasive species had been "a form of biological pollution," according to TNC,[124] in 2002 each invasive species required its own "rapid-response team"; plans were drawn up for "Exotic Plant Eradication Strike Teams" in Florida, which would consist of "specially trained corps of volunteers who adopt exotic species control projects for initial attack and long-term follow-up."[125]

Into the twenty-first century, ecological restoration would continue to professionalize. Today the Society for Ecological Restoration has nearly four thousand members and partners across eighty-five countries.[126] Restoration work remained deeply enmeshed in the politics of belonging, of deciding what is considered wild and authentic nature. With the establishment of regulations requiring ecological restoration as a form of "compensatory mitigation," this question would only become more complicated.

8

An Ecological Tomorrowland

The Disney Wilderness Preserve, one of the first and largest off-site wetland mitigation projects ever undertaken, encompasses 11,500 acres of Everglades headwaters and is home to more than one thousand native plant and animal species, including cypress trees, scrub jays, and alligators. Fifteen miles south of Disney World, it can be visited easily after a day at Disney's Animal Kingdom Theme Park, whose attractions include the Primeval Whirl (a rollercoaster), the Affection Section (a petting zoo), and the Conservation Station (a veterinary facility).

Theme parks and wildernesses may seem antithetical, but Florida is a place of telling juxtapositions. When, in 1989, Disney pursued plans to build on top of wetlands in Osceola County, Florida, it was required by Section 404 of the Clean Water Act to obtain permission from the U.S. Army Corps of Engineers. The very agency that had drained millions of acres of wetlands in the name of national expansion was now responsible for protecting and restoring wetlands nationwide. The Disney Wilderness Preserve resulted from a complex and innovative agreement between the Corps, the Walt Disney Company, The Nature Conservancy, the Florida Department of Environmental Regulation, and five other entities. The so-called win-win agreement enabled Disney to expand its parks while also funding ecological restoration, and in 2014, The Nature Conservancy declared that the Disney Wilderness Preserve ecosystem had been "restored to very near to its original state." Reflecting restoration ecology's emphasis on historical fidelity at the time, The Nature Conservancy boasted that the site's species assemblage once again resembled "the descriptions left by the area's first Spanish missionaries."[1]

Ultimately, the Disney Wilderness Preserve would prove to be much more than an early example of ecosystem restoration. The project was larger in scale than any previously approved by the Corps under Section 404, and its many-partied negotiations and implementation, involving government entities, corporations, and environmental NGOs, would serve as a template for future restoration megaprojects in the United States and abroad. During the 1990s, ecological restoration became commercialized and consolidated, enacted by large networks of agency managers, academic scientists, and private consultants. Crucially, the Disney Wilderness Preserve was also one of the first off-site "compensatory mitigation" projects in the world. Disney compensated for environmental destruction on their property by funding ecological care miles away. That regulators would allow such compensation was not obvious at the time, and the agreement set important precedents for wetlands restoration in the United States, and, ultimately, for compensatory mitigation worldwide.

The emergence of off-site mitigation as a concept and a practice had profound ecological and political consequences. By decoupling the site of ecological damage from the site of ecological care, the practice of off-site mitigation changed the geographical distribution of wetlands, forests, and other ecosystems at the regional and global scale. Further, the practice of off-site mitigation reconfigured the attitudes of industrialized nations toward developing nations, positioning the latter not only as sources of inexpensive labor or raw materials, but also as compensatory "natural" areas, as with today's carbon offsetting schemes.

Draining Florida's Wetlands

Walt Disney began scouting sites for an eastern resort to complement Disneyland, California, in 1959. Central Florida's new freeways and temperate weather made it a promising theme park site, and a few years later, under conditions of strict secrecy, real estate agents began purchasing land for Disney in Orange and Osceola Counties. Local residents speculated that NASA was buying land to develop space flight facilities, but in October, the plans for the Disney World theme park were leaked to the *Orlando Sentinel*. Development on the 27,443 purchased acres (43 square miles), much of it wetlands, began in 1969. To transform the swampy property into Disney World, construction crews razed forests, built more than 50 miles of levees and canals, and moved seven million cubic yards of dirt. In doing so they erased thousands of acres of wetlands.[2]

Such a project was not anomalous for the times; the federal government had incentivized wetland drainage and development for nearly two hundred years. European colonists typically viewed wetlands as wastelands that bred disease, restricted travel, and impeded the production of food and fiber. In 1849, Congress passed the first of three Swamp Land Acts, giving 9.5 million acres of federal wetlands in Louisiana to the state to drain and "reclaim" as agricultural land. In total, more than sixty-four million acres of land were given to multiple states in this manner. Universities began teaching soil-drainage science in the late 1800s, and if wetlands were not drained and filled, they might instead be dredged or channelized to create ports and improve navigation. In 1899, Congress charged the U.S. Army Corps of Engineers with maintaining navigation by regulating the dredging and filling of navigable waters. This federal sponsorship of wetland destruction continued through the 1960s.[3]

When Disney World looked to build on wetlands in 1989, however, it faced a new regulatory regime. Concern about wetlands loss had mounted as public awareness of industrial pollutants aligned with ecologists' efforts to portray wetlands as efficient waste recycling systems. Wetlands species, too, were newly valued. In 1920 a popular magazine had described alligators as "good for nothing except to furnish the makings of traveling bags," but in 1964 the FWS Committee on Rare and Endangered Wildlife Species paved the way for federal protection of the alligator, putting it on the first federal list of endangered native wildlife.[4] Wetlands, once threatening to settlement and development, were now threatened. An influential 1971 Fish and Wildlife Service (FWS) report estimated that at least 35 percent of the wetlands of the continental United States had been destroyed.[5] Later studies would put that number at 53 percent.[6] Ecologists began to argue that the Everglades itself, as an ecosystem, was endangered.

The Everglades are part of a large watershed that originates near Orlando, Florida, and drains into Lake Okeechobee, the hydrology of which had been drastically altered since the early twentieth century. Prior to widespread water control, water from Lake Okeechobee flowed slowly to the Atlantic during the wet season, forming a shallow and wide "river" some 60 miles wide. This "sheet flow" might take more than a year to reach the ocean. When Florida gained statehood in 1845, much of the southern portion of the state was underwater for portions of the year, and much of it was owned by the federal government. Soon, though, the Swamp Land Act of 1850 transferred the majority of the Everglades to the state of Florida, on the condition that it undertake the draining and development of those lands.

8.1 Disney World plans unveiled in 1970. From left to right, Domm Tatum, Governor Claude Kirk, Roy Disney, and Lt. Governor Ray Osborne. Orlando Sentinel / Television Critics Association

Thus early draining projects were motivated by territorial claims. The success of the first large-scale drainage project, completed in 1913, and the promise of further water control prompted interest from railroad companies and fruit producers, who dreamed of cultivating tropical produce at a large-scale domestically. Demand for food during World War I also sparked a boom in the Everglades, precipitating the clearing of hundreds of acres of mangrove forests and pine scrublands. Incoming settlers, following jobs in agriculture and related businesses, hunted alligators, panthers, otters, egrets, spoonbills, and herons.[7]

The threat of natural disaster also inspired large-scale changes to Florida's hydrology. Devastating hurricanes in 1926 and 1928 spurred the federal government to construct a massive levee, the Hoover Dike, around the southern shore of Lake Okeechobee. The Army Corps of Engineers completed the project in 1938. A decade later, the Corps initiated an extensive reclamation plan known as the Central & Southern Florida Project (C&SF Project) that incorporated the Kissimmee River basin. The plan was supported by agriculturists hoping to produce winter vegetables, sugar, oranges, and other tropical fruit. From 1949 to 1969, the Corps and the C&SF Flood Control District constructed more than 1,000 miles of canals and hundreds of pumping stations and levees. Into the 1990s, this system diverted approximately 1.7 billion gallons of freshwater per day east to the Atlantic Ocean, and nearly half of the Everglades had been converted into plantations, roads, and cities.[8]

Some of the environmental effects of the C&SF Project were immediately apparent. Draining led to muck fires, soil subsidence, and saltwater intrusion

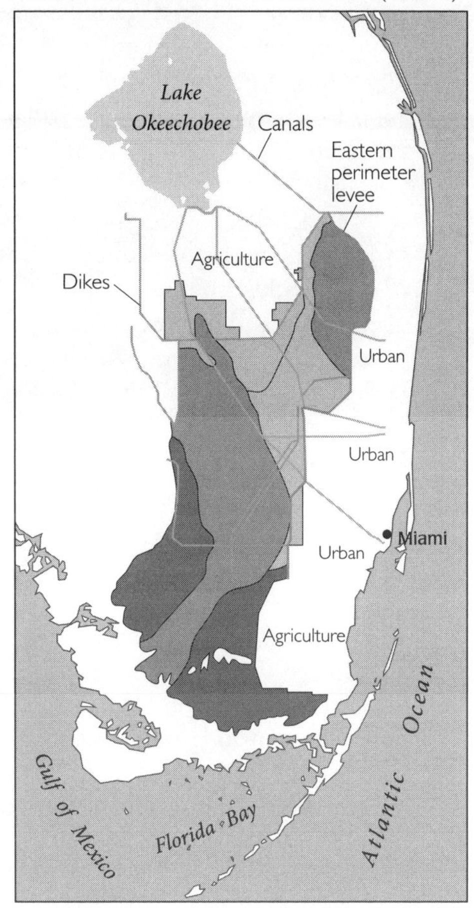

8.2 **The construction of extensive dams and levees and subsequent draining and development changed vegetation patterns throughout the Everglades.** Devin L. Galloway, David R. Jones, and S. E. Ingebritsen, "Land Subsidence in the United States," U.S. Geological Survey 1182 (Reston, VA: U.S. Geological Survey, 1999), 100.

into freshwater aquifers. The massive amounts of water diverted east inundated the estuaries on the Atlantic Coast with freshwater, while the reduced flow to southern estuaries caused hypersalinity there. Meanwhile, development in south Florida skyrocketed; the population of Miami-Dade County more than tripled between 1940 and 1960. National environmental controversy finally erupted when, in fall 1968, the Dade County Port Authority broke ground for a new supersonic jetport, planned to be the largest airport in the world, just six miles north of Everglades National Park. Responding to pressure from local environmental organizations, the Department of the Interior organized a six-scientist committee to consider the jetport's environmental implications, and the committee produced one of the nation's first environmental impact assessments.[9] Very soon such assessments would be mandated by the National Environmental Policy Act of 1969 (NEPA). The jetport assessment, authored by Luna Leopold, a hydrologist with the United States Geological Survey and another of Aldo Leopold's children, garnered national attention, and its assessment of the proposed airport project was bluntly critical.[10] The first sentence proclaimed that the proposed jetport would "inexorably destroy the south Florida ecosystem and thus the Everglades National Park."[11] Following the report, President Nixon weighed in against the project, and in 1970 work was halted.[12] The metaphorical tides had turned; once a wasteland in need of development, the Everglades were now an ecosystem in need of protection.

Section 404 Permitting

Of the various federal laws that enabled the rise of compensatory mitigation, which include the National Environmental Policy Act of 1969 and the Endangered Species Act of 1973, the Clean Water Act of 1972 played the greatest role in producing a commercial market for ecological restoration. In October 1972, Congress amended the Federal Water Pollution Control Act (renamed the Clean Water Act in 1977) with the stated purpose of restoring the chemical, physical, and biological integrity of the nation's waters.[13] The Clean Water Act established a structure for regulating pollutant discharges and for setting quality standards for surface waters.

The best-known section of this Act is Section 301, which prohibits the discharge of pollutants from point sources, like factory drainpipes, into navigable waters. Equally important, though less studied by environmental scholars, is Section 404, which established a permit program to regulate the discharge of dredged or fill material into "waters of the United States," defined to include wetlands. The U.S. Army Corps of Engineers has administered the permit program since its inception, while the Environmental Protection Agency (EPA) is responsible for establishing environmental guidelines that the Corps must use to assess proposed projects when it makes its permitting decisions. The EPA also has the authority to veto permits approved by the Corps, and the Fish and Wildlife Service, the National Marine

Fisheries Service, and the Natural Resources Conservation Service have the opportunity to review and comment on Corps decisions, too.

Despite these possible checks on Corps' approvals, the Corps has been the major player in permitting decisions since the passage of the Clean Water Act, and in 1977, by administrative fiat, the Corps greatly expanded the scope of its regulatory jurisdiction to encompass bogs, vernal pools, and other types of water-saturated lands that are not what we might think of as "waters." In doing so, they joined three other agencies that had legislative or administrative authority to define "wetlands" according to their respective mandates: the EPA, the FWS, and the Department of Agriculture's Soil Conservation Service. Because the amount of water varies seasonally in many types of wetlands, as it once did in central Florida, attempts to define wetlands by hydrology alone failed. Regulators and ecologists in turn defined and delineated wetlands by their biological components: hydric soils (soils characterized by their anaerobic conditions) and hydrophytes (plants adapted to water-saturated soils).[14] Ecology programs began training students to delineate wetlands according to federal guidelines, and by the 1990s, both academic restoration ecologists and private restoration consultants were working in the field of wetlands delineation.

Until 1990, there was no comprehensive federal policy regarding enforcement under the Clean Water Act, and the administering of Section 404 was highly irregular, as each agency adopted its own regime of policies, guidelines, and practices. The concept and practice of compensatory mitigation emerged unevenly, sometimes inconsistently, across federal agencies in the 1980s. The National Environmental Policy Act required all federal agencies to identify the potential adverse environmental impacts of the major actions they proposed to undertake and to consider reasonable alternatives to those actions. President Carter's Council on Environmental Quality established a "mitigation sequence" in its 1978 clarifications of the NEPA regulations, an ordered preference for how projects should be modified. Avoiding adverse impacts was the first choice. Minimizing them was second, if avoidance was not deemed reasonable. The third option was to compensate for impacts if they could not be minimized.[15]

The FWS was the first federal agency to implement a formal mitigation policy, in 1981. It adopted the definition of mitigation instituted by NEPA, which included not only "rectifying the impact by repairing, rehabilitating, or restoring the affected environment," but also "compensating for the impact by replacing or providing substitute resources or environments."[16] According to FWS policy, the second option, compensatory mitigation, would be permissible so long as the damaged habitat was not deemed of "high value" and "unique and irreplaceable on a national basis or in the ecoregion section." Under the FWS's compensatory mitigation policy, a developer could mitigate habitat loss by "restoration or rehabilitation of previously altered habitat" or through "increased management of similar replacement habitat so that the in-kind value of the lost habitat is replaced." Notably, the FWS policy did not mandate replacement of "acre for acre loss," but rather a replacement of "habitat value."[17]

Meanwhile, different mitigation policies developed at the U.S. Army Corps of Engineers and the EPA. Notably, when the EPA issued its Section 404(b)(1) guidelines in 1975, they did not include compensatory mitigation.[18] The guidelines stressed the avoidance of harm and made no mention of mechanisms either to mitigate or compensate for environmental damage. Indeed, the EPA assumed that permits for work that significantly damaged wetlands would either be denied by the Corps or vetoed by the EPA under its Section 404(c) powers. But into the early 1980s, the EPA rarely exercised its veto powers over permits approved by the Corps, and the Corps, for its part, tended to resolve the most difficult permit decisions by seeking substantial mitigation rather than by threatening to withhold permits.[19] The different approaches of the Corps and EPA played roles in numerous lawsuits and constituted what one analyst called "fullblown, institutional schizophrenia" regarding Section 404 implementation.[20]

The proliferation of (sometimes contradictory) regulations and court actions surrounding Section 404 presented clear opportunities for practitioners of ecological restoration. Once a volunteer activity, ecological restoration became professionalized in the 1980s, as developers seeking Section 404 permits employed an increasing number of wetlands restoration consultants.[21] Under the emerging federal mitigation regulatory regime, a 1983 article in *Restoration & Management Notes* speculated, "restoration technology is likely to assume new significance as a lever and bargaining chip in planning as well as litigation having to do with the environment."[22] The Department of the Interior advertised in the journal that they sought information on the location, extent, protocols, and cost of restoration projects across the country.[23] Section 404 requirements thus also bolstered the status of academics in the new discipline of restoration ecology, and this led to a set of conflicts between restoration practitioners, whose job it was to build and manage mitigation projects, and academic restoration ecologists, who questioned the "functional equivalency" between created and "natural" wetlands.

Compensatory Mitigation, On-Site and Off

Amid this administrative and legal uncertainty, three quite different mechanisms emerged in the 1980s for providing "compensatory mitigation" under the Section 404 permitting process: permittee-responsible mitigation, mitigation banks, and in-lieu fee mitigation. In the 1980s, the Corps showed a preference for the first: small, on-site wetlands, with the permittee or a contractor—often a private ecological restoration company—responsible for restoring or creating them.[24] Between 1983 and 1989, more than a thousand wetlands were created in Massachusetts alone through the Section 404 permitting process.[25] Indeed, on-site compensatory mitigation projects of this sort were common enough that, by the mid-1980s, the Corps claimed to have halted the trend of wetland loss in the United States while continuing to permit development.[26]

However, a push by contractors to make the Section 404 permitting system more streamlined eventually led to the widespread adoption of the second mechanism, wetland mitigation banks.[27] Upon establishing a "mitigation bank" by creating or restoring wetlands, the creator ("banker") can sell wetlands "credits" to developers, which the developer can use to meet Section 404 requirements. The first commercial sale of wetland credits occurred in February 1986 at the Fina LaTerre Bank in southern Louisiana, which had initially been established to provide in-house credits for the Tenneco Oil Company.[28] Commercial wetland banking aligned with President George H. W. Bush's "No Net Loss of Wetlands" agenda, which promoted market-based incentives. As a candidate in 1988, struggling to combat Democratic candidate Michael Dukakis's environmentalist platform, Bush adopted the "no net loss" slogan and made wetland advocacy one of his campaign's central themes. After winning the election, Bush followed through by creating a "net" accounting for wetlands loss, which enshrined compensatory mitigation within the Section 404 permit program. The Bush administration also influenced the 1990 Memorandum of Agreement between the Corps and the EPA that formally clarified the procedures to be followed in determining what mitigation is necessary for Section 404 compliance. The memorandum embraced the "no net loss of wetlands" goal and defined it to mean "no overall net loss of values and functions."[29] Thus it refrained from requiring acre-for-acre replacement of wetlands damaged or destroyed by development.

Federal promotion of commercial wetland mitigation banks began in earnest with a 1990 EPA workshop on the future of mitigation banking policy.[30] Within a few years, entrepreneurial wetland banks were permitted and selling credits.[31] In a fairly early and influential example, a contractor in the Chicago area, in 1991, searching for a way to meet Section 404 compliance, floated the idea of a wetlands bank to the project manager at the Chicago Army Corps of Engineers. As the contractor later recalled in an interview with geographer Morgan Robertson, "I just said, 'Maybe I'll build some big-ass wetlands somewhere, somewhere out there, and build some really good ones, and that ought to make these agencies really happy.' "[32] The project manager approved the idea, and the contractor began constructing a wetland on a former agricultural site. The set of guidelines developed by the Chicago office for this project would become the foundation for the 1995 federal guidance document on wetlands banking policy.[33]

The number of commercial wetland mitigation banks increased rapidly, from 46 active banks in 1992 to 330 in 2005.[34] Section 404 permittees could now purchase credits whose value resided not in work to be done in the future, but in wetlands that had already been constructed or restored by one of the banks. So, in 1989, when the Walt Disney World Company revealed its plan to develop 11,000 acres of the company's central Florida landholdings, including a proposed 1.5-million-square-foot shopping mall and an exclusive residential community, it confronted a vastly different set of laws and an altered public sentiment from its initial Florida build-out twenty years prior. The development would destroy hundreds of acres of wetlands.[35] But this did not necessarily mean the new project would not be permitted. Perhaps more surprisingly, it also did not mean that environmental organizations would object to the build-out. Rather, with off-site mitigation on the rise, organizations like the Florida Audubon Society and The Nature Conservancy (TNC) saw Disney's proposal as a real opportunity. They would use the permitting process to push for the acquisition and restoration of substantial off-site lands. Florida Audubon wrote to Disney executives in Florida and California in 1990, suggesting the company consider purchasing the Walker Ranch, a former cattle ranch 15 miles south of their Florida theme parks, as a possible mitigation site. The Audubon Society had become aware of the property the previous year, when its owners had proposed the development of 5,700 resort residences. Testifying to the rise of conservation mapping and planning, TNC had separately identified the area as one of twelve potential "megasites" for natural areas acquisition in Florida.[36]

The Disney mitigation proposal required the input, collaboration, and cooperation of numerous entities, and TNC led talks among the Army Corps, the Walt Disney Company, the Florida Game and Freshwater Fish Commission, the South Florida Water Management District, the U.S. Environmental Protection Agency, the U.S. Fish and Wildlife Service, and Florida Audubon. TNC and Florida Audubon maintained that such a project would be of higher "ecological value" than any on-site wetlands construction Disney would otherwise undertake. At least some regulatory agency representatives expressed concern that if they permitted off-site mitigation for Disney, they would have to do so for other proposals. But TNC prevailed, and the Disney Wilderness Preserve mitigation project was permitted in late 1992.[37]

Back in 1971, Robert Jenkins had speculated that new environmental legislation provided an opportunity for TNC: protected areas, he imagined, could serve as "baselines" for environmental impact statements.[38] As it turned out, however, the implementation of the Clean Water Act would provide TNC and other environmental NGOs with opportunities on a different scale and of a different kind. Playing a role in the assessment phase of new development was one thing, but TNC and similar organizations played an expanding role in the post-permit phase as well. Through mitigation projects, they acquired new natural areas and new management responsibilities, developed new practices, and reshaped environmental policy and practice.

In 1992, the finalized Section 404 permit agreement between Disney and the Corps specified that Disney purchase 8,350 acres at Walker Ranch, deed it to TNC, and provide funding for restoration and management for the next twenty years. The restoration and management cost was projected at around $45 million.[39] Less than five years after the founding of the Society for Ecological Restoration, restoration had become big business.

With the Disney Wilderness Preserve deal in place, Disney had immediate permission to build the town of Celebration, and the agreement streamlined the permitting process for Disney going forward, ensuring twenty years of development rights on the company's landholdings. As a director of business development later recounted, "At that time, the pendulum had swung over onto the side of the environmental community, and each year landowners throughout the United States were watching developable land get taken away to save one more plant species that got added to the list that year. [. . .] Well, by having that permit, it locked our long-term development rights in place, and that was of enormous value to the company."[40]

This paved the way for Disney to build a fourth theme park. Disney's Wild Kingdom—renamed Animal Kingdom because of a trademark dispute with the Mutual of Omaha insurance company, which owned a syndicated TV show of the same name—opened on Earth Day in 1998. At more than 500 acres, it was Disney's largest theme park, encompassing exhibits of African and Asian wildlife, as well as extinct animals and imaginary ones.[41] Tellingly, Animal Kingdom is also the only place where visitors can meet Disney's Pocahontas, the Native American princess of Disney's 1995 animated musical. The "nature" constructed by Animal Kingdom and by the Disney Wilderness Preserve might not be so different after all.

Restoring the Disney Wilderness Preserve

The property that would become the Disney Wilderness Preserve had been constantly inhabited since 500 A.D. When the Spanish arrived in Florida in 1513, there were more than twenty thousand people living in the greater Everglades area. In the decades that followed, war, disease, and cultural disruption devastated communities. In the early eighteenth century, Seminoles moved from Creek Confederacy towns in Georgia and Alabama into northern Florida. There they, along with descendants of enslaved African Americans, kept cattle, until many of them were forcibly removed to Indian Territory in Oklahoma in the mid-1800s. In 1881, white industrialist Hamilton Diston bought four million acres around the tiny town of Kissimmee and attempted to drain the property. Other developers soon moved in, and by the early 1900s, around 50,000 acres of Everglades headwater wetlands had been drained. In 1925, the Everglades Cypress and Candler Lumber Co. began logging the area. Around the same time, companies set up facilities to extract longleaf pine resin, which was used in soap, varnish, and boat repair. And as part of the C&SF Project described above, the Army Corps of Engineers altered the flow of the Kissimmee River in the 1960s, dropping the water level substantially. After this point cattle ranches proliferated, and ranchers planted thousands of acres of nonnative grass for forage.

The Nature Conservancy's restoration goal was to return the Disney Wilderness Preserve property to a pine flatwoods ecosystem, an ecosystem that encompassed a variety of habitats, including cypress dome, wet prairie, pine forest, and pine scrub. By 1989, the southeastern pine flatwoods ecosystem had been reduced to 3 percent of its original extent.[42] To restore the hydrology of the site, TNC compared the current patterns to photographs from 1941 and 1944 that pre-dated the most extensive drainage. Meanwhile, the plan for the ecological community was to reestablish plant and animal species found prior to colonization. The management plan blamed the degradation of the local ecosystem on "European man's introduction of cattle and hogs, combined with his use of fire and fencing that confined livestock to specific tracts of land."[43] Thus the managers hoped to create a historical vista, a settler scene. A 1993 magazine article proclaimed, "By the time a traveler reaches the [entrance] road's end at an oak hammock overlooking the lake's cypress shore, the illusion of moving back 100 or more years in time is firmly impressed on the mind's eye."[44]

To promote the reestablishment of native plants, TNC began to fill drainage ditches, kill nonnative plants like bahiagrass, tropical soda apple, and Brazilian peppertree, and burn swaths of land to simulate lightning fires, which some species require to germinate. They also paid hunters to shoot feral hogs.[45] The initial restoration plans deemed it more desirable to take nonnative species out of the landscape than to reintroduce native ones. The hope was that by removing hogs and restoring hydrological and fire cycles, natural vegetation would regenerate itself, making the area attractive once again to animal species like sandhill cranes and otters. One project ecologist explained, "Initially, we will not introduce plants artificially. Since each wetland is different, we want the plants to come back naturally to maintain diversity."[46]

This hope was not borne out, however. During the 1990s, natural areas organizations like TNC practiced increasingly interventionist restoration in their pursuit of preservation (as chapter 7 details), and the Disney Wilderness Preserve project managers regularly attended the annual Society for Ecological Restoration meeting, where they learned about changing restoration practices across the country.[47] By the late 1990s, TNC had planted many thousands of seeds and seedlings in the preserve, and they sometimes called for unusual forms of volunteer assistance. In 1999, for example, "volunteers dug into their single sock collections and contributed more than 3,000 socks." These socks were used to cradle root masses of wetland plants thrown into deep water.[48]

TNC not only provided staff for the site and recruited hundreds of volunteers, but also contracted restoration work as commercial restoration companies emerged. They hired Dames & Moore for "wetland restoration engineering," Lake Doctors for "exotic/nuisance weed control," The Natives for plant identification, and the University of Central Florida for monitoring Florida scrub-jays and gopher tortoises.[49] Thus the project served as a nexus for the academic, activist, and research communities, just as it did for the various governmental, corporate, and private organizations that planned and funded the work.

On April 23, 1993, the governor of Florida, state and federal environmental officials, and representatives of Disney and TNC gathered at the base of a live oak tree to announce the establishment of the Disney Wilderness Preserve. Carol Browner, administrator of the U.S. Environmental Protection Agency, proclaimed that she would like to see consolidated off-site mitigation pursued nationwide. Indeed, it was she who had advocated for the mitigation approach during the permit negotiation phase in her previous role as secretary of the Florida Department of Environmental Regulation. "This is the future of environmental protection," she said.[50] Browner's prediction was accurate. Later that year, the Greater Orlando Aviation Authority paid for TNC to expand the preserve by 3,000 acres when it sought permits to build an access road and two runways for Orlando International Airport.[51] This expansion of the preserve was soon followed by 270 additional acres funded by Universal Studios Florida and 114 acres from the owner of the Arabian Nights Dinner Theatre in Kissimmee. TNC would manage these additional acreages, and the land would be owned by the South Florida Water Management District.[52] These connections to suburbs, theme parks, airports, and television studios would be lost on the casual visitor to the wilderness preserve. The site was designed to appear timeless and pristine, to hide even the fact of its restoration.

In 1992 there were few precedents for a multipartner compensatory mitigation project. TNC had pitched the Disney Wilderness Preserve as "an innovative experiment in the ecosystem approach to environmental mitigation" and as a "living laboratory" in which to test "innovative biological restoration and management techniques" to "keep the area a thriving example of wild Florida." Promotional materials noted that while many different permits made up the mitigation project area, the project was "designed to be one large ecosystem conservation, restoration and management project."[53] By 2012, however, when Florida and federal regulators deemed the project a success, such multipartner projects had become common practice. Subscribing to the idea that restoration ecologists and environmental NGOs were better qualified to create off-site wetlands and ensure their quality than corporate landowners were on-site, natural areas organizations argued that off-site compensatory mitigation led to the creation of larger, higher-quality wetlands. One TNC representative described the Disney Wilderness Preserve as a "win-win partnership," in which Disney was "able to proceed with its development, on lands that were already marginal from an ecological perspective," while also "setting aside a large, intact parcel of land of very high ecological significance."[54]

This win-win language was, of course, widely appealing. As one reporter wrote in 1993, the practice of off-site compensatory mitigation "promised a way to have your K-mart and your wetland, too."[55] TNC promoted the Disney project as pioneering a new approach to balancing environmental concerns with the need for economic growth, and in the 1990s, TNC would dramatically increase its corporate partnerships.[56] For example, in 1999, TNC and the aluminum producer Alcoa entered a joint-management agreement for 1,400 acres near Alcoa's bauxite mines in central Arkansas.[57] Such corporate partnerships would prove lasting. As forestry-based carbon offsetting emerged as a new form of off-site compensatory mitigation in the early 2000s, some of the same companies that had partnered with TNC on wetland mitigation and forest management projects would team up with TNC again. In 2009, the Walt Disney Company announced that to offset its greenhouse gas emissions, it would partner with TNC and Conservation International to invest $7 million in forest conservation and restoration projects in the Peruvian Amazon, the Congo Basin, and the lower Missis-

8.3 A University of Idaho student participates in a controlled burn at Disney Wilderness Preserve in Florida. Leslie Fowler

sippi River. At the time, that was the largest single corporate contribution ever made to carbon offsetting.[58] In 2017, Alcoa reported that TNC, with Alcoa Foundation support, was working in Australia, Brazil, and Canada "to strengthen the role of indigenous and local communities in managing lands to help mitigate global climate change."[59]

In addition to influencing the future of off-site compensatory mitigation, the Disney Wilderness Preserve also set important precedents for multi-partner restoration megaprojects like the Comprehensive Everglades Restoration Plan (CERP). The CERP, approved by the U.S. Congress in December 2000, was at the time the most expensive and expansive restoration project ever attempted, and it in turn would serve as a model for other wetlands restoration megaprojects like those in the lower Mississippi River System, the San Francisco Bay, the Mekong Delta, and the marshes of Mesopotamia. Covering 18,000 square miles, the CERP would cost $7.8 billion and take thirty-six years to complete.

The Comprehensive Everglades Restoration Plan

President Harry S. Truman dedicated the Everglades National Park in December 1947, only a few months before Congress authorized the C&SF Project. A vision of South Florida as a wild place, teeming with autonomous life, was thus juxtaposed with a massive water control project, and the two were built at the same time. The embedding of the national park within the development-oriented drainage project would eventually catalyze Everglades restoration in the 1980s. In fact, it was the federal government that in 1988 filed suit against the State of Florida, alleging that the state had failed to prevent discharge of phosphorus-polluted waters into the Loxahatchee National Wildlife Refuge and Everglades National Park. Governor Lawton Chiles agreed to reach a settlement in 1991, and mediation began.[60] A resulting accord directed the U.S. Army Corps of Engineers to conduct a comprehensive review of the C&SF District for the purpose of restoring the hydrology of South Florida, and in 1998, the Corps' four-thousand-page review became the framework for the Comprehensive Everglades Restoration Plan.

Despite the increasingly interventionist approach to restoration spearheaded by TNC and other natural areas organizations, the CERP rested on the assumption that the Everglades ecosystem would recover if natural hydrologic patterns were restored. The focus was on "getting the water right," including the quantity, timing, and distribution of the flow. The plan listed indicators of the degree of ecosystem restoration occurring; these included the recovery of endangered species, the return of large nesting rookeries, and the improvement of water quality in estuaries.[61] The Everglades National Park science team, however, was critical of the Corps' proposal, arguing that the plan focused primarily on water supply for urban and agricultural users and not on ecosystem restoration. Park staff shared their concerns with conservation groups, and top officials in the Department of the Interior and the Department of the Army were displeased when newspapers reported the National Park Service and FWS's critiques. Among environmental NGOs, the National Audubon Society and its Florida affiliate emerged as the strongest supporters of the CERP. Other groups, like the Sierra Club and the Friends of the Everglades, along with a number of prominent ecologists, remained vocal critics.[62] Facing pressure from environmental and scientific communities, the Department of the Interior put in place the Committee on Restoration of the Greater Everglades Ecosystem, contracted though the National Research Council.[63]

Without substantially addressing ecologists' objections, the Comprehensive Everglades Restoration Plan was ultimately approved by the U.S. Congress and signed by President Bill Clinton in December 2000. Congressional negotiations were almost derailed, however, by a last-minute proposal to allow 20 percent more water flow into Everglades National Park. A group of ecologists argued that, without that increased flow, animal species that rely on marl prairies, such as the endangered Cape Sable seaside sparrow, would not recover. But the increased flow would mean flooding in the central Everglades, which included crucial roosting grounds for endangered snail kites. Thus the proposed amendment pitted endangered species advocates against one another. This proposal was also opposed by the Miccosukee tribe, who argued that it would lead to the flooding and destruction of their homes and property. The tribal water resources manager testified that it was an effort to sacrifice the people of the central Everglades to the national park. Eager to pass the legislation, the Senate abandoned the proposal to increase the amount of water for the park.[64]

Through the CERP, the Everglades came to be defined by the boundaries of the water management district, the administrative agency in charge of the daily management of the C&SF Project.[65] But the water management district spanned several Florida counties, the Miccosukee Indian Reservation, the Cypress Seminole Indian Reservation, public and private land, sixteen national wildlife refuges, and four national park units. It was also home to more than six million people and significant agricultural lands. In this area, the jurisdictional boundaries of dozens of federal, state, and local agencies, including the FWS, the National Park Service, the Florida Game and Freshwater Fish Commission, and the Florida Department of Environmental Protection overlapped. Further, many environmental groups expressed interest in shaping Everglades restoration, ranging from international groups like The Nature Conservancy, the Sierra Club, and the Audubon Society to state and local organizations, among them the Friends of the Everglades and the Everglades Coordinating Council.[66] The CERP was thus truly a megaproject, and, like the Disney Wilderness Preserve, it would set precedents for other restoration work around the world. The final plan approved sixty-eight water management projects, including the removal of 240 miles of levees and canals.[67] The federal government paid half the plan's costs, and state, tribal, and local agencies were responsible for the other half.[68] Although the Corps and the C&SF District received the majority of the funds, the National Park Service and the Fish and Wildlife Service also had a considerable financial stake.[69]

Apart from its massive scale, the CERP is significant as an early manifestation of the ecosystem management paradigm that defined ecological restoration in the 1990s.[70] Vice President Al Gore established the Interagency Ecosystem Management Task Force in 1993, in response to the northern spotted owl controversy, an enormously complex political and legal battle that pitted the logging industry against environmentalists in the Pacific Northwest.[71] The task force defined ecosystem management as "a method for sustaining or restoring natural systems and their functions and values" that was also "a mechanism for resolving conflicts that protects our national economy and the resources on which it is based."[72] Ecosystem management, in other words, was not just a mode of environmental management but a mode of interagency management, too. Advocates of ecosystem management as a "mechanism for resolving conflicts" promised that it would integrate federal environmental management across jurisdictional boundaries, and that it would resolve tensions between ecological protection and resource development through inclusive deliberation. According to historian James Skillen, changing theories of public administration were as important to the development of ecosystem management as changes in ecological science, and the vagueness of ecosystem management's definition was considered its strength. Federal agencies hoped to use ecosystem management as a process to increase direct public participation and interagency collaboration in natural resources decision-making.[73] In December 1994, all executive branch agencies signed a Memorandum of Understanding agreeing to "foster the ecosystem approach."[74] Thus, twenty years after its popularization, the ecosystem idea was fully enshrined in federal environmental regulation.

"Faking Nature" and Wetland Units

While many bureaucratic matters of ecological management were negotiated through projects like the Disney Wilderness Preserve and the CERP, there were also matters of value, of aesthetics, of wildness and nativity, or, as Australian philosopher Robert Elliott might have put it, of nature. Not all ecologists and not all environmental organizations approved of off-site compensatory mitigation or of megaprojects like CERP. Elliott argued in his 1982 paper, "Faking Nature," that wild nature is valued in part because it results from "natural processes" and that nature restored by humans could not possess this property. Just as faked art was less valuable than authentic art, he argued, human-created nature was less valuable than autonomous nature. While acknowledging as plausible the argument that "there is no longer any such thing as 'natural' wilderness, since the preservation of those bits of it which remain is achievable only by deliberate policy," Elliott responded that "what is significant about wilderness is its causal continuity with the past," an articulation of a preservationist, rather than a restorationist, mindset.[75] In a similar vein, William Jordan III wrote in 1989, "however *accurate* it may be, the restored community can never be *authentic*. It is not just technically inferior, it is somehow less *real*."[76]

Along with process, many ecologists sought authenticity in species composition. In another 1989 article, titled "Disneyland or Native Ecosystem," ecologists Constance Millar and William Libby argued that if a restoration project did not replicate "native genetic structure," reintroducing genotypes that were already adapted to the local environment, then restorationists were creating a mere "simulation" of a native ecosystem, a "tangible fantasy," like Disneyland itself.[77] The question remained whether reversing ecological damage on the ground (or in the water), even if technically feasible, addressed and reversed the full harms of environmental damage. Whereas Edith Roberts and Elsa Rehmann, writing in the 1930s, had been confident in their ability to re-create wild scenes, not all ecologists of the 1980s were so sure.

Elliott's thought experiments focused on in situ restoration; for example, sand dunes mined for rutile and then reconstructed down to the last detail. The real and increasingly prevalent practice of off-site mitigation only further complicated the quest for ecological authenticity, technically as well as philosophically. A 1988 *Newsweek* article noted that the Sierra Club worried that "if a developer promises to build a new wetland, he will be able to destroy the original with impunity—a zero-sum game for the environment."[78] Further, at the practical level, the idea that wetlands could be restored or created de novo ran counter to claims of ecosystem vulnerability and irreplaceability that had become prominent in the 1970s: Would re-created ecosystems really function like "natural" wetlands? The first wetland creation experiments were undertaken with the expectation that wetland plant communities could be restored simply by reestablishing hydrological regimes. The "efficient community hypothesis" contended that constructed wetlands would be quickly regenerated from relict seed banks, nearby refugial populations, and propagules dispersed by waterfowl.[79] But many constructed wetlands were quickly overtaken by unwanted species, in a process ecologist Bill Niering jokingly called "cattailization."[80]

Philosophical and practical distinctions between natural and reconstructed ecosystems put ecologists and pro-wetland regulators in a bind: on the one hand, it was in their interest to advocate for a capacious definition of wetlands for the sake of delineating and protecting them. As when the Army Corps of Engineers had expanded its definition of waters, it expanded its oversight, so the widest possible definition of wetlands was useful for putting more sites, more development projects, under some kind of ecological and regulatory review. Wetlands, by the broadest definitions, contained hydric soils *or* hydrophytic plants, not necessarily both. But on the other hand, a too-broad definition of wetlands only made it easier for constructed wetlands to come up short, ecologically, and many ecologists already felt that constructed wetlands failed to replace "natural" or "true" ones. They wanted restoration sites to be counted as wetlands only if they had hydric soils *and* hydrophytic plants, as well as diverse animal communities, high rates of primary productivity, stable sediments, and so on.

Such questions about the replacement value of restored and constructed wetlands would come to define a subfield of restoration ecology. In the late 1980s, an expansive literature emerged on the idea of "functional equivalency" among field sites, and wetlands ecologists, many of whom joined the newly created Society for Ecological Restoration, began to publish data indicating that mitigation wetlands were not functionally equivalent to the wetlands they were meant to replace.[81] One study found, for example, that while 71 percent of impacted wetlands in Massachusetts were forested wetlands, only 25 percent of designated replacement wetlands were forested.[82] Meanwhile, a widely circulated 1991 report on mitigation projects in Florida found that only half of wetlands required to be created under Section 404 permitting had actually been built. Even when projects were completed, the study found, the success of wetlands restoration was questionable: colonization by "undesirable plant species" such as cattail and melaleuca occurred in thirty-two of forty projects.[83]

Comparing created wetlands to protected wetlands, ecologists found differences in soil chemistry, plant diversity, and hydrological regimes.[84] A 2001 National Academy of Sciences report concluded that while some types of wetlands could be restored and/or created (e.g., freshwater marshes), others could not be (e.g., fens and bogs).[85] Somewhat perversely, however, regulators would use such findings to justify the shift away from on-site wetlands creation to off-site compensatory mitigation and wetland mitigation banks. Consolidated, higher-quality wetlands, the regulators reasoned, would be more successful restoration projects. The wetland unit emerged as a buyable, fungible entity, a crucial element of the ecological management regime. In fiscal year 2003, one-third of Section 404 permits required investment in mitigation banks.[86] By 2005, private entrepreneurs or companies sponsored the most banks (72 percent), followed by state agencies (14 percent), local governments (7 percent), and nonprofit conservation organizations like TNC (5 percent).[87] The Environmental Law Institute estimates that, by 2006, private and public expenditures for compensatory mitigation totaled approximately $3.8 billion annually. The vast majority of this expenditure—over 77 percent—was generated through the mitigation requirements of Section 404 of the Clean Water Act.[88]

The economization and commodification of wetlands was a significant development in U.S. environmental governance, but the spatial redistribution of wetlands mattered equally. Mitigation banking changed the spatial distribution of wetlands at a national scale. Wetland banks consolidated mitigation acreage into a few large sites that were overseen by third parties (neither the permit-seeking corporation nor the permit-granting agency). Information on acreage is spotty, but the Army Corps of Engineers reported that from 2000 to 2006, an average annual area of 20,620 existing wetland acres were permitted for adverse impacts under the Section 404 program. Over the same period, the area of required wetlands compensation projects averaged about 47,384 acres per year.[89]

In addition to reshaping the material environment, wetland banks normalized off-site mitigation: by concept and design they were not located on the same parcel as the development project. Off-site wetland mitigation and the system of mitigation banks that allowed wetland units to be traded among corporate entities paved the way conceptually and procedurally for other forms of commodified mitigation that decoupled the site of environmental damage from the site of its restoration. With carbon offsetting, entities as various as municipalities, universities, for-profit enterprises, and even individuals wishing to offset their air travel now pay for projects elsewhere to compensate for greenhouse gas emissions. This was a new way of relating ecological damage and ecological restoration. Off-site mitigation projects, the earliest among them the Disney Wilderness Preserve, untethered sites of destruction from sites of care.

From Wetland Mitigation Banks to Carbon Sequestration

The idea that forests could absorb excess carbon dioxide was first proposed by physicist Freeman Dyson in 1977. He speculated that even if rising carbon dioxide emissions were on a trajectory to cause "acute ecological disaster," it should be possible "to plant enough trees and other fast-growing plants to absorb the excess CO_2" and avert a worldwide emergency. At the time, Dyson wrote that it was "highly unlikely that the particular emergency program here proposed will ever be implemented."[90] But only twenty years later, environmental organizations would be protecting and planting trees in the name of carbon offsetting around the world. The idea behind carbon offsetting is that emissions generated in one location can be "offset" by removing greenhouse gases from the atmosphere somewhere else, through tree-planting, say, or underground carbon storage.

That the Disney Wilderness Preserve was an important precedent for this new type of consolidated off-site mitigation can be seen in the role The Nature Conservancy played in promoting the carbon market. In fact, TNC began exploring forest-based carbon sequestration as a conservation finance mechanism in 1990, when an independent power producer, AES/Barbers Point Co., offered to fund the protection of a property in Paraguay to offset carbon dioxide emissions from its power plant in Hawai'i. The offer resulted in the establishment of the Mbaracayú Forest Nature Reserve. By 1996, TNC had also initiated a carbon mitigation project in Belize for the Wisconsin Electric Power Company and one at the Noel Kempff Mercado National Park in Bolivia for American Electric Power. The latter was at the time the largest forest carbon sequestration project in the world.[91]

Just as TNC had recognized Disney's 1989 Florida build-out plan as a way to secure long-term funding and increase the area under its management, it recognized carbon mitigation's enormous potential as a financial mechanism for its sites; in no other situation had TNC, as one internal memo put it, "had the ability to raise this type of capital from one single donor."[92] TNC therefore worked to promote carbon offset markets internationally, and in 1997, TNC paid the costs for its Latin American partners to attend COP 3, the third Conference of the Parties to the UN Framework Convention on Climate Change, in Kyoto, Japan. At COP 3, the meeting where the Kyoto Protocol was negotiated, TNC and their collaborators argued for the protocol's inclusion of forest-based projects as a greenhouse gas mitigation strategy. Those that allied with TNC included both NGOs and for-profit entities, including the Union of Concerned Scientists, Conservation International, the World Resources Institute, the Wisconsin Electric Power Co., International Bamboo Development Co., and PacifiCorp. On the other side were Greenpeace and the World Wildlife Fund, who took the position that mitigation would become a "loophole" for industry to escape emissions reduction commitments.[93] Countries were also divided on the question: Paraguay, Ecuador, and Mexico, for example, supported forest-based mitigation, whereas Brazil and Peru opposed it.[94]

Ultimately, TNC's position prevailed, and the final Kyoto Protocol allowed for forest-based emissions reductions. Through the treaty, most developed nations agreed to legally binding targets for their emissions of six major greenhouse gases, and the Kyoto Protocol defined several mechanisms that allowed nations to meet their reduction commitments, including the Joint Implementation mechanism, which allows a country to invest in a project to reduce greenhouse gas emissions in another country. The guiding idea was that it would be cheaper to fund a reforestation project in Costa Rica than an energy efficiency project at a U.S. utility, while either option would lead to similar total greenhouse gas reductions.

Having spearheaded the Kyoto negotiations, TNC successfully positioned itself as a leader in the development and implementation of forest-based carbon sequestration projects.[95] The Rio Bravo Carbon Sequestration Pilot Project in Belize and the Noel Kempff Climate Action Project in Bolivia were among the first projects approved under the U.S. Initiative on Joint Implementation.[96] Investors in the Rio Bravo Carbon Sequestration Pilot Project in Belize included Wisconsin Electric Power, Cinergy, PacifiCorp, Detroit Edison, UtiliTree, and Suncor.[97] By 2001, TNC was conducting carbon sequestration feasibility studies in Belize, Guatemala, Peru, Panama, Costa Rica, Mexico, Ecuador, the Dominican Republic, Papua New Guinea, Louisiana, and Indiana.[98] TNC recognized, though, that many of their members would want recognition that forests had value beyond carbon—and assurance that resources were going toward biodiversity protection, too. A 1998 white paper suggested that TNC could "affirm this position in a statement with each [mitigation] agreement."[99] At the same time, however, TNC sought to make sure carbon sequestration projects were attractive to U.S. corporations and to "inform U.S. decision-makers of the cost effective role of forests in sequestering carbon."[100]

Since TNC first spearheaded forest-based carbon offsetting, it has become widespread. It was, for example, a major point of negotiation during the 2007 and 2008 UN Climate Change Conferences, and the resulting REDD+ program instituted incentives for reducing deforestation and also, crucially, for restoring degraded forests. This secured the place of forest-based carbon offsetting in international environmental governance.[101] Forest-based and land-use carbon offsets also represented 56.4 percent of transactions in the 2019 voluntary carbon offset market (followed by renewable energy projects at 21.3 percent and household device projects at 8.8 percent).[102] Today forest-based offsetting projects include restoration, as well as forest protection, afforestation (planting trees on lands that weren't previously forested), and timber plantations. Given this widespread buy-in to forest-based carbon offsetting, many ecologists have lauded carbon markets as a new funding mechanism for forest restoration and preservation, especially for tropical ecosystems.[103] One paper estimated that "natural climate solutions," or the storage of carbon in natural ecosystems like forests, wetlands, and grasslands, could provide up to 37 percent of the carbon mitigation needed by 2030 to keep global warming under 2°C; another recent study concluded that global-scale tree restoration was the "most effective climate change solution to date."[104]

Other ecologists, however, have argued for caution: planting trees in historic grasslands can harm grassland-adapted species; tree-planting programs can displace people and agriculture; and tree-planting programs are bound to fail if they are not actively maintained.[105] Some carbon offsetting projects focus exclusively on maximizing carbon sequestration, and their gains in carbon storage can come at the expense of local biodiversity. Scientists have used the term "bio-perversity" to refer to outcomes in which projects ostensibly meant to reduce harm to the environment in one dimension—climate change—end up causing further harm in another dimension, such as habitat for endemic bird species, say, if a native forest is cleared to make way for a carbon plantation. Plantation projects aimed at carbon offsetting may also introduce invasive species and otherwise alter key ecosystem processes such as water tables and fire regimes.[106]

And then there is the mismatch between where emissions offset credits are purchased and where they are produced. In a global market, purchasers of carbon offsets are able to appreciate the benefits of offsetting while continuing to emit greenhouse gasses and outsourcing the possible negative social and ecological impacts of mitigation. At present the greatest number of voluntary carbon credit buyers are the United States, France, the United Kingdom, Germany, and Switzerland. The major offset producing countries include Peru, Brazil, Kenya, Zimbabwe, Bolivia, Indonesia, and Cambodia.[107] In 2010, Europe purchased 10.6 metric tons of carbon dioxide equivalent ($MtCO_2e$) in the carbon markets and contracted to supply only 0.2; in contrast, Latin America contracted to supply 16.9 $MtCO_2e$ and purchased just 4.5.[108] Indeed, the global carbon market incentivizes the privatization and commodification of land and forest resources in developing regions, and it is not difficult to see forest- and land-use-based carbon offsetting as a mode of appropriating land in the Global South for the alleged "universal" environmental end of solving climate change. In the emerging carbon economy, the Global North continues to pollute, while communities in the Global South lose land and sovereignty.

Indeed, the establishment of carbon offsetting sites routinely involves the exclusion of local inhabitants from land and resources that were previously under public or shared jurisdiction. A 2011 study by Oxfam, for example, estimates that at least 22,500 people were evicted from their homes in the creation of timber plantations in Uganda that the UK-registered New Forests Company refashioned to generate carbon offsets.[109] Carbon offsetting projects can also erode food and resource security. Sociologists found this to be the case when a private Norwegian company planted stands of pine and eucalyptus in the shrubland of the Kachung Forest Reserve, also in Uganda. Prior to the establishment of the plantation, community members had possessed long-standing access and use rights, including for animal grazing, fishing, and the collection of water, firewood, and medicinal herbs. Since the establishment of the plantation, however, villagers have been vilified and sometimes criminalized; some have been arrested as "illegal encroachers" and "trespassers" on license areas. For their part, since the beginning of the company's plantation activities, community members have reported the destruction of crops, housing, and trading centers by a collection of state, police, and private sector actors.[110] Those purchasing carbon offsets seek guarantees that the carbon stock will remain stable into the future, and this incentivizes consolidated single-entity ownership. For these reasons, critics describe the international carbon offset market as "carbon colonialism."[111]

The term "carbon colonialism" is apt: as off-site mitigation redistributes ecosystems, it also redistributes power. Under its logic, wetlands, forests, and other ecosystems are destroyed in one location and restored elsewhere, but elsewhere is always a political and economic matter in addition to an ecological one. By decoupling the site of damage from the site of restoration, off-site mitigation redistributes benefits and harms. It thus raises fundamental questions about who should shoulder the responsibility of ecological repair. It is a matter of human rights.

Sites of Ecological Harm

In 2009, Disneynature film company released *Earth,* a documentary film that highlights the threats of rapid environmental change to the survival of three species: the polar bear, the African bush elephant, and the humpback whale. As part of the film's promotion, Disney promised to fund The Nature Conservancy to plant one tree in Brazil's endangered Atlantic rain forest for every viewer who saw the movie during its first week: 2.7 million trees. That same year, Walt Disney Co. announced that it would fund $7 million of forest restoration in partnership with The Nature Conservancy, the Conservation Fund, and Conservation International at sites that included California's northern coast, the Amazon basin of Peru, and the eastern Democratic Republic of Congo. Peter Seligmann, chairman of Conservation International, said the commitment represented "the largest single corporate contribution ever made to reduce [greenhouse gas] emissions from deforestation."[112] Since 2009, Disney has also funded three large-scale TNC forest carbon projects in China. In a 2019 retrospective, The Nature Conservancy heralded the Disney Wilderness Preserve as the beginning of this corporate partnership, a "historic beginning of a new philosophy in land conservation and mitigation," one that resulted in more than fifty mitigation banks in the state of Florida alone, along with methods used in Everglades wetlands restoration and, ultimately, carbon sequestration.[113]

Today ecological restoration is big business, in large part because wetland mitigation created a demand for ecological expertise that TNC and other environmental organizations amplified and helped meet. A 2015 study estimated that, in the United States, the ecological restoration sector directly employed 126,000 workers and generated approximately $9.5 billion in sales annually.[114] Where restoration was once accomplished through voluntary donations of money and labor, many large-scale restoration projects are now paid for by general taxes or international corporations.[115]

The business of off-site wetland mitigation created the institutional links, regulatory mechanisms, and modes of ecological inquiry that formed the foundations of the international carbon offsetting market. One of the striking consequences is that invasive species, first framed as a threat to native species and to the integrity of ecosystems, are today increasingly framed as threats to carbon storage. As of October 2020, TNC's website states that invasive species "crowd out and can kill important tree species that provide shade, carbon storage and habitat for native wildlife."[116] Restoration has become a global practice not only because the threat of climate change is global in nature, but also because corporations and governments of the Global North have had incentives to move their mitigation practices to the Global South. Today, ecological restoration is mandated and regulated through international treaties, and it is pursued by international corporations and environmental NGOs as well as governments.

With ecological restoration framed as a climate change solution, off-site mitigation is on the rise. It remains to be seen, though, how the practice of off-site mitigation will change understandings of the permanence of material damage. We should be asking how off-site mitigation has reconfigured power relations between and within nations and how it has refigured material environments, communities, and individual organisms. Crucially, we must remember to look away from the restoration site and toward the site of harm.

Epilogue

DESIGNING THE FUTURE

The question of how to reverse human-caused ecological damage is the most pressing of our century. Confronted with climate change, ocean acidification, persistent pollution, and other extraordinary challenges, species are struggling. Between 2015 and 2017, bleaching killed almost one-third of the shallow-water corals on the Great Barrier Reef. In northern Florida, populations of *Torreya taxifolia,* an endangered conifer, are rapidly dwindling. Who will take responsibility for the harms done to wild species across the globe? Which individuals, communities, coalitions, corporations, and governments will attempt to repair them? When we fail to reverse particular ecological harms, is that because of technological constraints, political intractability, or sheer scale? And how are we to imagine a future in which people and other species coexist harmoniously?

It is increasingly clear that climate change necessitates new ways of caring for species.[1] Climate change complicates place-based restoration, as many species cease to be suited to the climatic conditions in their historical ranges. The Florida torreya, for example, finds its habitat shifting northward, while changing ocean temperatures and chemistries are making the Great Barrier Reef inhospitable to the coral species that built it up over thousands of years.[2] While migratory species can sometimes find new territories that suit them, other species face substantial barriers to movement: newly suitable ranges might be hundreds of miles away from their current ranges, or beyond an impassable obstacle, like a city or a mountain range; further, species that

9.1 Hawai'i Institute of Marine Biology researcher Jen Davidson places a tray of enhanced coral onto a reef during a practice run for future transplants off the island of O'ahu. Courtesy of Vulcan Inc.

depend on one another may shift their ranges at different rates, which may disrupt ecological relationships that have evolved over thousands of years. Indeed, studying the effects of climate change shows just how entangled life is. One study in central Arizona found that climate-related declines in vegetation cover led to decreased abundance of bird habitat and increased nest predation rates, resulting in local extinction of some previously common bird species.[3]

Because some species may not be able to move fast enough on their own, ecological restoration's best hope may be to move species, or even whole communities of species, to new ranges where they can thrive in future climate conditions.[4] Proponents of this approach call it "managed relocation" or "assisted migration," and they are already pursuing it on a small scale. The loosely organized Torreya Guardians, for instance, have planted seedlings of Florida torreya as far north as Vermont. Recently, British Columbia extended seed transfer zones 200 meters higher in elevation for most tree species and introduced a new policy allowing the planting of western larch outside of its previous range.[5]

In parallel, proponents of the emerging concept of "assisted evolution" argue that, without help from humans, organisms will not be able to evolve quickly enough to adapt to climate change. Scientists at the Hawai'i

Institute of Marine Biology are attempting to produce strains of "super coral." Corals are colonies of tiny organisms called polyps that live in a symbiotic relationship with zooxanthellae, a type of alga. When corals become stressed—because of rising ocean temperatures, say—they expel the algae, losing the nutrients the algae produce through photosynthesis and appearing "bleached." Researchers are identifying bleaching-resistant corals, breeding them in the laboratory, and transplanting them to places where reefs are struggling.[6] Elsewhere, laboratories are using CRISPR-Cas9 genome editing to introduce beneficial mutations to coral genes toward the same end: altering species so that they can survive in a warming world.[7]

Thus far, both assisted migration and assisted evolution have proven intensely controversial among scientists, managers, and the public. Opponents of assisted evolution argue that human-modified organisms might have a competitive advantage over unmodified ones and that this threatens to reduce a species' overall genetic diversity.[8] Ecologists concerned with the genetic "purity" of native species worry, relatedly, that assisted migration will lead to increased hybridization. Opponents of assisted migration also argue that intentionally moved species are no different from unintentionally introduced species: they could compete with native species or irreversibly disrupt food webs at translocation sites. Those in favor of assisted evolution and assisted migration believe, however, that the consequences of doing nothing—species decline and extinction—are far worse than the potential risks of human efforts to design the wild.[9] These debates are unresolved, but some governmental agencies and environmental organizations are already recommending—and beginning to implement—forms of assisted evolution and migration as climate change adaptation strategies.[10]

Driving the controversies around assisted evolution and assisted migration are two questions that restorationists have grappled with since the early twentieth century: Does human intervention necessarily threaten wildness? And who gets to decide where and how restoration occurs? However unorthodox assisted evolution seems, it has important precedents in the work of groups like the American Bison Society and the Peregrine Fund, which sought to breed species while maintaining their wildness. The idea of assisted migration, meanwhile, represents a new turn in the long-standing debate over ecological baselines. Today, many ecologists argue that historical fidelity is not an achievable restoration goal and that ecosystem functioning is a better and more realistic one.[11] In line with this argument, the Society for Ecological Restoration has defined "ecological restoration" broadly enough to include assisted evolution and assisted migration, practices that aim at an improved future instead of a reconstructed past. The organization first defined ecological restoration in 1990 as "the goal of intentionally altering a site to establish a defined, indigenous, historic ecosystem," but members revised the official definition in 2002 to be "the process of assisting the recovery of an ecosystem that has been degraded, damaged, or destroyed"; in other words, they removed any mention of a historical reference point for restoration.[12] Indeed, the history of restoration ecology in the United States reveals that the goal of historical fidelity only really took hold in the late 1980s and began to fall out of favor by the early 2000s. Previous restorationists had pursued such disparate goals as cultivating a "laissez-faire" aesthetic in the garden, saving individual species on the brink of extinction, and enabling game to "farm itself." What united these diverse projects was a desire to intervene in a limited way, to help species while honoring their autonomy. Today a designed species or ecosystem might strike some as the opposite of a wild one. But restorationists have been designing the wild for more than a century.

Working outside the goal of historical fidelity, some restoration ecologists are also questioning the practicality and wisdom of attempting to turn back the clock on "novel ecosystems"—new, resilient ecological communities that have resulted from human activity. These ecologists recognize the millennia long history of human ecological intervention, and they argue that some ecosystems have been so profoundly transformed (by climate change, nutrient loads, changes in land use, nonnative species invasions, and so on) that it would be impossible to return them to a historical state.[13] The core idea of the "novel ecosystem" is that an ecological threshold has been irreversibly crossed, as is plausibly the case with the turkey oak. The turkey oak was native to the British Isles prior to the last glaciation, following which it disappeared from the landscape. Sometime during the past three hundred years, however, it was reintroduced to the British Isles, and it is now a food source for blue tits and great tits, bird species that are laying eggs earlier in the year in response to climate change. The oak is also a host for gall wasps, many of which are nonnative and which provide food for the tit populations earlier in the season than other food sources. The nonnative gall wasps also interact with native gall wasps.[14] Should environmental managers intervene in this ecosystem? If so, how? Proponents of novel ecosystems consider the turkey oak and its interacting insect species to be sufficiently established and resilient and intertwined with native bird species that it would be impossible to eradicate them and ecologically counterproductive to try.

Recognizing that such novel assemblages are everywhere, some ecologists argue that restoration efforts should steer ecosystems toward desired "functions" rather than toward the recomposition of historical species communities. Instead of focusing on the presence or absence of particular species, in other words, these efforts aim to rehabilitate processes like soil stabilization and nutrient cycling. Critics of novel ecosystems worry, however, that accepting human-altered ecosystems constitutes a "license to trash," a free pass for business-as-usual pollution and natural resources depletion. They worry, too, that managing novel ecosystems means domesticating nature and operating with a hubristic, managerial mindset that scorns nature's autonomy. In their view, nature should not be framed as a source of services for humans, and a highly managed novel ecosystem is no different from a garden or a zoo. Those who reject novel ecosystems hope to maintain and restore a different nature, the nature that novel ecosystems displace.

Assisted migration, assisted evolution, and novel ecosystems management represent the leading edge of ecological interventionism in the twenty-first century, but passive habitat preservation also has vocal advocates. While many people share the belief that it is more praiseworthy *to do* good actively than merely *to allow* a good thing to occur, a narrow view of what constitutes wildness can invert this intuition. Some ecologists thus maintain that allowing a wild species to survive is far preferable to intervening on its behalf. Such ecologists might advocate for the creation of "conservation corridors" to allow butterfly species to move their ranges poleward in response to climate change, but oppose doing the good of actually moving butterflies from Costa Rica to Nicaragua. It is this narrower view of wildness, one that sharply limits human collaboration with the wild, that guides the most ambitious recent calls to expand protected areas. In this vein, E. O. Wilson proposes in his 2017 book *Half-Earth* to set aside an entire half the world in nature reservations. Similarly, the 2019 Global Deal for Nature, endorsed by a broad coalition of environmental organizations, calls for 30 percent of Earth to be formally protected and an additional 20 percent to be designated as "climate stabilization areas."[15]

Troublingly, although such anti-interventionist proposals have the support of many environmentalists and policy makers, they neglect to describe where this preservation would occur and which human activities this preservation would limit. One recent paper estimates that if E. O. Wilson's half-earth plan were implemented, more than one billion people, primarily in middle-income countries, would find themselves living in newly protected areas.[16] Many of these people would join the already large number of conservation refugees who have been banished from their homes and histories in the name of protecting other species.[17] Restricting human access to enormous territories in the name of ecological care presumes that human presence and activity necessarily diminish wildness, or destroy it completely.

Half-earth advocates assume that overconsumption, or overpopulation, or climate change is inevitable, and they claim that separating humans from other species is all that we can do to care for the wild. Theirs is a call not for coexistence, but for segregation.

Wilderness preservation presumes that the best way to protect other species from human domination and violence is to keep humans out of nature. As historian William Cronon points out, the trouble with wilderness is that it offers no solution to environmental problems, only the false hope of an escape from responsibility.[18] Preservation is not about relating: it explicitly demands disengagement in designated protected areas, and it ignores the ecological value of every unprotected place. It says nothing about how to slow or stop environmental degradation in the places where we live, work, and extract resources. Meanwhile conservation, in the sense of sustainable use of natural resources, concerns only a narrowly circumscribed set of relations between humans and some of the species they consume. Restoration, in contrast to both preservation and conservation, is about repairing the broadest set of relationships among humans and the species that our economies affect, whether directly and indirectly, nearby or at a distance anywhere in the world.

Ecological restoration represents a hopeful path. It has the potential to ameliorate people's alienation from nature at a time when few people know the names of more than a handful of species in their neighborhoods or where their food comes from. Care and repair counter the helplessness that so many people feel when reading another headline about climate change, marine plastics, or biodiversity loss. There is no better way to understand, and begin to address, how our distributed systems of extraction and consumption harm other species than by attempting to work with those species directly. Restoration acknowledges damage, makes it visible, and attempts repair. It is not about where to be, but how to act.

Restorationists acknowledge and embrace the human ability to co-design the wild and to choose among many possible natures. They study and promote actions that people can take to make the shared world more hospitable for wild species. In 2014, governments at the United Nations Climate Summit chose this more hopeful path and committed to restoring a staggering 350 million hectares by 2030, an action that would remove an estimated thirteen to twenty-six gigatons of greenhouse gases from the atmosphere.[19] And, in an effort to massively scale up the restoration of degraded ecosystems, on March 1, 2019, the United Nations General Assembly declared 2021–2030 the UN Decade on Ecosystem Restoration.[20] Once the pursuit of a handful of enthusiasts, ecological restoration is now a centerpiece of global environmental policy.

Ecological restoration is not without its hazards, however, as this book has made clear. The practices of mitigation and offsetting rely on the ideas that ecosystems are fully replicable and that ecological restoration at one site can compensate, politically and biologically, for damage inflicted elsewhere. But ample evidence from studies of wetlands mitigation demonstrates that restored sites are rarely ecologically equivalent to lost sites.[21] And the expectation that environmental degradation can be offset by restoration elsewhere does not hold decision-makers accountable for environmental harms. This redistribution of care has clear implications for human communities as well as for nonhuman ones. It is quite likely that compensatory mitigation enables some forms of hazardous development and thus reinforces or amplifies environmental racism. Those who defend global offset markets argue that it is best and most efficient to offset ecological harm where it is cheapest to do so—in the Global South—rather than in proximity to local ecological damage. But local offsetting projects offer two significant benefits. First, they allow oversight and accountability. Second, they maintain the geographic link between environmental harm and remediation. As chapter 8 details, carbon offsetting can incentivize governments and corporations to push people off of land deemed ecologically valuable. But we should be willing to pay the cost of local remediation, especially if the higher price prevents land dispossession and other human rights violations. As environmentalists around the world embrace ecological restoration, they must also center human justice.

The history of ecological restoration in the United States reveals that caring for wild species has often gone hand in hand with harming marginalized people. Ecological restoration is a powerful response to environmental degradation, but it is not a merely technical solution; restoration targets are moral and political matters as well as logistical and scientific ones. Everyone involved with ecological restoration, from policy makers and land managers to ecologists and other scholars, has the obligation to scrutinize which processes and events restoration attempts to undo and which cultural practices it criticizes, implicitly and explicitly. Attempting to offset wetland filling from the construction of a Walmart differs conceptually and ethically from attempting to reverse the population effects of federal predator eradication. And yet it is easy to conflate importantly different causes of ecological damage, because many of the practices of ecological restoration—captive breeding, transplanting, digging, burning, sowing—are shared among enormously varied projects.

Since its inception, ecological restoration has been pursued in both protected areas and unprotected areas, in mountains and streams, oceans and farmlands and cities, at large scales and on tiny patches of land. Along with megaprojects like the Comprehensive Everglades Restoration Plan, there are projects like pollinator gardens in people's backyards. Whatever the scale of restoration, we must keep in mind that what seems possible in ecological management today depends on what we have learned to perceive, from climate patterns to ecosystems to genes. It depends, too, on how we understand the limits of both ecological transformation and social change. Lauren Donaldson and his collaborators, for example, believed it was easier to change trout bodies than the trajectory of river industrialization. And today, to many, it seems entirely possible to genetically engineer coral and fund tree planting in Costa Rica, but virtually impossible to reduce carbon emissions in the United States. Perhaps restoring a plant community to what it looked like in 1492 seems easier than grappling with the ongoing violence of colonialism. Whatever paths restorationists choose, restoration must happen in tandem with other changes in human behavior. If we don't reduce the ongoing harms of racism, fossil fuel burning, overconsumption by the wealthy, and toxic industrial chemicals, restoration will offer no more than a temporary repair, a way to move a problem to some other place or time.

In this book I have sought to denaturalize ecological restoration by questioning how restoration practices were constructed, why they were proposed in the first place, and what values or judgments they encoded. I suggest that we conceive of restoration as an optimistic collaboration with nonhuman species, a practice of co-designing the wild with them. But we still have the responsibility to collaborate with one another, too. Otherwise, ecological restoration, like the designation of protected areas, can be patently unjust, even when it appears that all the human stakeholders have been invited to the negotiating table (which is itself rare). An instructive example is the case of humpback chub restoration. The humpback chub is a federally listed endangered species found only in the Colorado River basin, and it faces multiple human-caused threats. The National Park Service began stocking nonnative fish in the Grand Canyon in 1920, and these competed with the humpback chub for food and space.[22] Then, in 1964, the completion of the Glen Canyon Dam created Lake Powell upstream of Grand Canyon and transformed the silty, warm Colorado River into a cold, clear-flowing river below the dam. Subsequently the Arizona Game and Fish Department and various federal agencies managed the section below the dam as a fishery for rainbow trout, a species native to the Pacific Coast.[23]

In the early 2000s, biologists from the U.S. Fish and Wildlife Service concluded that nonnative trout were one of the main threats to the humpback chub. At the confluence of the Colorado River and the Little Colorado River, in what is now Grand Canyon National Park, rainbow trout were competing for food with the endangered chub and even preying on them. The operator of the Glen Canyon Dam, the U.S. Bureau of Reclamation, is responsible under the Endangered Species Act for protecting the chub, and in 2009 the bureau and the Fish and Wildlife Service unveiled a plan to "mechanically eliminate" tens of thousands of trout. Thus, eighty-nine years after the fledgling National Park Service first packed trout eggs onto the backs of mules and hauled them into Grand Canyon, federal environmental managers planned to take even greater pains to undo the ecological effects of that decision.

The Bureau of Reclamation consulted with five tribal governments as it prepared its environmental assessment of an electrofishing proposal. The idea was to stun all the fish in sections of the Colorado River, then extract and kill the nonnative trout—more than twenty thousand trout annually. In the consultation, the Zuni made clear their opposition to the plan. They explained that the place of Zuni emergence into this current world, Chimik'yana'kya dey'a, is in Grand Canyon, and that the location selected for trout removal is sacred: the place simultaneously expresses and represents the fertility of nature and is connected to the prosperity of all life, including the life of nonnative trout. In the Pueblo of Zuni, no life is taken casually, and moreover, the Zuni maintain a familial relationship with all aquatic life. Further, the Zuni were unconvinced that scientific evidence supported removing trout to benefit humpback chub. Yet the Bureau of Reclamation had taken other restoration options off the table, options like changing water releases from the Glen Canyon Dam or not continuing to stock nonnative trout upstream.[24]

For two years, federal agencies and the Pueblo of Zuni conferred on the Zuni objection to the planned killing of nonnative trout. Ultimately, however, the Bureau of Reclamation moved forward over Zuni objections. The Department of the Interior announced in a press release that after "extensive government-to-government tribal consultations" and "multi-criteria decision analysis," nonnative fish control would be "implemented in a way that respects tribal perspectives."[25] Various federal agencies maintained that they were relying on scientific evidence and a technically rigorous means of deciding how to manage species, but Kurt Dongoske, tribal historic preservation officer, put things differently: "With its uncritical reliance on science to provide the only answer to a perceived ecosystem problem, even to the detriment of Zuni traditional cultural values, the Bureau of Reclamation imposed a cultural bias in favor of Western scientific materialism on the Zuni, thereby further subjecting the Zuni people to the ongoing detrimental effects of colonialism. This NEPA process did not equitably weigh or consider Zuni traditional perspectives in its environmental analysis;

rather, it favored hegemonic scientism over Zuni traditional perspectives as the only valid form of understanding the Grand Canyon's ecosystem."[26]

Ecologists often argue that more and better science equals better environmental management.[27] But when ecologists look only to nature for their answers, they disavow their own roles—and the roles of people more generally—in constituting that nature. Prioritizing historical baselines when we decide how the ecological world should look erases the human actions that built those baselines and made them visible. Restorationists must contest such erasure; they must not naturalize what are in fact ethical, aesthetic, and political decisions about which species and which people are to be cared for and how.

Aldo Leopold, the oft-quoted author of *A Sand County Almanac,* suggested we enlarge the boundaries of our ethical community to include not only humans, but also "soils, waters, plants, and animals, or collectively: the land." He speculated that such an ethic could not "prevent the alteration, management, and use of these 'resources,'" but could nevertheless "affirm their right to continued existence."[28] Leopold asks us to consider wild species when making decisions about how we live, but of course, communities differ vastly in their ethical systems, in their ways of deliberating, and in how they apportion power. A commitment to wildness by design therefore requires deliberate ethical thinking.

The concepts of distributive justice and procedural justice, as delineated by environmental justice scholars and activists, should guide the ethics of collaboration in ecological restoration.[29] Distributive justice depends on our making visible the distribution of burdens and benefits among members of a society. We can ask these questions of ecological restoration proposals and practices: How are sites of ecological repair distributed in relation to sites of ecological harm? Who benefits from restoration? Who is harmed? Who does the work of care, and who is cared for? Procedural justice, in turn, is about how decisions are made, who is included in decision-making processes, and what principles are used to make normative claims about just or unjust procedures. With procedural justice in mind, we ask: Who decides where restoration happens? Who decides which species, ecosystems, or other entities are restored? Whose vision of wildness is acted on? These questions are particularly urgent given the ascendence of off-site mitigation and forestry-based carbon offsetting, both of which uncouple sites of harm from sites of restoration. Ecological restoration is increasingly corporatized and consolidated, rather than democratic and locally focused.

It is possible, though, to align social justice and ecological restoration. This book opened with the carving out of what would become the National Wildlife Refuge System from Indian reservation lands appropriated under the Dawes Act. On January 15, 2021, after 113 years and a history that includes three rounds of failed agreements between the United States and tribal governments, numerous lawsuits, a federal investigation, and a massive public education campaign to combat racist resistance, the National Bison Range—the second bison reservation founded by the American Bison Society—was transferred to the Confederated Salish and Kootenai Tribes. The Tribal Council is working with the U.S. Fish and Wildlife Service on the transition from federal to tribal management of the land and bison herd.[30] To the northeast, leaders of the four tribes that make up the Blackfoot Confederacy (Blackfeet Nation, Kainai Nation, Piikani Nation, and Siksika Nation) launched the Iinnii Initiative in 2009 to conserve traditional lands, protect Blackfeet culture, and create a home for bison to return to. Today the Blackfeet Nation manages about eight hundred bison, descendants of the Hornaday reserves, and the Iinnii Initiative helped to negotiate an intergovernmental agreement called the Buffalo Treaty, signed in 2014. The Buffalo Treaty's signatories—now more than thirty tribes and First Nations in the United States and Canada—commit to active pursuit of bison restoration and its ecological and cultural benefits.[31] Their vision combines ecological restoration and historical reparation.

Such examples of justice-oriented restoration are rare. Large-scale restoration projects typically have top-down and technocratic approaches that threaten to sideline local customs, local land use practices, and land tenure. And ecological restoration is still largely confined to protected areas: parks and forests and wetlands owned by land trusts or state and national governments.[32] Nevertheless, restoration has enormous promise. Restorationists see a human role in the making of nature, and this allows for reconciling past antagonisms between humans and wild species. Ecological restoration sits squarely at the intersection of scientific and humanistic work: on the one hand, technical knowledge is crucial to handling species without harming them and to helping them thrive; on the other hand, deciding which species to help, and where, is a political and ethical matter.

Recently scholars have argued that despite climate catastrophe and infrastructure breakdown, maintenance and repair are undervalued and understudied in a world obsessed with perpetual innovation.[33] A lesson from the history of ecological restoration is that repair—and care—are ongoing tasks. Scholars of feminist science studies have argued that even though households, economies, and the flourishing of life on earth all depend on care, we continue to value self-sufficiency and independence from others more highly.[34] To care is to attach and commit to something, and caring is also a practice. But care is not an uncomplicated practice; indeed, care is notorious for the problems that it raises when it is standardized, commodified, prescribed, and evaluated.[35] Ecological restoration will always be ethically complex, in the way that all care is.

Practices of ecological restoration will continue to evolve as we change both how we care and what we know. History reveals that ecological restoration was often designed to achieve goals other than justice; indeed, restoration projects often perpetuated injustices against people in the name of caring for nature. But restoration remains a hopeful practice, endeavoring to undo harm and to help heal. To care the best we can for the worlds we live in and the species we live with, we must design the wild with an eye toward justice. We must design places in which humans and nonhumans not only coexist, but flourish.

ABBREVIATIONS

NOTES

ACKNOWLEDGMENTS

INDEX

ABBREVIATIONS

ABS	American Bison Society
AEC	U.S. Atomic Energy Commission
AFL	University of Washington Applied Fisheries Laboratory
BFER	Biographical File Edith Roberts, Catherine Pelton Durrell Archives & Special Collections Library, Poughkeepsie, New York
BRAU	Annette and E. Lucy Braun Papers, Cincinnati Museum, Ohio
BUTP	Eloise Butler Papers, Minnesota Historical Society, Minneapolis, Minnesota
CARES	Center for Applied Research in Environmental Sciences
CCC	Civilian Conservation Corps
CERP	Comprehensive Everglades Restoration Plan
COP 3	Conference of the Parties to Kyoto International Conference
CREWS	U.S. Fish and Wildlife Service Committee on Rare and Endangered Wildlife Species
C&SF	Central and Southern Florida Project
DARPA	Defense Advanced Research Projects Agency of the U.S. Department of Defense
DCOEL	Dutchess County Outdoor Ecological Laboratory, Catherine Pelton Durrell Archives & Special Collections Library, Poughkeepsie, New York
EPA	Environmental Protection Agency
EPOP	Eugene P. Odum Papers, Hargrett Rare Book & Manuscript Library, University of Georgia, Athens, Georgia
ESA	Ecological Society of America
FWS	U.S. Fish and Wildlife Service
GEHP	G. Evelyn Hutchinson Papers, Manuscripts and Archives, Yale Sterling Memorial Library, New Haven, Connecticut

IBP	International Biological Programme
IUCN	International Union for Conservation of Nature
JCPL	Jimmy Carter Presidential Library, Atlanta, Georgia
LEOP	Aldo Leopold Papers, University of Wisconsin–Madison Libraries, Madison, Wisconsin
LRBR	Laboratory of Radiation Biology Records, 1944–1970, University of Washington Special Collections, Seattle, Washington
LRDP	Lauren R. Donaldson Papers, University of Washington Libraries Special Collections, Seattle, Washington
LRER	Laboratory of Radiation Ecology Records, 1948–1984, University of Washington Special Collections, Seattle, Washington
MED	Manhattan Engineer District
NARA II	National Archives Record Administration II, College Park, Maryland
NEPA	National Environmental Policy Act of 1969
NOAA	National Oceanic and Atmospheric Administration
NOHP	Neal O. Hines Papers, University of Washington Special Collections, Seattle, Washington
NPS	U.S. National Park Service
NWF	National Wildlife Federation
OSRD	Office of Scientific Research and Development
REDD	UN Programme on Reducing Emissions from Deforestation and Forest Degradation
SCOPE	Scientific Committee on Problems of the Environment
SERM	Society for Ecological Restoration & Management
SHEL	Victor E. Shelford Papers, University of Illinois Archives, Urbana-Champaign, Illinois
TNC	The Nature Conservancy
TNCR	The Nature Conservancy Records, Denver Public Library, Denver, Colorado
WFPS	Wild Flower Preservation Society
WFPSA	Wild Flower Preservation Society of America Records, The LuEsther T. Mertz Library, New York Botanical Garden, Bronx, New York
WTH	William T. Hornaday Papers, Wildlife Conservation Society Archives, Bronx, New York

NOTES

Introduction

1. My account derives from George Archibald, *My Life with Cranes: A Collection of Stories* (Baraboo, WI: The International Crane Foundation, 2016).
2. Ray Erickson, "Propagation Studies of Endangered Wildlife at the Patuxent Center," *International Zoo Yearbook* 20 (1980): 40–47; Matthew Perry, *Patuxent History—65th Anniversary* (Laurel, MD: USGS Patuxent Wildlife Research Center, 2004).
3. George Archibald, "Methods for Breeding and Rearing Cranes in Captivity," *International Zoo Yearbook* 14 (1974): 147–155; Ronald Sauey and C. Barbara Brown, "The Captive Management of Cranes," *International Zoo Yearbook* 17 (1977): 89–92; Faith McNulty, "A Reporter at Large: The Thread Remains Very Thin," *New Yorker,* August 6, 1966, pp. 31–82; John R. Cannon, "Whooping Crane Recovery: A Case Study in Public and Private Cooperation in the Conservation of Endangered Species," *Conservation Biology* 10 (1996): 813–821.
4. As quoted in Rene Ebersole, "The Man Who Saves Cranes," *Audubon,* January 18, 2013.
5. Canadian Wildlife Service and U.S. Fish and Wildlife Service, *International Recovery Plan for the Whooping Crane (Grus americana), Third Revision,* March 2007. Ottawa: Recovery of Nationally Endangered Wildlife (RENEW), and U.S. Fish and Wildlife Service, Albuquerque, New Mexico.
6. "First Whooping Crane Hatched at International Crane Foundation in Baraboo, Wisconsin, Dies," *Chicago Sun Times,* March 11, 2021, https://chicago.suntimes.com/2021/3/11/22325239/whooping-crane-baraboo-international-crane-foundation-gee-whiz.
7. Canadian Wildlife Service and U.S. Fish and Wildlife Service, *International Recovery Plan for the Whooping Crane (Grus americana), Third Revision.*
8. "Operation Migration to End 25-Year Mission to Help Save Whooping Cranes," *Lakeland Times,* August 24, 2018; USFWS, "Proposal to Establish a Nonessential Experimental Population of Whooping Cranes in the Eastern United States," *Federal Register* 66, no. 47 (March 9, 2001): 14107–14119; USFWS, "Establishment of a Nonessential Experimental Population of Endangered Whooping Cranes in Southwestern Louisiana," *Federal Register* 76, no. 63 (February 3, 2011): 6066–6082.

9. Pete Fasbender et al., *The Eastern Migratory Population of Whooping Cranes: FWS Vision for the Next 5-Year Strategic Plan,* June 23, 2015.
10. Karl Etters, "Whooping Cranes May Lose Ultralight Assistance," *Tallahassee Democrat,* October 31, 2015; Becca Cudmore, "In New Plan, Baby Whooping Cranes to Be Led by Parents, Not Planes," *Audubon,* February 12, 2016; Wade Harrell and Mark Bidwell, *Report on Whooping Crane Recovery Activities (2015 Breeding Season–2016 Spring Migration),* October 2016, http://www.thewheelerreport.com/wheeler_docs/files/0817om.pdf; Operation Migration, press release, "Operation Migration to End 25-Year Mission to Help Save Whooping Cranes," August 17, 2018.
11. Donna Haraway, *The Companion Species Manifesto: Dogs, People, and Significant Otherness* (Chicago: Prickly Paradigm Press, 2003).
12. For overview, see Bradley Cantrell, Laura J. Martin, and Erle C. Ellis, "Designing Autonomy: Opportunities for New Wildness in the Anthropocene," *Trends in Ecology and Evolution* 32 (2017): 156–166.
13. See, for example, J. Michael Scott et al., "Conservation-Reliant Species and the Future of Conservation," *Conservation Letters* 3 (2010): 91–97.
14. Society for Ecological Restoration International Science & Policy Working Group, *The SER Primer* (Tucson, AZ: SER, 2004).
15. "Gorongosa Restoration Project," https://www.usaid.gov/mozambique/fact-sheets/gorongosa-project; "Coral Restoration Foundation," https://www.coralrestoration.org/.
16. Because so many entities are involved in restoration work, from environmental consulting agencies to plant nurseries, it is difficult to estimate the scale of the global restoration economy. A recent study estimated U.S. federal appropriations for restoration-related programs at $1.9 billion per year (2011–2013). Estimates of U.S. private sector investments range from $1.3 billion to $9.47 billion per year. See Environmental Law Institute, *Mitigation of Impacts to Fish and Wildlife Habitat: Estimating Costs and Identifying Opportunities* (Washington, DC: Environmental Law Institute, 2007), https://www.eli.org/sites/default/files/eli-pubs/d17_16.pdf; Todd BenDor et al., "Defining and Evaluating the Ecological Restoration Economy," *Restoration Ecology* 23 (2015): 209–219; Todd BenDor et al., "Estimating the Size and Impact of the Ecological Restoration Economy," *PLOS One* 10 (2015): e0128339.
17. Intergovernmental Science-Policy Platform on Biodiversity and Ecosystem Services, The Assessment Report on Land Degradation and Restoration (Bonn: IPBES Secretariat, 2018); El Salvador Ministerio de Medio Ambiente y Recursos Naturales, *UN Decade of Ecosystem Restoration 2021–2030: Initiative Proposed by El Salvador with the Support of Countries from the Central American Integration System,* Concept Note, 2019, https://wedocs.unep.org/handle/20.500.11822/26027.
18. My work builds on that of Marcus Hall and Ian Tyrrell, who root ecological restoration in histories of forestry and botanical science, and that of Thomas Dunlap, Mark Barrow, Peter Alagona, and Miles Powell, who each explore histories of concern about extinction. In *Earth Repair,* a study of nineteenth-

century reforestation projects in the Alps and the Rocky Mountains, Marcus Hall roots both restoration and landscape architecture in national gardening traditions. He shows that Italian restorationists, building on their cultural model of ecological health, aimed at "returning order to an unkempt garden," whereas American restorationists "simulated samples of untouched nature." Ian Tyrell, meanwhile, writes about environmental reform movements in California and Australia that worked to "renovate" post-mining landscapes into idealized gardens. Marcus Hall, *Earth Repair: A Transatlantic History of Environmental Restoration* (Charlottesville: University of Virginia Press, 2005); Ian Tyrrell, *True Gardens of the Gods: California-Australian Environmental Reform, 1860–1930* (Berkeley: University of California Press, 1999). See also Thomas R. Dunlap, *Saving America's Wildlife: Ecology and the American Mind, 1850–1990* (Princeton, NJ: Princeton University Press, 1988); Mark V. Barrow Jr., *Nature's Ghosts: Confronting Extinction from the Age of Jefferson to the Age of Ecology* (Chicago: University of Chicago Press, 2009); Peter Alagona, *After the Grizzly: Endangered Species and the Politics of Place in California* (Berkeley: University of California Press, 2013); Miles A. Powell, *Vanishing America: Species Extinction, Racial Peril, and the Origins of Conservation* (Cambridge, MA: Harvard University Press, 2016).

19. William Jordan III and George Lubick, for example, describe the history of ecological restoration as a case of "arrested development," in which Aldo Leopold's ideas were lost for half a century—"a lull during which Americans concluded a war, embarked on a cold war, moved to the suburbs, and went shopping"—before they were taken up again by the founders of the SER. Jordan and Lubick, *Making Nature Whole: A History of Ecological Restoration* (Washington, DC: Island Press, 2011), 105. See also T. A. Pickett and V. Thomas Parker, "Avoiding the Old Pitfalls: Opportunities in a New Discipline," *Restoration Ecology* 2 (1994): 75–79; William R. Jordan III, *The Sunflower Forest: Ecological Restoration and the New Communion with Nature* (Berkeley: University of California Press, 2003).

20. To dive into the vast literature on the American wilderness preservation movement, see Roderick Frazier Nash, *Wilderness and the American Mind* (New Haven, CT: Yale University Press, 1967); William Cronon, "The Trouble with Wilderness; or, Getting Back to the Wrong Nature," in *Uncommon Ground: Rethinking the Human Place in Nature*, ed. William Cronon (New York: W. W. Norton, 1995), 69–90; J. Baird Callicott and Michael P. Nelson, eds., *The Great New Wilderness Debate* (Athens: University of Georgia Press, 1998); Paul Sutter, *Driven Wild: How the Fight against Automobiles Launched the Modern Wilderness Movement* (Seattle: University of Washington Press, 2002). On the American conservation movement, see Samuel P. Hays, *Conservation and the Gospel of Efficiency: The Progressive Conservation Movement, 1890–1920* (Cambridge, MA: Harvard University Press, 1959); John Reiger, *American Sportsmen and the Origins of Conservation* (Norman: University of Oklahoma Press, 1986); Louis Warren, *The Hunter's Game: Poachers and Conservationists in Twentieth-Century America* (New Haven, CT: Yale University Press, 1997); Karl Jacoby, *Crimes against Nature:*

Squatters, Poachers, Thieves and the Hidden History of American Conservation (Berkeley: University of California Press, 2003); Sarah Phillips, *This Land, This Nation: Conservation, Rural America, and the New Deal* (Cambridge, UK: Cambridge University Press, 2007); Neil Maher, *Nature's New Deal: The Civilian Conservation Corps and the Roots of the American Environmental Movement* (New York: Oxford University Press, 2008).

21. Mark David Spence, *Dispossessing the Wilderness: Indian Removal and the Making of the National Parks* (New York: Oxford University Press, 2000).
22. United Nations Environment Programme, Protected Planet, https://www.protectedplanet.net/en (accessed April 26, 2021).
23. See also Ben Minteer and Stephen Pyne, *After Preservation: Saving American Nature in the Age of Humans* (Chicago: University of Chicago Press, 2015); Paul Sutter, "What Can U.S. Environmental Historians Learn from Non-U.S. Environmental Historiography?" *Environmental History* 8 (2003): 109–129.
24. William Jordan III and George Lubick distinguish two types of restoration, "meliorative land management" and "ecocentric restoration." According to their definition, meliorative land management is "conservation" intended to make an environment "better for someone," while ecocentric restoration is "a self-conscious encounter with nature as other in the form of ecosystems that were there when we got there and owe nothing to us." Whereas Jordan and Lubick distinguish these two modes of land management by their intended beneficiaries—meliorative restoration is anthropocentric, ecocentric restoration is biocentric—historian Marcus Hall identifies three modes of restoration by what they attempt to ameliorate. "Maintenance gardening," he writes, is when humans work to maintain "ideal domesticated forms" in the face of pests, droughts, and other natural damaging agents. "Reparative gardening" is when humans work to address environmental damage caused not by nature, but by humans: logging, invading exotics, overgrazing. And "reparative naturalizing" is when humans, seeing themselves as more "directors" than gardeners, attempt to address damage caused by culture by mimicking "nature's healing processes" (Hall, *Earth Repair,* 212–217). The examples of ecological restoration I consider in this book are motivated, sometimes simultaneously, by anthropocentric and biocentric concerns—thus they do not fit neatly in Jordan and Lubick's framework. My use of ecological restoration encompasses Hall's "reparative gardening" and "reparative naturalizing." See Hall, *Earth Repair;* Jordan and Lubick, *Making Nature Whole.* I exclude efforts to restore hydrology, soil nutrients, and other environmental processes, not because they are not rich areas of inquiry, but because I am interested in efforts to manage the wildness of species. Works that touch on the history of soil restoration include Hall, *Earth Repair;* Neil Maher, *Nature's New Deal: The Civilian Conservation Corps and the Roots of the American Environmental Movement* (New York: Oxford University Press, 2007); Paul Sutter, *Let Us Now Praise Famous Gullies: Providence Canyon and the Soils of the South* (Athens: University of Georgia, 2015). Geographer Rebecca Lave interrogates the political economy of stream restoration in the United States, analyzing how river scientists shift their work in response

to the demands of funding agencies, in *Field and Streams: Stream Restoration, Neoliberalism, and the Future of Environmental Science* (Athens: University of Georgia Press, 2012).

25. National Research Council, *Compensating for Wetland Losses under the Clean Water Act* (Washington, DC: National Academy Press, 2001); USFWS, "Guidance for the Establishment, Use, and Operation of Conservation Banks," 2003, https://www.fws.gov/endangered/esa-library/pdf/Conservation_Banking_Guidance.pdf.
26. Louisiana Coastal Protection and Restoration Authority, *Louisiana's Comprehensive Master Plan for a Sustainable Coast* (Baton Rouge: CPRA, 2012), https://issuu.com/coastalmasterplan/docs/coastal_master_plan-v2.
27. As Robert Wilson notes, the FWS is the primary federal agency responsible for wild animals in the United States, yet it has received remarkably little attention from environmental historians. Wilson, "The Ugly Duckling," *Environmental History* 16 (2011): 439–444.
28. A smaller literature concerns the history of ecological ideas, but few environmental histories pursue the intersection of ideational and material change. For a classic exchange on this topic, see Donald Worster, "Transformations of the Earth: Toward an Agroecological Perspective in History," *Journal of American History* 76 (1990): 1087–1106; William Cronon, "Modes of Prophecy and Production: Placing Nature in History," *Journal of American History* 76 (1990): 1122–1131. For histories of ecological ideas, see, for example, Donald Worster, *Nature's Economy: A History of Ecological Ideas* (Cambridge, UK: Cambridge University Press, 1977); Gregg Mitman, *The State of Nature: Ecology, Community, and American Social Thought, 1900–1950* (Chicago: University of Chicago Press, 1992); Sharon E. Kingsland, *Modeling Nature* (Chicago: University of Chicago Press, 1995), and *The Evolution of American Ecology, 1890–2000* (Baltimore, MD: Johns Hopkins University Press, 2005).
29. On the intersection of environmental history and science and technology studies, see Sara Pritchard, Dolly Jørgensen, and Finn Arne Jørgensen, eds., *New Natures: Joining Environmental History with Science and Technology Studies* (Pittsburgh: University of Pittsburgh Press, 2013). For work at this intersection, see, for example, Michelle Murphy, *Sick Building Syndrome and the Problem of Uncertainty: Environmental Politics, Technoscience, and Women Workers* (Durham, NC: Duke University Press, 2006); Linda Nash, *Inescapable Ecologies: A History of Environment, Disease, and Knowledge* (Berkeley: University of California Press, 2006); Jeremy Vetter, ed., *Knowing Global Environments: New Historical Perspectives on the Field Sciences* (New Brunswick, NJ: Rutgers University Press, 2010). Meanwhile, work under the banners of actor-network theory, posthumanism, multispecies ethnography, and new materialism strives to expand the range of actors involved in history-making, a project that has been central to the discipline of environmental history since its inception. For an introduction to these concerns, see Bruno Latour, *Reassembling the Social: An Introduction to Actor-Network Theory* (New York: Oxford University Press, 2005); S. Eben Kirksey

and Stefan Helmreich, "The Emergence of Multispecies Ethnography," *Cultural Anthropology* 25 (2010): 545–576. Recent environmental histories that engage this literature include Murphy, *Sick Building Syndrome;* Paul Sutter, "Nature's Agents or Agents of Empire? Entomological Workers and Environmental Change during the Construction of the Panama Canal," *Isis* 98 (2007): 724–754; Edmund Russell, *Evolutionary History: Uniting History and Biology to Understand Life on Earth* (Cambridge, UK: Cambridge University Press, 2011); Donna J. Haraway, *Staying with the Trouble: Making Kin in the Chthulucene* (Durham, NC: Duke University Press, 2016); Timothy LeCain, *The Matter of History: How Things Create the Past* (Cambridge, UK: Cambridge University Press, 2017).

30. For STS approaches to what counts as environmental knowledge and an environmental problem, see Peter Taylor and Frederick Buttel, "How Do We Know We Have Global Environmental Problems? Science and the Globalization of Environmental Discourse," *Geoforum* 23 (1992): 405–416; Kim Fortun, *Advocacy after Bhopal: Environmentalism, Disaster, New Global Orders* (Chicago: University of Chicago Press, 2001); Michelle Murphy, *Sick Building Syndrome;* Sheila Jasanoff, *Designs on Nature: Science and Democracy in Europe and the United States* (Princeton, NJ: Princeton University Press, 2007); Paul Edwards, *A Vast Machine: Computer Models, Climate Data, and the Politics of Global Warming* (Cambridge, MA: MIT Press, 2010).
31. As is the case with so many environmental practices, recently, the work of restoring whooping cranes has shifted from the federal government to private organizations. In 2017, the fifty-one-year-old Patuxent whooping crane program was dismantled after the Trump administration reduced the budget of the U.S. Geological Survey, which administered the program. As I write, the captive cranes are in the process of being relocated to zoos and nonprofits. Karen Brulliard, "A 50-Year Effort to Raise Endangered Whooping Cranes Comes to an End," *Washington Post,* September 18, 2017; U.S. Geological Survey, *Budget Justifications and Performance Information Fiscal Year 2019* (Washington, DC, 2019), https://www.doi.gov/sites/doi.gov/files/uploads/fy2019_usgs_budget_justification.pdf.
32. On the Society for Ecological Restoration's changing definitions of ecological restoration, see Eric Higgs, *Nature by Design: People, Natural Process, and Ecological Restoration* (Cambridge, MA: MIT Press, 2003), 107–110. Higgs argues that restoration is concerned with two primary concepts: ecological integrity and historical fidelity.
33. Stephen T. Jackson and Richard J. Hobbs, "Ecological Restoration in the Light of Ecological History," *Science* 31 (2009): 567–569.
34. Marcus Hall, ed., *Restoration and History: The Search for a Usable Environmental Past* (New York: Routledge, 2009).
35. Jean-Luc Dupouey et al., "Irreversible Impact of Past Land Use on Forest Soils and Biodiversity," *Ecology* 83 (2002): 2978–2984; Nanci J. Ross, "Modern Tree Species Composition Reflects Ancient Maya 'Forest Gardens' in Northwest Belize," *Ecological Applications* 21 (2011): 75–84; Susan C. Cook-Patton et al., "Ancient Experiments: Forest Biodiversity and Soil Nu-

trients Enhanced by Native American Middens," *Landscape Ecology* 29 (2014): 979–987.

36. See, for example, William Denevan, "The Pristine Myth: The Landscape of the Americas in 1492," *Annals of the Association of American Geographers* 82 (1992): 369–385; Cronon, "The Trouble with Wilderness," in *Uncommon Ground,* 69–90; J. Baird Callicott and Michael P. Nelson, eds., *The Great New Wilderness Debate* (Athens: University of Georgia Press, 1998); Shepard Krech III, *The Ecological Indian: Myth and History* (New York: W. W. Norton, 1999).
37. Marcus Hall, ed., *Restoration and History: The Search for a Usable Environmental Past* (New York: Routledge, 2010); Peter Alagona, John Sandlos, and Yolanda Wiersma, "Past Imperfect: Using Historical Ecology and Baseline Data for Contemporary Conservation and Restoration Projects," *Environmental Philosophy* 9 (2012): 49–70.
38. Susan Solomon et al., "Irreversible Climate Change Due to Carbon Dioxide Emissions," *PNAS* 106 (2009): 1704–1709.
39. Philip Kiefer, "Iconic Joshua Trees May Disappear—But Scientists Are Fighting Back," *National Geographic,* October 15, 2018. See, for example, Camille Parmesan and Gary Yohe, "A Globally Coherent Fingerprint of Climate Change Impacts across Natural Systems," *Nature* 421 (2003): 37–42; Camille Parmesan, "Biotic Response: Range and Abundance Changes," in *Climate Change and Biodiversity,* ed. Thomas Lovejoy and Lee Hannah (New Haven, CT: Yale University Press, 2005), 41–55; Craig Moritz et al., "Impact of a Century of Climate Change on Small-Mammal Communities in Yosemite National Park, USA," *Science* 322 (2008): 261–264.
40. Jurriaan M. De Vos et al., "Estimating the Normal Background Rate of Species Extinction," *Conservation Biology* 29 (2014): 452–462; Stuart L. Pimm et al., "The Biodiversity of Species and Their Rates of Extinction, Distribution, and Protection," *Science* 344 (2014): 987; Elizabeth Kolbert, *The Sixth Extinction: An Unnatural History* (New York: Henry Holt, 2014); Gerardo Ceballos and Paul R. Ehrlich, "The Misunderstood Sixth Mass Extinction," *Science* 360 (2018): 1080–1081.
41. Intergovernmental Science-Policy Platform on Biodiversity and Ecosystem Services, *Media Release: Nature's Dangerous Decline "Unprecedented"; Species Extinction Rates "Accelerating,"* May 6, 2019, https://www.ipbes.net/news/Media-Release-Global-Assessment. For humanistic approaches to extinction, see Thom van Dooren, *Flight Ways: Life and Loss at the Edge of Extinction* (New York: Columbia University Press, 2014); Ursula K. Heise, *Imagining Extinction: The Cultural Meanings of Endangered Species* (Chicago: University of Chicago Press, 2016); Dolly Jørgensen, *Recovering Lost Species in the Modern Age: Histories of Longing and Belonging* (Cambridge, MA: MIT Press, 2019).
42. G. D. Gann et al., "International Principles and Standards for the Practice of Ecological Restoration," 2nd ed., *Restoration Ecology* 27 (2019): S1–S46.
43. I have reviewed this literature elsewhere. See Laura J. Martin et al., "Conservation Opportunities across the World's Anthromes," *Diversity and Distributions*

20 (2014): 745–755. In 2009, two prominent restoration ecologists, Steven Jackson and Richard Hobbs, summarized why historical baselines were under scrutiny in "Ecological Restoration in the Light of Ecological History," *Science* 325 (2009): 567–569. See also Thomas W. Swetnam, Craig D. Allen, and Julio L. Betancourt, "Applied Historical Ecology: Using the Past to Manage the Future," *Ecological Applications* 9 (1999): 1189–1206; Young D. Choi, "Restoration Ecology to the Future: A Call for a New Paradigm," *Restoration Ecology* 15 (2007): 351–353; Emma Marris, *Rambunctious Garden: Saving Nature in a Post-Wild World* (New York: Bloomsbury, 2011); Eric Higgs et al., "The Changing Role of History in Restoration Ecology," *Frontiers in Ecology and the Environment* 12 (2014): 499–506.

44. Paul J. Crutzen, "Geology of Mankind," *Nature* 415 (2002): 23–32; Will Steffen et al., "The Anthropocene: Conceptual and Historical Perspectives," *Philosophical Transactions of the Royal Society A* 369 (2011): 842–867.
45. Simon L. Lewis and Mark A. Maslin, "Defining the Anthropocene," *Nature* 519 (2015): 171–180.
46. Bruno Latour, "Agency at the Time of the Anthropocene," *New Literary History* 45 (2014): 1–18; Dipesh Chakrabarty, "The Climate of History: Four Theses," *Critical Inquiry* 35 (2009): 197–222.
47. See, for example, Haraway, *Staying with the Trouble;* Jason W. Moore, ed., *Anthropocene or Capitalocene?: Nature, History, and the Crisis of Capitalism* (Oakland, CA: PM Press, 2016).
48. In "The Politics of Care in Technoscience," Aryn Martin and colleagues challenge the idea that care is always innocent and good. Focusing on health care, they write, "Care is a selective mode of attention: it circumscribes and cherishes some things, lives, or phenomena. In the process, it excludes others. Further, practices of care are always shot through with asymmetrical power relations: Who has the power to define what counts as care and how it should be administered?" Aryn Martin, Natasha Myers, and Ana Viseu, "The Politics of Care in Technoscience," *Social Studies of Science* 45 (2015): 625–641.
49. Mark Dowie, "Conservation Refugees," *Orion,* Nov/Dec 2005.

1. Uncle Sam's Reservations

1. Lukas Rieppel, "Museums and Botanical Gardens," in *The Wiley Blackwell Companion to the History of Science,* ed. Bernard Lightman (New York: Wiley-Blackwell, 2016): 238–251. On Hornaday's life, see Stefan Bechtel, *Mr. Hornaday's War: How a Peculiar Victorian Zookeeper Waged a Lonely Crusade for Wildlife That Changed the World* (New York: Beacon Press, 2012); Gregory J. Dehler, *The Most Defiant Devil: William Temple Hornaday and His Controversial Crusade to Save American Wildlife* (Charlottesville: University of Virginia Press, 2013).
2. Society of American Taxidermists, *Third Annual Report* (Washington, DC: Gibson Brothers Printers, 1884). See also Mark V. Barrow Jr., "The Specimen Dealer: Entrepreneurial Natural History in America's Gilded Age," *Journal of the History of Biology* 33 (2000): 493–534.

3. Report of Professor Baird, in *Annual Report of the Board of Regents of the Smithsonian Institution for the Year Ending June 30, 1887,* Part I (Washington, DC: Government Printing Office, 1889), 6.
4. William Temple Hornaday, "The Passing of the Buffalo," *The Cosmopolitan,* October 1887, p. 9. See also Hanna Rose Shell, "Introduction: Finding the Soul in the Skin," in William Temple Hornaday, *The Extermination of the American Bison* (Washington, DC: Smithsonian Institution Press, 2002), viii–xxiii; Mark Barrow, *Nature's Ghosts: Confronting Extinction from the Age of Jefferson to the Age of Ecology* (Chicago: University of Chicago Press, 2009), chapter 4.
5. William T. Hornaday, *Taxidermy and Zoological Collecting* (New York: C. Scribner's Sons, 1891), as quoted in Shell, "Introduction," in Hornaday, *Extermination of the American Bison,* xix.
6. Andrew Isenberg details the interaction of economic, cultural, and ecological circumstances, including drought and disease, that brought bison to the brink of extinction in *The Destruction of the Bison: An Environmental History, 1750–1920* (Cambridge, UK: Cambridge University Press, 2000).
7. Megan Black, *The Global Interior: Mineral Frontiers and American Power* (Cambridge, MA: Harvard University Press, 2018).
8. Columbus Delano, *Annual Report of the Secretary of the Interior, 1873,* as quoted in Isenberg, *Destruction of the Bison,* 152.
9. R. C. McCormick, "Restricting the Killing of the Buffalo, Speech of R. C. McCormick of Arizona, House of Representatives, April 6, 1872," *Congressional Globe,* Appendix 42nd Congress, 2nd Sess., 179–180. On bison numbers, see David Smits, "The Frontier Army and the Destruction of the Buffalo: 1865–1883," *Western Historical Quarterly* 25 (1994): 312–338.
10. Isenberg, *Destruction of the Bison,* 145.
11. Extract from the *New Mexican,* reprinted in *Annual Report of the Board of Regents of the Smithsonian,* p. 516.
12. Hornaday, "The Passing of the Buffalo," 9.
13. *Annual Report of the Board of Regents of the Smithsonian,* 464.
14. Brown, as quoted in Bechtel, *Mr. Hornaday's War,* 134. William T. Hornaday, "The Zoological Park of Our Day," *Zoological Society Bulletin* 35 (1909): 543.
15. On the rise of natural history displays and zoological gardens, see Elizabeth Hanson, *Animal Attractions: Nature on Display in American Zoos* (Princeton, NJ: Princeton University Press, 2002); Lynn Nyhart, *Modern Nature* (Chicago: University of Chicago Press, 2009); Carin Berkowitz and Bernard Lightman, eds., *Science Museums in Transition: Cultures of Display in Nineteenth-Century Britain and America* (Pittsburgh: University of Pittsburgh Press, 2017).
16. "Department of Living Animals: Annual and Monthly Reports, 1887–88," SIA RU000158, United States National Museum (Record Unit 158), Smithsonian Institution Archives, Washington, DC.
17. As recounted in Dehler, *Most Defiant Devil,* chapters 3 and 4. See also William Bridges, *Gathering of Animals: An Unconventional History of the New York Zoological Society* (New York: Harper and Row, 1974).

18. Article II, The Constitution of the Boone and Crockett Club, Founded December 1887; Jonathan Spiro, *Defending the Master Race: Conservation, Eugenics, and the Legacy of Madison Grant* (Burlington: University of Vermont Press, 2008), 74.
19. Roderick Nash first distinguished the Progressive Era conservation movement from the wilderness movement in *Wilderness and the American Mind* (New Haven, CT: Yale University Press, 1967). On the American conservation movement, see also Samuel P. Hays, *Conservation and the Gospel of Efficiency: The Progressive Conservation Movement, 1890–1920* (Cambridge, MA: Harvard University Press, 1959); John Reiger, *American Sportsmen and the Origins of Conservation* (New York: Winchester Press, 1975); Thomas Dunlap, *Saving America's Wildlife: Ecology and the American Mind, 1850–1990* (Princeton, NJ: Princeton University Press, 1988); Louis Warren, *The Hunter's Game: Poachers and Conservationists in Twentieth-Century America* (New Haven, CT: Yale University Press, 1997); Karl Jacoby, *Crimes against Nature: Squatters, Poachers, Thieves, and the Hidden History of American Conservation* (Berkeley: University of California Press, 2001); Miles Powell, *Vanishing America: Species Extinction, Racial Peril, and the Origins of Conservation* (Cambridge, MA: Harvard University Press, 2016). On women's Progressive Era conservation organizations, see Jennifer Price, *Flight Maps: Adventures with Modern America* (New York: Basic Books, 1999); Carolyn Merchant, "Women of the Progressive Conservation Movement: 1900–1916," *Environmental Review* 8 (1984): 57–85; on the founding of the Audubon Society, see Mark V. Barrow Jr., *A Passion for Birds: American Ornithology after Audubon* (Princeton, NJ: Princeton University Press, 1998), chapters 5 and 6.
20. Elizabeth Britton, "Going, Going, Almost Gone! Our Wild Flowers," *New York Times,* May 4, 1913.
21. Mary Perle Anderson, "The Protection of Our Native Plants," *Plant World* 7 (1904): 123–129; William T. Hornaday, *Our Vanishing Wild Life: Its Extermination and Preservation* (New York: Charles Scribner's Sons, 1913), 105.
22. Warren, *Hunter's Game;* Jacoby, *Crimes against Nature.*
23. Redfield Proctor, Preservation of the Buffalo, 35th Cong., 1st sess., *Congressional Record* 35, pt. 2 (January 30, 1902): 1902.
24. Department of the Interior, *The American Bison in the United States and Canada* (Washington, DC: Government Printing Office, 1902).
25. Philip Buffalo Ranch, South Dakota, c. 1910, American Bison Society Papers, as quoted in Isenberg, *Destruction of the Bison,* 176.
26. "How to Preserve the Once Mighty Monarch of the Plains," *Washington Times,* May 4, 1902.
27. On private parks, see Jacoby, *Crimes against Nature,* 39; Edward Comstock and Mark Webster, *The Adirondack League Club, 1890–1990* (Old Forge, NY: Adirondack League Club, 1990).
28. Ernest Harold Baynes, *War Whoop and Tomahawk: The Story of Two Buffalo Calves* (New York: MacMillan Co., 1929). As early as 1887, private herd

owners had appealed to Washington for land and money with which to continue their herds, to no avail. See George Coder, "The National Movement to Preserve the American Buffalo in the United States and Canada between 1880 and 1920" (PhD diss., Ohio State University, 1975), chapter 3.

29. William T. Hornaday, Wildlife Conservation Scrapbook, volume 2, "The Founding of Two National Bison Herds, 1906–1911," series 4, William T. Hornaday Papers (Collection 1007), Wildlife Conservation Society Archives, Bronx, NY (hereafter WTH Papers).
30. Ernest Harold Baynes, "Save the Buffalo: An Appeal for this Greatest of American Animals," *Bottineau Courant,* January 27, 1905. Hornaday to John Lacy, March 24, 1906, as quoted in Dehler, *Most Defiant Devil,* chapter 4.
31. Ernest Harold Baynes, "The American Bison Society," *Western Field* 8 (1906): 138–140.
32. Ernest Harold Baynes, "The Bison Society Organization," *Perth Amboy Evening News,* January 19, 1906.
33. Hornaday, *Extermination of the American Bison,* 526.
34. Robert C. Auld, "A Means of Preserving the Purity and Establishing a Career for the American Bison of the Future," *American Naturalist* 24 (1890): 787–796; "Saving the Buffalo: How the Few Remaining Bison Are Being Preserved," *Essex County Herald,* August 22, 1902; "New Hybrid Is Valuable," *Billings Gazette,* January 22, 1907.
35. Letter from Charles B. Davenport in U.S. Congress, Senate, *To Establish a Permanent Bison Range (Report to accompany S. 6159),* April 6, 1908, 60th Cong., 1st sess., Report No. 467.
36. Barbara A. Kimmelman, "The American Breeders' Association: Genetics and Eugenics in an Agricultural Context, 1903–13," *Social Studies of Science* 13 (1983): 163–204; Kathy Cooke, "The Limits of Heredity: Nature and Nurture in American Eugenics before 1915," *Journal of the History of Biology* 31 (1998): 263–278; Garland Allen, Daniel Kevles, and Peter J. Bowler, *The Mendelian Revolution: The Emergence of Hereditarian Concepts in Modern Science and Society* (London: Bloomsbury Publishing, 2001).
37. Spiro, *Defending the Master Race,* xii.
38. Madison Grant, "A Canadian Moose Hunt," 1895, cited in Spiro, *Defending the Master Race,* 41.
39. American Breeders Association, *Proceedings of the Meeting Held at Lincoln, Nebraska, January 17–19, 1906,* vol. 2, p. 15; D. E. Lantz, "Report of the Committee on Breeding Wild Mammals" *Journal of Heredity* 4 (1908): 184–192.
40. Garland Allen, "Eugenics and American Social History 1880–1950," *Genome* 31 (1989): 885–889. See also Michael Freeden, "Eugenics and Progressive Thought: A Study in Ideological Affinity," *Historical Journal* 22 (1979): 645–671. Lukas Rieppel expands on these insights in *Fossil Hunters, Tycoons, and the Making of a Spectacle* (Cambridge, MA: Harvard University Press, 2019), chapter 5.
41. Hornaday, *Extermination of the American Bison,* 394.

42. "How to Preserve the Once Mighty Monarch of the Plains," *Washington Times,* May 4, 1902.
43. As quoted in Spiro, *Defending the Master Race,* 40.
44. William T. Hornaday, "The Founding of the Wichita National Bison Herd," in U.S. Congress, Senate, *To Establish a Permanent Bison Range (Report to accompany S. 6159),* April 6, 1908, 60th Cong., 1st sess., Report No. 467, p. 19.
45. The dichotomy between the wild and the domesticated, as Harriet Ritvo points out, has operated powerfully in scientific and general culture but has not been much reflected on. See "Calling the Wild: Selection, Domestication, and Species," *Clio Medica* 93 (2015): 262–280. See also Harriet Ritvo, "Race, Breed, and Myths of Origin: Chillingham Cattle as Ancient Britons," *Representations* 39 (1992): 1–22. On the history of the concept of "wildlife," see the compelling set of essays in *Environmental History* 16 (2011): 391–445.
46. Andrew Isenberg has argued that the preservation of bison was not an end in itself, but a means to preserve an idealized, masculinized vision of frontier life. See Isenberg, *Destruction of the Bison.*
47. "How to Preserve the Once Mighty Monarch of the Plains," *Washington Times,* May 4, 1902.
48. National Park Service, *Homesteading by the Numbers,* https://www.nps.gov/home/learn/historyculture/bynumbers.htm.
49. On the role the Department of the Interior played in this reorientation, see Megan Black, *Global Interior,* chapter 1. For other works that center the settler colonial project in the history of natural resources management (although not restoration), see Bruce Schulman, "Governing Nature, Nurturing Government: Resource Management and the Development of the American State, 1900–1912," *Journal of Policy History* 7 (2005): 375–403; Ian Tyrell, *The Crisis of the Wasteful Nation: Conservation and Empire in Teddy Roosevelt's America* (Chicago: University of Chicago Press, 2015).
50. Sadiah Qureshi, "Dying Americans: Race, Extinction, and Conservation in the New World," in *From Plunder to Preservation: Britain and the Heritage of Empire, c. 1800–1940,* ed. Astrid Swenson and Peter Mandler (Oxford: Oxford University Press, 2013), 269–288.
51. Roosevelt, as quoted in Kevin Bruyneel, *The Third Space of Sovereignty: The Postcolonial Politics of U.S.-Indigenous Relations* (Minneapolis: University of Minnesota Press, 2007), 94. See also Henry E. Fritz, *The Movement for Indian Assimilation, 1860—1890* (Philadelphia: University of Pennsylvania Press, 1963); Frederick E. Hoxie, *A Final Promise: The Campaign to Assimilate the Indians, 1880–1920* (Lincoln: University of Nebraska Press, 1984); Klaus Frantz, *Indian Reservations in the United States* (Chicago: University of Chicago Press, 1999).
52. Along with wildlife reservations, the federal government continued to reserve lands through the creation of forest reserves and national parks. The implications of these programs for Native American sovereignty have been much better studied than those of wildlife refuges. See Peter Nabokov, *Restoring a Presence: American Indians and Yellowstone National Park* (Norman: Uni-

versity of Oklahoma Press, 2004); Mark David Spence, *Dispossessing the Wilderness: Indian Removal and the Making of the National Parks* (New York: Oxford University Press, 2000).

53. U.S. Congress, House of Representatives, *Protection of Game, Etc., In the Forest Reserves of California,* June 9, 1906, 59th Cong., 1st sess., Report No. 4907, p. 6. For a detailed account, see Charles Wilkinson and H. Michael Anderson, "Land and Resource Planning in the National Forests," *Oregon Law Review* 61 (1985): 273–282.
54. The U.S. Fish and Wildlife Service has received much less attention from environmental historians than the U.S. National Park Service or the U.S. Forest Service. On the history of the National Wildlife Refuge System, see Ira N. Gabrielson, *Wildlife Refuges* (New York: Macmillan Company, 1943); Robert Wilson, *Seeking Refuge: Birds and Landscapes of the Pacific Flyway* (Seattle: University of Washington Press, 2010).
55. *An Act for the Protection of Wild Animals and Birds in the Wichita Forest Reserve,* Public Law 23, *U.S. Statues at Large* 33 (1905): 614. See also Hornaday, "The Founding of the Wichita National Bison Herd," *Annual Report of the American Bison Society* 1 (1908): 55–69.
56. "The Wichita National Bison Herd," *Zoological Society Bulletin* 35 (1909): 556; Martin S. Garretson, *The American Bison* (New York: New York Zoological Society, 1938).
57. Elwin R. Sanborn, "The National Bison Herd: An Account of the Transportation of the Bison from the Zoological Park to the Wichita Range," *Zoological Society Bulletin* 28 (1908): 400–412.
58. Hornaday, "Founding of the Wichita National Bison Herd"; "An Object Lesson in Bison Preservation: The Wichita National Bison Herd after Five Years," *Zoological Society Bulletin* 16 (1913): 990–993; Jack Haley, "A History of the Establishment of the Wichita National Forest and Game Preserve, 1901–1908" (PhD diss., University of Oklahoma, 1973).
59. "For a Buffalo Preserve," *Saturday Evening Post,* March 6, 1908; *Second Annual Report of the American Bison Society, 1908–1909* (New York: American Bison Society, 1909); William T. Hornaday, "Report of the President on the Founding of the Montana National Bison Herd," Annual Report of the ABS, 1908–1909, clipping in Hornaday Wildlife Scrapbook Collection, Wildlife Conservation Scrapbook, volume 2, series 4, WTH Papers.
60. M. J. Elrod, "Report on Flathead Buffalo Range," *Annual Report of the American Bison Society, 1905–1907* (1908): 15–49. U.S. Congress, Senate, *To Establish a Permanent Bison Range (Report to accompany S. 6159),* April 6, 1908, 60th Cong., 1st sess., Report No. 467. "Another Reservation for Buffalo," *Washington Star,* March 4, 1908, clipping in Hornaday, Wildlife Conservation Scrapbook, volume 2, series 4, WTH Papers.
61. "Starting a Bison Herd," *Wellsboro Gazette,* August 17, 1910; *Third Annual Report of the American Bison Society, 1909–1910* (New York: ABS, 1910).
62. *Fourth Annual Report of the American Bison Society* (New York: ABS, 1911); *Fifth Annual Report of the American Bison Society* (New York: ABS, 1912); *Seventh Annual Report of the American Bison Society* (New York:

ABS, 1914); Coder, "National Movement to Preserve the American Buffalo," chapter 5.

63. Fred M. Dille, "The Niobrara Reservation," in *Sixth Annual Report of the American Bison Society* (1913): 33–39.
64. Series P147: Field Workers' Wildlife Refuges Reports 1915–1937, Record Group 22, National Archives Record Administration II, College Park, MD (hereafter NARA II).
65. "To Stock National Reservations," *Forest and Stream,* February 25, 1911; U.S. Forest Service, *The Wichita National Forest and Game Preserve,* USDA Miscellaneous Circular 36 (Washington, DC: U.S. Government Printing Office, 1925, revised 1928); Haley, "A History of the Establishment of the Wichita National Forest and Game Preserve."
66. Permanent Wild Life Protection Fund, "Game in the National Forests: Common Sense Views of a Practical Man," extracts from a paper by Mr. Smith Riley, U.S. District Forester of Denver, 1915, clippings in William T. Hornaday, Wildlife Conservation Scrapbook, volume 7, "Documentary History of Making Game Sanctuaries in National Forests, 1915–1928," series 4, WTH Papers.
67. P. Chalmers Mitchell, "Zoological Gardens and the Preservation of Fauna," *Science* 36 (1912): 353–365.
68. Robert Sterling Yard, *Our Federal Lands: A Romance of American Development* (New York: Charles Scribner's Sons, 1928), 319.
69. "Restoring the Buffalo," *El Paso Herald,* July 15, 1911, p. 23.
70. "The Proposed New York State Bison Herd," *Annual Report of the American Bison Society, 1905–1907* (1908), 50.
71. Hornaday, draft of "Basic Requirements of a General Scheme for National Forest Sanctuaries," c. 1915, William T. Hornaday Wildlife Conservation Scrapbook, volume 7, series 4, WTH Papers; American Bison Society, *Annual Report 1905–1907.*
72. Hornaday, *Our Vanishing Wild Life,* 329.
73. "Address on Developments in Federal Big-Game Refuges," May 1937, RG 22, series P146: Records Concerning Wildlife Refuges 1892–1939, box 7, NARA II. The Division of Biological Survey was formed out of the Division of Economic Ornithology and Mammalogy in 1896. It was replaced in 1905 by the Bureau of Biological Survey in the Department of Agriculture. In 1940 the Fish and Wildlife Service was created by combining the Bureau of Biological Survey and the Bureau of Fisheries within the Department of the Interior.
74. Each refuge has a different administrative history. The Wichita reserve, for example, was designated a game preserve by presidential proclamation in 1905. It was placed under the administration of the U.S. Forest Service until it was transferred to the Biological Survey in 1935, "in recognition of the primary importance of the area as a wildlife preserve and out-of-door laboratory for wildlife research," and renamed the Wichita Mountains Wildlife Refuge. The U.S. Fish and Wildlife Service maintains a useful list at https://www.fws.gov/laws/lawsdigest/nwrsact.html. Refuges were consolidated into the

National Wildlife Refuge System Administration Act of 1966 (16 U.S. Code § 668dd).

75. Gabrielson, *Wildlife Refuges,* 104; T. S. Palmer, *National Reservations for the Protection of Wild Life,* U.S. Department of Agriculture, Circular 87 (Washington, DC: U.S. Government Printing Office, 1912).

76. *Twentieth Census of Living American Bison* (New York: American Bison Society, 1934); C. Gordon Hewitt, "The Coming Back of the Bison: Under Government and Private Protection Bison Have Increased," *Natural History,* December 1919; Victor H. Cahalane, "Restoration of the Wild Bison," in *Transactions of the Ninth North American Wildlife Conference* (Washington, DC: American Wildlife Institute, 1944).

77. Raf De Bont, "Extinct in the Wild: Finding a Place for the European Bison, 1919–1932," in *Spatializing the History of Ecology: Sites, Journeys, Mappings,* ed. Raf De Bont and Jens Lachmund (London: Routledge, 2017), 165–184.

78. David A. Nesheim, "Profit, Preservation, and Shifting Definitions of Bison in America," *Environmental History* 17 (2012): 1–31.

79. Martin S. Garretson, "Report of the Secretary," *Report of the American Bison Society, 1919–1920* (1920): 18.

80. Garretson, *American Bison,* 153.

81. Taft, as quoted in Russel Lawrence Barsh, "An American Heart of Darkness: The 1913 Expedition for American Indian Citizenship," *Great Plains Quarterly* 13 (1993): 91–115, 99.

2. Ecology in the Public Service

1. Henry Chandler Cowles, "The Work of the Year 1903 in Ecology," *Science* 19 (1904): 879–885.

2. Walter P. Taylor, "What Is Ecology and What Good Is It?" *Ecology* 17 (1936): 333–346.

3. Account loosely based on a letter from Caroline Henderson to Henry A. Wallace, July 26, 1935, as quoted in Ronald Reis, *The Dust Bowl* (New York: Chelsea House Publishers, 2008), 65. See also Studs Terkel, *Hard Times: An Oral History of the Great Depression* (New York: Pantheon Books, 1970); Donald Worster, *Dust Bowl: The Southern Plains in the 1930s* (New York: Oxford University Press, 1979).

4. Environmental historians continue to see new stories in the "dirty thirties," stories of justice, reform, and catastrophe. See William Cronon, "A Place for Stories: Nature, History, and Narrative," *Journal of American History* 78 (1992): 1347–1376.

5. On erosion, see J. Donald Hughes, *Pan's Travail: Environmental Problems of the Ancient Greeks and Romans* (Baltimore, MD: Johns Hopkins University Press, 1996); Paul Sutter, *Let Us Now Praise Famous Gullies: Providence Canyon and the Soils of the South* (Athens: University of Georgia Press, 2015).

6. Christophe Masutti makes a similar observation in "Frederic Clements, Climatology, and Conservation in the 1930s," *Historical Studies in the Physical and*

Biological Sciences 37 (2006): 27–48. For some influential examples, see Donald Worster, *Nature's Economy: A History of Ecological Ideas* (Cambridge, UK: Cambridge University Press, 1977); Robert Croker, *Pioneer Ecologist: The Life and Work of Victor Ernest Shelford, 1877–1968* (Washington, DC: Smithsonian Institution Press, 1991); Gregg Mitman, *The State of Nature: Ecology, Community, and American Social Thought, 1900–1950* (Chicago: University of Chicago Press, 1992); Joel Hagen, "Clementsian Ecologists: The Internal Dynamics of a Research School," *Osiris* 8 (1993): 178–195; Michael Barbour, "Ecological Fragmentation in the Fifties," in *Uncommon Ground,* ed. William Cronon (New York: W.W. Norton & Co.), 233–255.

7. I write about the influence of the Dust Bowl crisis on statistical practices in Laura J. Martin, "Mathematizing Nature's Messiness: Graphical Representations of Variation in Ecology, 1930–present," *Environmental Humanities* 7 (2015): 59–88. On the professionalization of science, see Sally Kohlstedt, *The Formation of the American Scientific Community* (Chicago: University of Illinois Press, 1976); Andrew Jewett, *Science, Democracy, and the American University: From the Civil War to the Cold War* (Cambridge, UK: Cambridge University Press, 2012).
8. A successful neologist, Haeckel is also credited with "phylogeny," "word riddle," and "the first world war." An English translation of Ernst Haeckel, *Generelle Morphologie der Organismen* (Berlin: G. Reimer, 1866) can be found in Robert C. Stauffer, "Haeckel, Darwin, and Ecology," *Quarterly Review of Biology* 32 (1957): 138–414.
9. Eugenius Warming, *Oecology of Plants: An Introduction to the Study of Plant Communities* (Oxford: Clarendon Press, 1909), 13 (emphasis original). On early American ecology and its connections to biogeography, see Frank N. Egerton, *Ecological Phytogeography in the Nineteenth Century* (New York: Arno Press, 1977); Janet Browne, *The Secular Ark: Studies in the History of Biogeography* (New Haven, CT: Yale University Press, 1983); Eugene Cittadino, *Nature as the Laboratory: Darwinian Plant Ecology in the German Empire, 1880–1900* (Cambridge, UK: Cambridge University Press, 1990); Peder Anker, *Imperial Ecology: Environmental Order in the British Empire, 1895–1945* (Cambridge, MA: Harvard University Press, 2002); Aaron Sachs, *The Humboldt Current: Nineteenth Century Exploration and the Roots of American Environmentalism* (New York: Viking, 2006).
10. Henry Cowles, "The Ecological Relations of the Vegetation on the Sand Dunes of Lake Michigan," *Botanical Gazette* 27 (1899): 95–117, 167–202. On Cowles, see Ronald Tobey, *Saving the Prairies: The Life Cycle of the Founding School of American Plant Ecology, 1895–1955* (Berkeley: University of California Press, 1981); Eugene Cittadino, "A 'Marvelous Cosmopolitan Preserve': The Dunes, Chicago, and the Dynamic Ecology of Henry Cowles," *Perspectives on Science* 1 (1993): 520–559; Victor Cassidy, *Henry Chandler Cowles: Pioneer Ecologist* (Chicago: Kedzie Sigel Press, 2007). On early succession theory see Robert Park, "Succession: An Ecological Concept," *American Sociological Review* 1 (1936): 171–179; Worster, *Nature's Economy,* chapter 10.

11. Tobey traces this connection in *Saving the Prairies,* chapter 3. On the Clementses, see also Edith Schwartz Clements, *Adventures in Ecology: Half a Million Miles from Mud to Macadam* (New York: Pageant Press, 1960); Hagen, "Clementsian Ecologists."
12. Frederic E. Clements and Roscoe Pound, *The Phytogeography of Nebraska I: General Survey* (Lincoln: University of Nebraska Botanical Seminar, 1897), 4.
13. Tobey, *Saving the Prairies,* 126–127; Frederic Clements, *Research Methods in Ecology* (Lincoln: University of Nebraska Publishing, 1905), chapter 3.
14. "Complex organism": Clements, *Research Methods in Ecology,* 5; "unmistakable impress": Frederic E. Clements, "Experimental Ecology in the Public Service," *Ecology* 16 (1935): 342–363.
15. Laura Cameron and David Matless, "Translocal Ecologies: The Norfolk Broads, the 'Natural,' and the International Phytogeographical Excursion, 1911," *Journal of the History of Biology* 44 (2011): 15–41.
16. Robb H. Wolcott to Victor Shelford, March 27, 1914, box 1, folder 1910–1919, series 15/24/20, Victor E. Shelford Papers, University of Illinois Archives, Urbana-Champaign (hereafter SHEL); Wolcott to Victor Shelford, April 24, 1914, box 1, folder 1910–1919, SHEL; "Handbook of the Ecological Society of America," *Bulletin of the Ecological Society of America* 1 (March 1917): 9–56. For ecologists' accounts of their early disciplinary history, see Barrington Moore, "The Scope of Ecology," *Ecology* 1 (1920): 3–5; Victor E. Shelford, "The Organization of the Ecological Society of America 1914–19," *Ecology* 19 (1938): 164–166; Norman Taylor, "The Beginnings of Ecology," *Ecology* 19 (1938): 352. On the history of ecology between 1890 and World War II, see Worster, *Nature's Economy,* chapters 10–11; Frank N. Edgerton, ed., *History of American Ecology* (New York: Arno Press, 1977); Tobey, *Saving the Prairies;* Mitman, *The State of Nature;* Robert E. Kohler, *Landscapes and Labscapes: Exploring the Lab-Field Border in Biology* (Chicago: University of Chicago Press, 2002); Sharon Kingsland, *The Evolution of American Ecology, 1890–2000* (Baltimore, MD: Johns Hopkins University Press, 2005), chapters 1–5.
17. Moore, "The Scope of Ecology"; Forrest Shreve, "Proceedings," *Ecology* 1 (1920): 61–70. On ecology's physiological roots see Kohler, *Landscapes and Labscapes.* For relevant histories of zoology see Everett Mendelsohn, *Heat and Life: The Development of the Theory of Animal Heat* (Cambridge, MA: Harvard University Press, 1964); Lynn Nyhart, *Biology Takes Form: Animal Morphology and the German Universities, 1800-1900* (Chicago: University of Chicago Press, 1995.
18. Federal Department of Agriculture, "Research Institutions in the United States," *Nature* 99 (1917): 274–277. On field stations, see also Ronald Rainger, Keith Benson, and Jane Maienschein, eds., *The American Development of Biology* (Philadelphia: University of Pennsylvania Press, 1988); Kohler, *Landscapes and Labscapes,* 41–55; Helen Rozwadowski, *Fathoming the Ocean: The Discovery and Exploration of the Deep Sea* (Cambridge, MA: Harvard University Press, 2005); Raf de Bont, *Stations in the Field: A History*

of Place-Based Animal Research, 1870–1930 (Chicago: University of Chicago Press, 2014).

19. Croker, *Pioneer Ecologist.*
20. Quotes from Lucile Durrell, "Memories of E. Lucy Braun," *Ohio Biological Survey Biol. Notes* 15 (1981): 37–39. On Braun's life see also Perry Peskin, "A Walk through Lucy Braun's Prairie," *Explorer: Bulletin of the Cleveland Museum of Natural History* 20 (1978): 15–21; Ronald Stuckey, *E. Lucy Braun (1889–1971): Ohio's Foremost Woman Botanist* (Columbus, OH: RLS Creations, 2001).
21. Francis Sumner, "The Responsibility of the Biologist in the Matter of Preserving Natural Conditions," *Science* 54 (1921): 39–43.
22. Willard Van Name, "Zoological Aims and Opportunities," *Science* 50 (1919): 81–84; E. Lucy Braun, "Preservation of Natural Conditions," *Ohio Journal of Science* 22 (1922): 99–100; Victor Shelford, "Nature's Sanctuaries," *Science* 6 (1932): 481–482.
23. In 1932 the committee was renamed the Committee for the Preservation of Natural Conditions in the United States. For clarity I use "preservation committee" throughout this chapter. On its history, see "Committee on the Preservation of Natural Conditions for Ecological Study," *Bulletin of the Ecological Society of America* 1, no. 6/9 (1917); Sara Tjossem, "Preservation of Nature and the Ecological Society of America, 1915–1979" (PhD diss., Cornell University, 1994); Mark Barrow, *Nature's Ghosts: Confronting Extinction from the Age of Jefferson to the Age of Ecology* (Chicago: University of Chicago Press, 2009), chapter 7; Abby J. Kinchy, "On the Borders of Post-War Ecology: Struggles over the Ecological Society of America's Preservation Committee, 1917–1946," *Science as Culture* 15 (2012): 23–44. As Gina Rumore points out, Russian ecologists were engaged in a parallel movement to preserve tracts of nature for scientific study, but there is no evidence that they influenced the ESA's efforts. See "Preservation for Science: The Ecological Society of America and the Campaign for Glacier Bay National Monument," *Journal of the History of Biology* 45 (2011): 613–650.
24. Charles C. Adams, "Ecological Conditions in National Forests and in National Parks," *Scientific Monthly* 20 (1925): 561–593; Victor Shelford, "The Preservation of Natural Biotic Communities," *Ecology* 14 (1933): 240–245.
25. Victor Shelford, "Conservation Versus Preservation," *Science* 77 (1933): 535.
26. See correspondence in box 1, folder 1924, SHEL; Victor Shelford, *Naturalist's Guide to the Americas* (Baltimore, MD: Williams & Wilkins Co., 1926); E. Lucy Braun, "The Present Emergency," series 2, subseries 8, box 27, folder 4, Annette and E. Lucy Braun Papers (MSS 1064), Cincinnati Museum, OH (hereafter BRAU); E. Lucy Braun, draft of "A National Monument of Every Type of Native Vegetation," series 2, subseries 8, box 27, folder 5, BRAU.
27. Rumore, "Preservation for Science."
28. Victor Shelford, "Suggested Program for the Establishment of Reservations for Ecological Research and Instruction Submitted to the Advisory Board," 1931, box 1, folder 1931–1935, series 15/24/20, SHEL.

29. Lizabeth Cohen, *Making a New Deal: Industrial Workers in Chicago, 1919–1939* (Cambridge, UK: Cambridge University Press, 1990); Anthony Badger, *FDR: The First Hundred Days* (New York: Hill and Wang, 2008).
30. Neil Maher, *Nature's New Deal: The Civilian Conservation Corps and the Roots of the American Environmental Movement* (New York: Oxford University Press, 2008); Sarah Phillips, *This Land, This Nation: Conservation, Rural America, and the New Deal* (Cambridge, UK: Cambridge University Press, 2007). Details from Maher, *Nature's New Deal,* 43.
31. Forrest Shreve to Shelford, March 4, 1916, box 1, folder 1910–1919, SHEL; Raphael Zon to Charles C. Adams, March 6, 1919, box 4, folder 17, Charles C. Adams Papers, New York State Library Archives, Albany, NY; Burgess, *Historical Data and Some Preliminary Analyses.* On the history of scientific forestry, see Henry Lowood, "The Calculating Forester: Quantification, Cameral Science, and the Emergence of Scientific Forestry Management in Germany," in *The Quantifying Spirit in the Eighteenth Century,* ed. Tore Frangsmyr, J. L. Heilbron, and Robin E. Rider (Berkeley: University of California Press, 1991), 315–342; Nancy Langston, *Forest Dreams, Forest Nightmares: The Paradox of Old Growth in the Inland West* (Seattle: University of Washington Press, 1995); Emily Brock, *Money Trees: The Douglas Fir and American Forestry, 1900–1944* (Corvallis: Oregon State University Press, 2015).
32. Taylor, "What Is Ecology and What Good Is It?," 342. See also Horace Albright, "Research in the National Parks," *Scientific Monthly* 36 (1933): 483–501; Walter P. Taylor, "'Man and Nature'—A Contemporary View," *Scientific Monthly* 41 (1935): 350–362.
33. Drury as quoted in Richard Sellars, *Preserving Nature in the National Parks: A History* (New Haven, CT: Yale University Press, 1997), 142–143. E. Lucy Braun, draft of "America's Wilderness—Where Can We Find It?" c. 1936, subseries 2, box 23, folder 1, BRAU; Forest Service to V. E. Shelford, March 4, 1936, box 1, folder 1936, SHEL; "Proceedings," *Ecology* 18 (1937): 307–309, reprint in box 2, folder 9, series 9/25/10-2.
34. Clements, "Experimental Ecology in the Public Service"; Frederic E. Clements and Ralph Chaney, *Environment and Life in the Great Plains* (Washington, DC: Carnegie Institution, 1936). See also Jonathan Mitchell, "Shelter Belt Realities," *New Republic* 90 (1934): 69; Frederic E. Clements, "The Relict Method in Dynamic Ecology," *Journal of Ecology* 22 (1934): 39–68; W. I. Joerg, "Geography and National Land Planning," *Geographical Review* 25 (1935): 177–208; Taylor, "What Is Ecology and What Good Is It?"
35. The issue of the *Journal of Forestry* 32 (1934): 927–1052; U.S. Forest Service, *Report of the Chief of the Forest Service* (Washington, DC: United States Forest Service, 1934); Mitchell, "Shelter Belt Realities"; "Fighting the Drouth," *Popular Mechanics Magazine* 62 (1934): 483–485; Raphael Zon, "Shelterbelts—Futile Dream or Workable Plan," *Science* 81 (1935): 391–394; U.S. Forest Service, *Possibilities of Shelterbelt Planting in the Plains Region* (Washington, DC: United States Forest Service, 1935); Paul B.

Sears, "The Great American Shelter-Belt: Review of Possibilities of Shelterbelt Planting in the Plains Region by R. Zon," *Ecology* 17 (1936): 683–684. For histories of the Shelter-Belt Program, see Thomas Wessel, "Roosevelt and the Great Plains Shelterbelt," *Great Plains Journal* 8 (1969): 57–74; Wilmon Henry Droze, *Trees, Prairies, and People: A History of Tree Planting in the Plains States* (PhD diss., Texas Woman's University, 1977); Joel Orth, "The Shelterbelt Project: Cooperative Conservation in 1930s America," *Agricultural History* 81 (2007): 333–357; Robert Gardner, "Trees as Technology: Planting Shelterbelts on the Great Plains," *History and Technology* 25 (2009): 325–341; Robert Gardner, "Constructing a Technological Forest: Nature, Culture, and Tree Planting in the Nebraska Sand Hills," *Environmental History* 14 (2009): 275–297.

36. Clements, "Experimental Ecology in the Public Service"; Ellsworth Huntington, "Marginal Land and the Shelter Belt," *Journal of Forestry* 32 (1934): 804–812; Sears, "Great American Shelter-Belt."
37. Poem quoted in David Moon, *The American Steppes: The Unexpected Russian Roots of Great Plains Agriculture, 1870s–1930s* (New York: Cambridge University Press, 2020), 402. R. S. Kellogg, "Proposed Tree Belt Regarded as Futile," *New York Times,* September 16, 1934, sec. 4, p. 5.
38. Richard Yahner, "Small Mammals in Farmstead Shelterbelts: Habitat Correlates of Seasonal Abundance and Community Structure," *Journal of Wildlife Management* 47 (1983): 74–84.
39. "The Nature Sanctuary Program," c. 1936, box 1, folder 1936, series 15/25/20, SHEL. See also "Proceedings: Business Meetings of the ESA at Richmond, Virginia, December 27 and 29, 1938," *Ecology* 20 (1939): 317–334.
40. Braun, draft of "A National Monument of Every Type of Native Vegetation," series 2, subseries 8, box 27, folder 5, BRAU.
41. On model organisms, see Angela Creager, *The Life of a Virus: Tobacco Mosaic Virus as an Experimental Model, 1930–1965* (Chicago: University of Chicago Press, 2002), 320–321; Robert Kohler, *Lords of the Fly: Drosophila Genetics and the Experimental Life* (Chicago: University of Chicago Press, 1994); Rachel A. Ankeny, "The Conqueror Worm: An Historical and Philosophical Examination of the Use of the Nematode *C. Elegans* as a Model Organism" (PhD diss., University of Pittsburgh, 1997); Karen Rader, *Making Mice: Standardizing Animals for American Biomedical Research, 1900–1955* (Princeton, NJ: Princeton University Press, 2004).
42. On the history of the "control group," see R. L. Solomon, "An Extension of the Control Group Design," *Psychological Bulletin* 46 (1949): 137–150; Edwin Boring, "The Nature and History of Experimental Control," *American Journal of Psychology* 67 (1954): 573–589.
43. Ronald Fisher, *Statistical Methods for Research Workers* (London: Oliver and Boyd, 1925); Ronald Fisher, *The Design of Experiments* (London: Oliver and Boyd, 1935); Joan Fisher Box, "R. A. Fisher and the Design of Experiments, 1922–1926," *American Statistician* 34 (1980): 1–7.
44. E. Lucy Braun, "Composition and Source of the Flora of the Cincinnati Region," *Ecology* 2 (1921): 161–180.

45. John E. Weaver and Frederic Clements, *Plant Ecology,* 2nd ed. (New York: McGraw-Hill Book Company, 1938), 30. On the history of ecological field methods, see Kohler, *Landscapes and Labscapes.*
46. Kohler notes this in *Landscapes and Labscapes,* 154–158.
47. Victor Shelford, "Faith in the Results of Controlled Laboratory Experiments as Applied in Nature," *Ecological Monographs* 4 (1933): 491–494.
48. Clements, *Research Methods in Ecology,* 306–314; Forrest Shreve, "The Influence of Low Temperatures on the Distribution of the Giant Cactus," *Plant World* 14 (1911): 136–146. See also Rexford F. Daubenmire, "Exclosure Technique in Ecology," *Ecology* 21 (1940): 514–515.
49. Clements, "Relict Method in Dynamic Ecology," 39–68.
50. John E. Weaver and Evan L. Flory, "Stability of the Climax Prairie and Some Environmental Changes Resulting from Breaking," *Ecology* 15 (1934): 333–347; John Weaver and William Noll, *Comparison of Runoff and Erosion in Prairie, Pasture, and Cultivated Land* (Lincoln: Conservation and Survey Division of the University of Nebraska, 1935); John E. Weaver, "Competition of Western Wheat Grass with Relict Vegetation of Prairie," *American Journal of Botany* 29 (1942): 366–372.
51. Weaver and Clements, *Plant Ecology,* chapters 2 and 10, 52–53, 37.
52. Clements, "Relict Method in Dynamic Ecology."
53. Weaver and Clements, *Plant Ecology,* 48–49.
54. Herbert C. Hanson, "Check-Areas as Controls in Land Use," *Scientific Monthly* 48 (1939): 130–146.
55. Michael Grant, *Down and Out on the Family Farm: Rural Rehabilitation in the Great Plains, 1929–1945* (Lincoln: University of Nebraska Press, 2002), 78. See also R. Douglas Hurt, "Federal Land Reclamation in the Dust Bowl," *Great Plains Quarterly* 6 (1986): 94–106; Dan L. Flores, "A Long Love Affair with an Uncommon Country: Environmental History and the Great Plains," in *Prairie Conservation: Preserving North America's Most Endangered Ecosystem,* ed. Fred Samson and Fritz Knopf (Washington, DC: Island Press, 1996).
56. Committee on the Preservation of Natural Conditions to the Biological Survey, January 15, 1936, box 1, folder 1936, series 15/25/20, SHEL.
57. Paul Sutter writes, "Indeed, ecological concerns were not a central causative agent or a major component in the [Wilderness Society] founders' definition of modern wilderness." Sutter, *Driven Wild: How the Fight against Automobiles Launched the Modern Wilderness Movement* (Seattle: University of Washington Press, 2002), 14. Here I am in disagreement with Abby J. Kinchy, who argues that the preservation committee was allied with the wilderness preservation movement, in Kinchy, "On the Borders of Post-War Ecology."
58. Committee on the Preservation of Natural Conditions to Members Past and Present [. . .], March 21, 1930, box 1, folder 1925–1930, SHEL.
59. Shelford, "Suggested Program for the Establishment of Reservations for Ecological Research and Instruction Submitted to the Advisory Board," 1931, box 1, folder 1931–1935, SHEL.
60. Herbert C. Hanson, "Check-Areas as Controls in Land Use," *Scientific Monthly* 48 (1939): 130–146. See also "Correspondence," *Journal of Forestry*

34 (1936): 1077–1078; Shelford, "Suggested Program for the Establishment of Reservations for Ecological Research and Instruction Submitted to the Advisory Board," 1931, box 1, folder 1931–1935, SHEL.

61. Committee on the Ecology of North American Grasslands, *Background Report for Western Nebraska Meeting* (Washington, DC: National Research Council, April 22, 1937); R. E. Coker, "Functions of an Ecological Society," *Science* 87 (1938): 309–315; "Great Plains National Monument Project," *Council Ring*, January 8, 1940; Shelford to Richard Pough (American Museum of Natural History), November 2, 1955, box 1, folder 1939–1958, SHEL; James Swint, *The Proposed Prairie National Park: A Case Study of the Controversial NPS* (MA thesis, Kansas State University, 1971).
62. Protected Planet database, https://www.protectedplanet.net/en, accessed October 1, 2020.
63. See, for example, E. Dinerstein et al., "A Global Deal for Nature: Guiding Principles, Milestones, and Targets," *Science Advances* (2019): eaaw2869; Edward O. Wilson, *Half Earth: Our Planet's Fight for Life* (New York: Liveright, 2017).
64. Surprisingly little has been written on the history of experimentation in ecology. See Kohler, *Landscapes and Labscapes;* Sharon E. Kingsland, "The Role of Place in the History of Ecology," in *The Ecology of Place: Contributions of Place-Based Research to Ecological Understanding*, ed. Ian Billick and Mary Price (Chicago: University of Chicago Press, 2010); Martin, "Mathematizing Nature's Messiness."
65. Clements, "Experimental Ecology in the Public Service," 345, 360, 344.
66. Clements, *Adventures in Ecology*, 231.
67. Clements, "Experimental Ecology in the Public Service"; Tobey, *Saving the Prairies*, 207; Robert Goodman, "The Regulation and Control of Land Use in Non-Urban Areas," *Journal of Land & Public Utility Economics* 9 (1933): 266–271. See also Arthur Sampson, "Plant Indicators—Concept and Status," *Botanical Review* 5 (1939): 155–206; Benton Mackaye, "Regional Planning and Ecology," *Ecological Monographs* 10 (1940): 349–353; Shelford, "Nature's Sanctuaries."
68. Walter P. Taylor, "Man and Nature—A Contemporary View," *Scientific Monthly* 41 (1935): 350–362.
69. Weaver and Flory, "Stability of the Climax Prairie and Some Environmental Changes Resulting from Breaking"; John Weaver and T. J. Fitzpatrick, "The Prairie," *Ecological Monographs* 4 (1934): 109–295; Weaver and Clements, *Plant Ecology*, 277.
70. Shelford, "The Preservation of Natural Biotic Communities"; Homer L. Shantz, "The Relation of Plant Ecology to Human Welfare," *Ecological Monographs* 10 (1940): 311–342.

3. An Outdoor Laboratory

1. Maria E. Carter of the Boston Society of Natural History formed the Society for the Preservation of Native Plants the following year. N. L. Britton, "The Preservation of Native Plants," *Plant World* 4 (1901): 230–231; "Wild Flower

Preservation Society," *Plant World* 5 (1902): 95–97. On the history of the New York Botanical Garden, see Sharon E. Kingsland, *The Evolution of American Ecology, 1890–2000* (Baltimore, MD: Johns Hopkins University Press, 2005), chapter 3; Peter Mickulas, *The New York Botanical Garden and American Botany, 1888–1929* (New York: NYBG Press, 2007).

2. Jennifer Price, *Flight Maps: Adventures with Modern America* (New York: Basic Books, 1999), 82. On the founding of the Audubon Society, see also Mark V. Barrow Jr., *A Passion for Birds: American Ornithology after Audubon* (Princeton, NJ: Princeton University Press, 1998), chapters 5 and 6. On the importance of women's civic clubs in propelling women into the public sphere, see Karen J. Blair, *The Clubwoman as Feminist: True Womanhood Redefined, 1868–1914* (New York: Holmes & Meier Publishers, 1980).
3. Margaret B. Harvey, "How to Arrange Wild-Flowers," *The Connoisseur* 1 (1887): 39–41; Ruth E. Messenger, "The Preservation of Our Native Plants," *Plant World* 6 (1903): 4–9.
4. Mary Perle Anderson, "The Protection of Our Native Plants," *Plant World* 7 (1904): 123–129.
5. Elizabeth Britton, "Going, Going, Almost Gone! Our Wild Flowers," *New York Times*, May 4, 1913, p. 8.
6. Jean Broadhurst, "The Protection of Our Native Plants (A Plea to Teachers)," *The Plant World* 7 (1904): 152–154.
7. Broadhurst, "Protection of Our Native Plants," 152–154; James Marten, ed., *Children and Youth during the Gilded Age and Progressive Era* (New York: New York University Press, 2014).
8. On automobiles and preservation, see Paul Sutter, *Driven Wild: How the Fight against Automobiles Launched the Modern Wilderness Movement* (Seattle: University of Washington Press, 2005); David Louter, *Windshield Wilderness: Cars, Roads, and Nature in Washington's National Parks* (Seattle: University of Washington Press, 2010).
9. Britton, "Going, Going, Almost Gone!"
10. John W. Harshberger, "Hemerecology: The Ecology of Cultivated Fields, Parks, and Gardens," *Ecology* 4 (1923): 297–306. Dock as quoted in Susan Rimby, *Mira Lloyd Dock and the Progressive Era Conservation Movement* (University Park: Penn State University Press, 2012), 115. For cartoons, see clippings of reports on vandals and cars from the 1920s in box 5, folders 2–4, Wild Flower Preservation Society of America Records, The LuEsther T. Mertz Library, New York Botanical Garden, Bronx, NY (hereafter WFPSA).
11. P. L. Ricker, "A Report to the Bureau of Plant Industry to the National Conference on Outdoor Recreation, May 28–30, 1925," The Wild Flower Preservation Society Inc., Washington, DC, Circular 9, 1925. On science and sentiment, see Jessica Riskin, *Science in the Age of Sensibility: The Sentimental Empiricists of the French Enlightenment* (Chicago: University of Chicago Press, 2002); Lorraine Daston and Peter Galison, *Objectivity* (Cambridge, MA: Zone Books, 2007), chapters 3 and 4.
12. Sally Gregory Kohlstedt, "In from the Periphery: American Women in Science, 1830–1880," *Signs* 4 (1978): 81–96; Margaret W. Rossiter, *Women Scientists*

in America: Struggles and Strategies to 1940 (Baltimore, MD: Johns Hopkins University Press, 1982); Elizabeth B. Keeney, *The Botanizers: Amateur Scientists in Nineteenth Century America* (Raleigh: University of North Carolina Press, 1992).

13. Emanuel Rudolph, "Women in Nineteenth Century American Botany: A Generally Unrecognized Constituency," *American Journal of Botany* 69 (1982): 1346–1355.
14. "Is Botany a Suitable Study for Young Men?" *Science* 9 (1887): 116–117.
15. Robert Kohler, *Landscapes and Labscapes: Exploring the Lab-Field Border in Ecology* (Chicago: University of Chicago Press, 2002), 7.
16. Nowhere in *Landscapes and Labscapes* does Kohler mention gender or the history of women in botany and ecology. This is especially notable in chapters 3 and 4.
17. Willard N. Clute, "What's the Use of Botany?" *School Science and Mathematics* 8 (1908): 470–472.
18. Ricker to Britton, May 8, 1924, box 2, folder 30, WFPSA; Henry Cowles to E. Britton, July 11, 1924, box 1, folder 20, WFPSA; Henry C. Cowles to E. Britton, April 27, 1924, box 1, folder 20, WFPSA; Elizabeth Britton to Henry Cowles, November 3, 1924, box 1, folder 20, WFPSA.
19. Ricker to Britton, September 25, 1925, box 2, folder 30, WFPSA.
20. P. L. Ricker, "The Protection of Our Native Flowers," *Scientific Monthly* 35 (1932): 273–275.
21. This difficulty was acknowledged early in the century. See F. H. Knowlton, "Suggestions for the Preservation of Our Native Plants," *Plant World* 5 (1902): 61–66. See also James A. Tober, *Who Owns the Wildlife?: The Political Economy of Conservation in Nineteenth Century America* (Westport, CT: Praeger, 1981); Michael J. Bean and Melanie Rowland, *The Evolution of National Wildlife Law,* 3rd ed. (Westport, CT: Praeger, 1997).
22. P. L. Ricker, "A Report to the Bureau of Plant Industry."
23. P. L. Ricker, "The Protection of Our Native Flowers."
24. Roderick Nash titled a chapter in his classic 1967 book, *Wilderness and the American Mind* (New Haven, CT: Yale University Press, 1967), simply "Aldo Leopold, Prophet." See also William Jordan II, Michael E. Gilpin, and John D. Aber, eds., *Restoration Ecology: A Synthetic Approach to Ecological Research* (Cambridge, UK: Cambridge University Press, 1990), 3. Curt Meine lists other examples of Leopold being described as a prophet in Curt Meine, *Correction Lines: Essays on Land, Leopold, and Conservation* (Washington, DC: Island Press, 2004), 172–183.
25. In "Some Reflections on Curtis Prairie," William R. Jordan III calls the arboretum the "Kitty Hawk" of ecological restoration. *Ecological Management & Restoration* 11 (2010): 99–107.
26. The text of this article is one of a number of short essays that Eloise Butler wrote between 1914 and 1920, which after her death were collected in a series titled "Annals of the Wild Life Reserve, 1914–1931," in box 2, Garden Records (Location 146.K.15.1B), Minnesota Historical Society, Minneapolis, MN.

27. Martha E. Hellander, *The Wild Gardener: The Life and Selected Writings of Eloise Butler* (St. Cloud: North Star Press of St. Cloud, 1992).
28. Unpublished typescript, March 1931, box 1, Eloise Butler Papers (Location 146.K.14.14F), Minnesota Historical Society, Minneapolis, MN (hereafter BUTP).
29. Emily Griswold, "The Origin and Development of Ecogeographic Displays in North American Botanic Gardens" (MA thesis, University of Washington, 2002). On botanical gardens, see Richard Drayton, *Nature's Government: Science, Imperial Britain, and the "Improvement" of the World* (New Haven, CT: Yale University Press, 2000); Londa Schiebinger and Claudia Swan, eds., *Colonial Botany: Science, Commerce, and Politics* (Philadelphia: University of Pennsylvania Press, 2004).
30. On acclimatization societies, see George Laycock, *The Alien Animals: The Story of Imported Wildlife* (New York: Doubleday, 1966); Thomas Dunlap, *Nature and the English Diaspora: Environment and History in the United States, Canada, Australia, and New Zealand* (Cambridge, UK: Cambridge University Press, 1999); Michael Osborne, "Acclimatizing the World: A History of the Paradigmatic Colonial Science," *Osiris* 15 (2000): 135–151.
31. Philip Pauly discusses this distinction in *Fruits and Plains: The Horticultural Transformation of America* (Cambridge, MA: Harvard University Press, 2008).
32. Frank Waugh, *The Natural Style in Landscape Gardening* (Boston: Richard G. Badger, 1917), 24. On naturalistic gardening, see Robert Grese, *Jens Jensen: Maker of Natural Parks and Gardens* (Baltimore, MD: Johns Hopkins University Press, 1998).
33. "Shy Wild Flowers to Be Given Hospice," *Minneapolis Journal*, May 5, 1907, in scrap book 1912–1919, box 1, BUTP; John Greer et al., Petition to the Board of Park Commissioners, Minneapolis, Minnesota, 1907. On the history of the Minneapolis municipal park system, see Aaron Sachs, in *Arcadian America: The Death and Life of an Environmental Tradition* (New Haven, CT: Yale University Press, 2013), chapter 6.
34. Greer et al., Petition to the Board of Park Commissioners; Eloise Butler, "The Wild Botanic Garden—Early History," from "Annals of the Wild Life Reserve, 1914–1931; Eloise Butler, "A Wild Botanic Garden," c. 1911, unpublished typescript, box 1, BUTP; Eloise Butler, "Native Wild Flowers of Minnesota To Be Shown to Hundreds of Visiting Florists This Week," *Minneapolis Tribune*, August 17, 1913, in scrap book 1912–1919, box 1, BUTP; Eloise Butler, "Early History of Eloise Butler Plant Reserve," 1926, unpublished book, box 2, Garden Records (Location 146.K.15.1B), Minnesota Historical Society, Minneapolis, MN.
35. Cora E. Pease, "A New Botanical Garden," *American Botanist* 13 (1907): 3–5, box 4, Garden Records (Location 146.K.15.1B), Minnesota Historical Society, Minneapolis, MN; Butler, "The Wild Botanic Garden—Early History."
36. The Wild Botanic Garden became the Native Plant Reserve and was then renamed the Eloise Butler Wildflower Garden in 1929. Butler, "Early History of Eloise Butler Plant Reserve"; Eloise Butler, "Native Plant Preserve, Glenwood

Park, Minneapolis," c. 1931, in correspondence files of Department of Botany (ua-00892), Elmer L. Andersen Library, University of Minnesota Archives and Special Collections, Minneapolis, MN.

37. Edith Roberts, "The Development of an Out-of-Door Botanical Laboratory for Experimental Ecology," *Ecology* 14 (1933): 163–223; Anderson, "The Protection of Our Native Plants." On forest nurseries, see R. Kasten Dumroese et al., "Forest Service Nurseries: 100 Years of Ecosystem Restoration," *Journal of Forestry* 103 (2005): 241–247; Emily K. Brock, "The Challenge of Reforestation: Ecological Experiments in the Douglas Fir Forest, 1920–1940," *Environmental History* 9 (2004): 57–79; Robert Gardner, "Constructing a Technological Forest: Nature, Culture, and Tree Planting in the Nebraska Sand Hills," *Environmental History* 14 (2009): 275–297.
38. Butler, "A Wild Botanic Garden"; Fannie Heath, "Flower Gardens on the Plains," *Plainswoman* 6 (1896): 7–9; Hellander, *Wild Gardener,* 77–84.
39. Hellander, *Wild Gardener,* 82, 85; Butler, "Early History of Eloise Butler Plant Reserve"; W. P. Kirkwood, "A Wild Botanic Garden," *The Bellman* 14 (1913): 559–562; "Botanists All Over U.S. Visit Glenwood Wildflower Garden," "Neighbors Don't Know She Exists, but Botanic Garden Curator Is Famous Over America," and "Glenwood Park Wants Wire Fence to Keep Out Spooners," clippings in Scrapbook III, box 1, BUTP.
40. "Roberts, Edith A," series 18, folder R, Vassar College Biographical Files, Archives and Special Collections Library, Vassar College Libraries, Poughkeepsie, NY.
41. Edith Roberts and Margaret Shaw, *Native Plants of Dutchess County* (New York: The Conservation Committee of the Garden Club of America, 1924); Roberts, "Development of an Out-of-Door Botanical Laboratory"; "Outdoor Ecological Laboratory at Vassar, First of Kind in U.S., Bears Most of Area's Native Plants," April 5, 1948, Biographical File Edith Roberts, Catherine Pelton Durrell Archives & Special Collections Library, Poughkeepsie, NY (hereafter BFER). Funding came largely from Frederic Newbold and the Garden Club of Orange and Dutchess Counties. See Conservation Committee of the Garden Club of America, "Sanctuary and Nature Trail Survey," 1939, Vassar Subject File 1.37, Dutchess County Outdoor Ecological Laboratory, Catherine Pelton Durrell Archives & Special Collections Library, Poughkeepsie, NY (hereafter DCOEL). The Vassar project is described briefly in William R. Jordan and George M. Lubick, *Making Nature Whole: A History of Ecological Restoration* (Washington, DC: Island Press, 2011), 65–67. Thank you to Meg Ronsheim for identifying further sources.
42. Conservation Committee of the Garden Club of America, "Sanctuary and Nature Trail Survey," DCOEL; "Outdoor Ecological Laboratory at Vassar," BFER; Roberts and Shaw, *Native Plants of Dutchess County;* Opal Davis, "Germination of Seeds of Certain Horticultural Plants," *Florists' Exchange* 63 (1926): 917–922; Esther Mitchell, "Germination of Seeds of Plants Native to Dutchess County, New York," *Botanical Gazette* 81 (1926): 108–112; Helen Hart, "Delayed Germination in Seeds of *Peltandra virginica* and *Celastrus scandens,*" *Puget Sound Biological Station Publication* 6 (1928):

255–261. Roberts and Shaw, *Native Plants of Dutchess County.* In 1926, the Boyd Thompson Institute of Plant Research and Vassar produced an exhibit of the project at the New York Flower Show.

43. Thaisa Way, *Unbounded Practice: Women and Landscape Architecture in the Early Twentieth Century* (Charlottesville: University of Virginia Press, 2009); Dorothy Wurman, "Elsa Rehmann, Ecological Pioneer: 'A Patch of Ground,'" in *Women in Landscape Architecture: Essays on History and Practice,* ed. Louise Mozingo and Linda Jewell (Jefferson: McFarland & Co., 2012), 113–128.
44. Wurman, "Elsa Rehmann, Ecological Pioneer"; Elsa Rehmann, "An Ecological Approach," *Landscape Architecture* 23 (1933): 240–241. Roberts and Rehmann's work was originally published in *House Beautiful,* as a series titled Plant Ecology, between June 1927 and May 1928. In 1929, these articles were collected in Edith Roberts and Elsa Rehmann, *American Plants for American Gardens* (New York: Macmillan Company, 1929).
45. Roberts and Rehmann, *American Plants for American Gardens,* 49.
46. Rehmann, "An Ecological Approach," 5, 8.
47. John W. Harshberger, "Hemerecology: The Ecology of Cultivated Fields, Parks, and Gardens," *Ecology* 4 (1923): 297–306.
48. On Leopold's biography, see Susan L. Flader, *Thinking Like a Mountain: Aldo Leopold and the Evolution of an Ecological Attitude toward Deer, Wolves, and Forests* (Madison: University of Wisconsin Press, 1974); Curt Meine, *Aldo Leopold: His Life and Work* (Madison: University of Wisconsin Press, 1988); Aldo Leopold, *The River of Mother God and Other Essays by Aldo Leopold,* ed. Susan Flader and J. Baird Callicott (Madison: University of Wisconsin Press, 1991); Marybeth Lorbiecki, *A Fierce Green Fire* (Helena, MT: Falcon, 2004); Julianne Lutz Newton, *Aldo Leopold's Odyssey: Rediscovering the Author of A Sand County Almanac* (New York: Shearwater, 2008).
49. According to Curt Meine, *Our Vanishing Wild Life* was the book that had the greatest effect on Leopold. Meine, *Aldo Leopold,* 128. Julianne Lutz Newton and Susan Flader downplay Hornaday's impact. See Newton, *Aldo Leopold's Odyssey,* 91; Flader, *Thinking Like a Mountain,* 12.
50. Karen Merrill examines the consequences of this shift for grazing practices in *Public Lands and Political Meaning: Ranchers, The Government, and the Property between Them* (Berkeley: University of California Press, 2002).
51. See Meine, *Aldo Leopold,* chapters 8–9.
52. The Office of Economic Ornithology and Mammalogy was established in 1885 under the U.S. Department of Agriculture. A year later the office changed its name to the Division of Economic Ornithology and Mammalogy. In 1905 it was renamed again, the Bureau of Biological Survey.
53. "Report on Experiments in the Use of War Gases as Bird Control Agencies, Conducted at the Edgewood, MD, Arsenal of the Chemical Warfare Service," 1922, box 4, series P177: Branch of Wild. Research, Research Reports 1912–1951, Record Group 22, National Archives II, College Park, MD (hereafter NARA II).

54. Report in Folder: Data for Report, box 1, series P144: Records Concerning the President's Committee on Wildlife 1/1934-7/1934, RG 22, NARA II. As the report notes, this is an underestimate—it is a count only of the animals for which the bureau retrieved and counted the skins and scalps.
55. On shifting U.S. federal policy toward large predators, see Thomas Dunlap, *Saving America's Wildlife* (Princeton, NJ: Princeton University Press, 1988); Michael J. Robinson, *Predatory Bureaucracy: The Extermination of Wolves and the Transformation of the West* (Boulder: University of Colorado Press, 2005); Peter Alagona, *After the Grizzly: Endangered Species and the Politics of Place in California* (Berkeley: University of California Press, 2013), chapter 4.
56. USDA Animal and Plant Health Inspection Service Wildlife Services, Program Data Report G-2018, Animals Killed or Euthanized, https://www.aphis.usda .gov/aphis/ourfocus/wildlifedamage/pdr/?file=PDR-G_Report&p =2018:INDEX:.
57. As quoted in Meine, *Aldo Leopold,* 155.
58. Lorbiecki, *A Fierce Green Fire,* 77.
59. Aldo Leopold, draft of "Southwestern Game Fields" (1927), folder 1, box 10, series 9/25/10-6, Aldo Leopold Papers, University of Wisconsin–Madison Libraries, Madison, WI (hereafter LEOP).
60. Aldo Leopold, draft of "Southwestern Game Fields" (1927), folder 1, box 10, series 9/25/10-6, LEOP.
61. Barrington Moore, "Review: Game Survey of the North Central States," *Ecology* 12 (1931): 748–749.
62. Folder: Publications—Reports Stoddard, box 33, series P254: Correspondence, 1934–1966, RG 22, NARA II. Aldo Leopold, *Report on a Game Survey of the North Central States* (Madison: Sporting Arms and Ammunition Manufacturers' Institute, 1931); Herbert L. Stoddard, *The Bobwhite Quail: Its Habits, Preservation, and Increase* (New York: Charles Scribner's Sons, 1931).
63. Herbert L. Stoddard to Aldo Leopold, April 3, 1930, box 003, folder 004, series 9/25/10-1, LEOP.
64. Robert E. Kohler, "Paul Errington, Aldo Leopold, and Wildlife Ecology: Residential Science," *Historical Studies in the Natural Sciences* 41 (2011): 216–254. Susan Flader emphasizes the importance to Leopold's thinking about game management of a 1935 trip to Germany and 1936 trip to northern Mexico in *Thinking Like a Mountain,* chapter 4.
65. Lorbiecki, *A Fierce Green Fire,* 123.
66. Folder "Minutes of Meeting with Sen. Wildlife Committee on Jan 9, 1934," box 1, series P144: Records Concerning the President's Committee on Wildlife 1/1934-7/1934, RG 22, NARA II. For criticism of the proposal, see "Recent Waterfowl Developments," December 4, 1931, box 001, folder 002, series 9/25/10-2, LEOP; A. Willis Robertson to Aldo Leopold, January 16, 1931, box 001, folder 002, series 9/25/10-2, LEOP. Leopold fell out with the More Game Birds Foundation in 1930 in private correspondence. Aldo

Leopold to Joseph P. Knapp, September 18, 1930, box 005, folder 004, series 9/25/10-2, LEOP.

67. See Meine, *Aldo Leopold,* chapter 15; Minutes of a Meeting with the Senate Wild Life Committee at the Capitol, Washington, DC, January 9, 1934, Bureau of Biological Survey Records, Office Files of J. N. "Ding" Darling, 1930–35, RG 22, National Archives, College Park, MD; Thomas Beck, "What President's Committee Intends To Do," in *American Game Conference, Transactions of the Twentieth American Game Conference* (Washington, DC: American Game Protective Association, 1934); Thomas H. Beck, Jay N. Darling, and Aldo Leopold, *Report of the President's Committee on Wild-Life Restoration* (Washington, DC: U.S. Government Printing Office, 1934).

68. Leopold worried about the "federalization of game," maintaining that states ought to administer federally purchased game lands. Flader, *Thinking Like a Mountain,* 131.

69. Jay N. Darling to Clarence Cottam, June 25, 1959, as cited in David Leonard Lendt, "Ding: The Life of Jay Norward Darling" (PhD diss., Iowa State University, 1978), 127–128.

70. "Minutes of a Meeting with the Senate Wild Life Committee at the Capitol, Washington D.C., 9 January 1934," Records Concerning the President's Committee on Wildlife, Records of Bureau of Biological Survey, entry no. 144, RG 22, NARA II; Beck, "What President's Committee Intends To Do," in *American Game Conference;* Theodore W. Cart, "'New Deal' for Wildlife: A Perspective on Federal Conservation Policy, 1933–40," *Pacific Northwest Quarterly* 63 (1972): 113–120; Michael W Giese, A Federal Foundation for Wildlife Conservation: The Evolution of the National Wildlife Refuge System, 1920–1968 (PhD diss., American University, 2008).

71. Jay N. Darling to Mr. President, July 26, 1935, box 11, entry no. 253, RG 22, National Archives II, College Park, MD.

72. Drawing in the margin of Jay N. Darling to Mr. President, July 26, 1935, box 11, entry no. 253, RG 22, NARA II.

73. "Memorandum for the Secretary, February 26, 1934, box 6, series P253: Office Files of Dr. Frederick C. Lincoln 1917–1960, RG 22, NARA II.

74. *An Act to Supplement and Support the Migratory Bird Conservation Act,* Public Law 124, *U.S. Statutes at Large* 48 (1934): 451–453; U.S. Congress, *Report No. 868,* July 6, 1937, 75th Congress, copy in box 1, folder 9, series 9/25/10-2, LEOP.

75. Thomas Allen, *Guardian of the Wild: The Story of the National Wildlife Federation, 1936–1986* (Indianapolis: Indiana University Press, 1987).

76. Victor Shelford, "The Preservation of Natural Biotic Communities," *Ecology* 14 (1933): 240–245; "Report of the Committee on Wild Life Studies, National Research Council, On the Proposed Wild Life Research Program of the U.S. Biological Survey," May 22, 1935, box 006, folder 001, series 9/25/10-2, LEOP.

77. Master Plan: Wildlife Research and Management, Patuxent Research Refuge, Bowie, MD, box 30, entry no. 253, RG 22, NARA II.

78. "New Friends of Wild Game: Firearms Companies Backing Protection and Propagation Society," *New York Times,* September 25, 1911; Gregory Dehler, *The Most Defiant Devil: William Temple Hornaday and His Controversial Crusade to Save American Wildlife* (Charlottesville: University of Virginia Press, 2013), chapter 6.
79. Box 5, series P144: Records Concerning the President's Committee on Wildlife 1/1934-7/1934, RG 22, NARA II.
80. H. P. Sheldon, Chief, Division of Public Relations, Bureau of Biological Survey, "History and Significance of American Wildlife: Trends from Exploitation to Restoration," Wildlife Research and Management Leaflet BS-126, February 1939, reprint in folder 12, box 7, series 9/25/10-4, LEOP.
81. A joint resolution of Congress on February 9, 1871 (16 Stat. 593) created an independent Commissioner of Fish and Fisheries to investigate the decline in food fish and to stock such fish. The position was reconstituted into a Bureau of Fisheries, Department of Commerce, in February 14, 1903 (32 Stat. 825). Reorganization Plan No. II, July 1, 1939 (53 Stat. 1433), transferred Bureau of Biological Survey and Bureau of Fisheries to the Department of the Interior. Reorganization Plan No. III, June 30, 1940 (54 Stat. 1232), consolidated the two bureaus into a Fish and Wildlife Service in the Department of the Interior under a Commissioner of Fish and Wildlife.
82. On the National Wildlife Refuge System, see Cart, "'New Deal' for Wildlife," 113–120; Giese, "A Federal Foundation for Wildlife Conservation."
83. Aldo Leopold to Mr. Thomas Beck, January 8, 1934, Folder: Aldo Leopold, box 2, series P144—Records Concerning the President's Committee on Wildlife 1/1934-7/1934, RG 22, NARA II.
84. The U.S. Department of Agriculture, Report of the President's Committee on Wild-life Restoration, 1934, Folder: Original Copy of President's Committee on Wild-life Restoration, box 1, series P144: Records Concerning the President's Committee on Wildlife 1/1934-7/1934, RG 22, NARA II.
85. Speech published as Aldo Leopold, "Conservation Economics," *Journal of Forestry* 32 (1934): 537–544. Leopold had already been thinking of how farmers could promote game on their lands in articles like "How the Country Boy or Girl Can Grow Quail," *Wisconsin Arbor and Bird Day Annual,* May 10, 1929, pp. 51–53.
86. Undated fragment, LP 10-6, p. 16, emphasis original, as quoted in Meine, chapter 15.
87. Aldo Leopold, Wildlife Conservation on the Farm, reprinted from *Wisconsin Agriculturalist and Farmer,* Racine, WI, reprint in folder 3, box 2, series 9/25/10-8, LEOP.
88. "University Arboretum Wildlife Management Plan," box 001, folder 002, series 9/25/10-5, LEOP; Henry P. Davis, "A Full Dinner Pail for Game Birds," *National Sportsman* 18–19 (Dec 1932): 32–33, in box 7, folder 8, series 9.25.10-4, LEOP.
89. Aldo Leopold, Wildlife Conservation on the Farm, Fall 1938, *Wisconsin Agriculturalist and Farmer,* Racine WI, reprint in folder 4, box 009, series 9/25/10-6: Writings, LEOP.

90. Philip J. Pauly discusses Leopold and Longenecker's debate in *Fruits and Plains,* 190–194. Thomas J. Blewett and Grant Cottam, "History of the University of Wisconsin Arboretum Prairies," *Transactions of the Wisconsin Academy of Sciences, Arts and Letters* 72 (1984): 130–144; J. W. Jackson, "Memorandum, Proposed Chair of Conservation," 1933, general correspondence, Arboretum Committee Records (Series 38/3/2); "Memorandum for President Dykstra on a Research Program for the University of Wisconsin Arboretum," July 10, 1938, General Files, box 12, folder "Research Arboretum," series 38/7/3, University of Wisconsin–Madison Archives, Madison, WI; Aldo Leopold, "University Arboretum Wild Life Management Plan," October 25, 1933, box 001, folder 002, series 9/25/10-5, LEOP. On the history of American arboretums, see also Emily Griswold, "The Origin and Development of Ecogeographic Displays in North American Botanic Gardens" (MA thesis, University of Washington, 2002).
91. Leopold wrote many versions of this talk. A draft written a few weeks after the dedication ceremony is quoted in Franklin Court, *Pioneers of Ecological Restoration: The People and the Legacy of the University of Wisconsin Arboretum* (Madison: University of Wisconsin Press, 2012), 75–76. For another version, see Aldo Leopold, "What Is the Arboretum?" Address to Nakoma Women's Club, September 20, 1934, series 38/7/2, LEOP.
92. Aldo Leopold, "Memorandum for President Dykstra"; Russ Pyre, "Hook, Line, and Sinker," *Madison Wisconsin State Journal,* December 26, 1943.
93. Court, *Pioneers of Ecological Restoration,* 79–85; Theodore Sperry, "Prairie Restoration on the University of Wisconsin Arboretum," 1939, box 1, Arboretum (series 38/7/2), University of Wisconsin–Madison Archives, Madison, WI; D. C. Peattie, Roger C. Anderson, *The Use of Fire as a Management Tool on the Curtis Prairie* (Madison: University of Wisconsin Arboretum, 1972); William R. Jordan III, "Some Reflections on Curtis Prairie and the Genesis of Ecological Restoration," *Ecological Management and Restoration* 11 (2010): 99–107.
94. Paul B. Riis, "Ecological Garden and Arboretum at the University of Wisconsin," *Parks & Recreation* 20 (1937): 382–389; Russell B. Pyre, "Clod by Clod, Historical Prairie Returns to Madison's Yard: CCC, Dr. Sperry Undo Plow's Work," *Madison Wisconsin State Journal,* November 12, 1939; T. J. Blewett and G. Cottam, "History of the University of Wisconsin Arboretum Prairies," *Transactions of the Wisconsin Academy of Science, Arts and Letters* 72 (1984): 130–144. The CCC remained at Camp Madison until November 1941. See Aldo Leopold to Roberts Mann, November 5, 1941, box 003, folder 003, Series 9/25/10-1, LEOP.
95. Here, I am in agreement with Peter Coates and Philip Pauly. In *Fruits and Plains,* Pauly argues that Americans began to worry about the introduction of crop-destroying insects, weeds, and plant diseases in the 1890s, leading to the establishment of legal and bureaucratic regulation of the movements of species across international and interstate boundaries in the 1910s. Pauly, *Fruits and Plains,* especially chapters 2 and 6; Peter Coates, *American*

Perceptions of Immigrant and Invasive Species: Strangers on the Land (Berkeley: University of California Press, 2007).

96. Aldo Leopold, "A Biotic View of Land," *Journal of Forestry* 37 (1939): 727–730. See also Paul B. Sears, "Report of the Committee on Summer Symposia," *Ecology* 20 (1939): 323–324.
97. Aldo Leopold, *A Sand County Almanac: And Sketches Here and There* (Oxford: Oxford University Press, 1949), 217, 197, 196.
98. Frederic Clements to Bernhard, April 5, 1924, quoted in Griswold, "The Origin and Development of Ecogeographic Displays," 57.
99. Donald C. Holden, "WPA Makes Possible Bird and Wild Flower Sanctuary at Petersburg," *The W.P.A. Record in Virginia* 1 (1937): 1–2; Nancy Kober, *With Paintbrush and Shovel: Preserving Virginia's Wildflowers* (Charlottesville: University of Virginia Press, 2000).
100. Franz Aust to F. Swingle, September 20, 1939, box 008, folder 023, series 9/25/10-4, LEOP.
101. V. E. Shelford to S. Charles Kendeigh, October 28, 1947, box 1, folder 1940-1946, Victor E. Shelford Papers, University of Illinois Archives, Urbana-Champaign, IL.
102. H. Greene and J. Curtis, "Germination Studies of Wisconsin Prairie Plants," *American Midland Naturalist* 39 (1950): 186–194; H. C. Greene and J. Curtis, "The Re-establishment of Prairie in the University of Wisconsin Arboretum," *Wild Flower* 29 (1953): 77–88.
103. "The Wild Flower Society," *Saturday Evening Post,* June 20, 1903.
104. Leopold, "The Conservation Ethic," 1933, emphasis original, in *The River of the Mother of God: And Other Essays by Aldo Leopold,* ed. Susan L. Flader and J. Baird Callicott (Madison: University of Wisconsin Press, 1991). See also Aldo Leopold, "Wildflower Management: A Virgin Field for Conservation Research," draft for *American Forestry,* 1941, box 8, folder 23, Series 9/25/10-4, LEOP.
105. Elizabeth Britton, "The Relations of Plants to Birds and Insects," *Plant World* 7 (1904): 69–70.
106. Aldo Leopold, "Exit Orchis," May 15, 1940, clipping in box 008, folder 023, series 9/25/10-4, LEOP.
107. Aldo Leopold, "Wildlife in American Culture," *Journal of Wildlife Management* 7 (1943): 1–6; Leopold, *A Sand County Almanac,* 285.
108. Meine, *Aldo Leopold,* 383. Leopold, *A Sand County Almanac,* 123, 226.

4. Atoms for Ecology

1. Neil O. Hines recounts in *Fish of Rare Breeding: Salmon and Trout of the Donaldson Strains* (Washington, DC: Smithsonian Institution Press, 1976) that Donaldson was driving from Seattle to New Westminster, British Columbia, when he received a telegram around August 15, 1943. I believe the date of the telegram to be August 18 based on a transcript of a phone call between Mr. Wensel of Knoxville, Tennessee (Clinton Engineering Works) and Mr. Hanford Thayer that can be found in box 9, folder 18, Lauren R. Don-

aldson Papers, Accession No. 2932-007, University of Washington Libraries Special Collections, Seattle, WA (hereafter LRDP).

2. Federal sponsorship of the AFL remained with the MED until the passage of the Atomic Energy Act in 1946, at which point the program was transferred to the Atomic Energy Commission. The lab was placed under the AEC's Division of Biology and Medicine, Environmental Sciences. See Neal O. Hines, *Proving Ground: An Account of the Radiobiological Studies in the Pacific, 1946–1961* (Seattle: University of Washington Press, 1962), chapter 1. For archival material on the establishment of the AFL, see boxes 3 and 9 of the Laboratory of Radiation Ecology Records, 1948–1984, University of Washington Special Collections, Seattle, WA (hereafter LRER); and Leslie R. Groves to Donaldson, March 10, 1961, box 12, folder 46, Laboratory of Radiation Biology Records, 1944–1970, University of Washington Special Collections, Seattle (hereafter LRBR). For histories of the Manhattan Project, and Hanford Works in particular, see Leslie Groves, *Now It Can Be Told: The Story of the Manhattan Project* (New York: Harper, 1962); John Findlay and Bruce Hevly, *Atomic Frontier Days: Hanford and the American West* (Seattle: University of Washington Press, 2011).
3. Kelshaw Bonham et al., "Lethal Effect of X-Rays on Marine Microplankton Organisms," *Science* 106 (1947): 245–246; Richard F. Foster and Lauren R. Donaldson, "The Effect on Embryos and Young of Rainbow Trout from Exposing the Parent Fish to X-Rays," *Growth* 13 (1949): 119–142. AFL reports on these experiments can be found in box 9, LRER.
4. See, for example, Peter Galison and Bruce Hevly, eds., *Big Science: The Growth of Large-Scale Research* (Stanford, CA: Stanford University Press, 1992); Stuart W. Leslie, *The Cold War and American Science: The Military-Industrial-Academic Complex at MIT and Stanford* (New York: Columbia University Press, 1993); Ronald E. Doel, "Constituting the Postwar Earth Sciences: The Military's Influence on the Environmental Sciences in the USA after 1945," *Social Studies of Science* 33 (2003): 635–666; Jacob Darwin Hamblin, *Oceanographers and the Cold War: Disciples of Marine Science* (Seattle: University of Washington Press, 2005).
5. U.S. Department of Energy, *United States Nuclear Tests: July 1945 through September 1992* (Las Vegas: USDOE, 2015), DOE/NV—209-REV 16. I include the weapons dropped on Hiroshima and Nagasaki.
6. Angela H. Creager, *Life Atomic: A History of Radioisotopes in Science and Medicine* (Chicago: University of Chicago Press, 2013), 5.
7. Creager, *Life Atomic.*
8. George Perkins Marsh, *Report Made under Authority of the Legislature of Vermont, on the Artificial Propagation of Fish* (Burlington, VT: Free Press, 1857), 9. See also T. Cumbler, "The Early Making of an Environmental Consciousness: Fish, Fisheries Commissions, and the Connecticut River," *Environmental History Review* 15 (1991): 73–91.
9. John Muir, "The Establishment on McCloud River—John Muir, the Naturalist, Gives a Graphic Description of What Is Being Done," *Daily Evening Bulletin* (San Francisco), October 29, 1874.

10. Fisheries biology and ecology were separate disciplines with their own societies and journals. But many biologists identified as both fisheries biologists and ecologists. As a result, the two disciplines deeply influenced each other. See Matthew Klingle, "Plying Atomic Waters: Lauren Donaldson and the 'Fern Lake Concept' of Fisheries Management," *Journal of the History of Biology* 31 (1998): 1–32. See also Tim D. Smith, *Scaling Fisheries: The Science of Measuring the Effects of Fishing, 1855–1955* (Cambridge, UK: Cambridge University Press, 1994); Joseph E. Taylor, *Making Salmon: An Environmental History of the Northeast Fisheries Crisis* (Seattle: University of Washington Press, 1999).
11. On the history of fisheries research centers, see N. G. Benson, ed., *A Century of Fisheries in North America* (Washington, DC: American Fisheries Society, 1970); Keith R. Benson, "Laboratories on the New England Shore: The 'Somewhat Different Decision' of American Marine Biology," *New England Quarterly* 56 (1988): 53–78; Dean Allard Jr., "The Fish Commission Laboratory and Its Influence on the Founding of the American Biological Laboratory," *Journal of the History of Biology* 23 (1990): 251–270; Arthur McEvoy, *The Fisherman's Problem: Ecology and Law in the California Fisheries, 1850–1980* (Cambridge, UK: Cambridge University Press, 1990).
12. Anthony Netboy, *The Columbia River Salmon and Steelhead Trout: Their Fight for Survival* (Seattle: University of Washington Press, 1981); Richard White, *The Organic Machine: The Remaking of the Columbia River* (New York: Hill and Wang, 1996).
13. Charles O. Hayford and George C. Embody, "Further Progress in the Selective Breeding of Brook Trout at the New Jersey State Hatchery," *Transactions of the American Fisheries Society* 60 (1930): 109–113.
14. See Hines, *Fish of Rare Breeding.*
15. "Notes May 20 1965," box 1, folder 37, LRDP.
16. Lauren R. Donaldson and Paul R. Olson, "Development of Rainbow Trout Brood Stock by Selective Breeding," *Transactions of the American Fisheries Society* 85 (1957): 93–101.
17. Lauren Donaldson and Deb Menasveta, "Selective Breeding of Chinook Salmon," *Transactions of the American Fisheries Society* 90 (1961): 160–164.
18. Angela N. H. Creager and María Santesmases, "Radiobiology in the Atomic Age: Changing Research Practices and Policies in Comparative Perspective," *Journal of the History of Biology* 39 (2006): 637–647. On the Atoms for Peace campaign, see Richard G. Hewlett and Jack M. Holl, *Atoms for Peace and War, 1953–1961: Eisenhower and the Atomic Energy Commission* (Berkeley: University of California Press, 1989); John Krige, "Atoms for Peace, Scientific Internationalism, and Scientific Intelligence," *Osiris* 21 (2006): 161–181.
19. "Atom Study Points to Food Plenty by Fast Development of New Plants," *New York Times*, January 31, 1952, as quoted in Helen Anne Curry, "Radiation and Restoration; or, How Best to Make a Blight-Resistant Chestnut Tree," *Environmental History* 19 (2014): 217–238, 223.
20. Curry, "Radiation and Restoration."
21. Lauren R. Donaldson and Kelshaw Bonham, "Irradiation of Chinook and Coho Salmon Eggs and Alevins," *Transactions of the American Fisheries Society* 93 (1964): 333–341.

22. Lauren R. Donaldson and Kelshaw Bonham, "Effects of Chronic Exposure of Chinook Salmon Eggs and Alevins to Gamma Irradiation," *Transactions of the American Fisheries Society* 99 (1970): 112–119.
23. Lauren R. Donaldson to Thomas Shipman, November 21, 1963, box 6, folder 16, LRBR.
24. See, for example, "The Island That Went Wrong," *Sunday Times,* May 28, 1962; Herb Caen, "The Voice of the Turtle," *San Francisco Chronicle,* June 17, 1962. James S. Jenkins to Allston Jenkins, May 18, 1962, box 6, folder 16, LRBR.
25. See box 6, folder 16, LRBR.
26. Harold Collidge to Lauren R. Donaldson, November 2, 1962, box 6, folder 16, LRBR.
27. Lauren R. Donaldson to John Devlin, August 3, 1962, box 6, folder 16, LRBR.
28. A growing literature in STS and animal studies that investigates the industrialization of organisms informs my analysis. See Edmund Russell, "Evolutionary History: Prospectus for a New Field," *Environmental History* 8 (2003): 204–228; Susan R. Schrepfer and Philip Scranton, *Industrializing Organisms: Introducing Evolutionary History* (New York: Routledge, 2004); Edmund Russell, *Evolutionary History* (Cambridge, UK: Cambridge University Press, 2011).
29. I elaborate on this argument in Laura J. Martin, "Proving Grounds: Ecological Fieldwork in the Pacific and the Materialization of Ecosystems," *Environmental History* 23 (2018): 567–592.
30. William A. Shurcliff, *Bombs at Bikini: The Official Report of Operation Crossroads* (New York: William H. Wise & Co., 1947); Jonathan M. Weisgall, *Operation Crossroads: The Atomic Tests at Bikini Atoll* (Annapolis, MD: Naval Institute Press, 1994); Scott Kirsch, "Watching the Bombs Go Off: Photography, Nuclear Landscapes, and Spectator Democracy," *Antipode* 29 (1997): 227–255.
31. Jeffrey S. Davis, "Representing Place: 'Deserted Isles' and the Reproduction of Bikini Atoll," *Annals of the Association of American Geographers* 95 (2005): 607–625. For anthropological perspectives on American nuclear colonialism, see Barbara R. Johnston and Holly M. Barker, *The Consequential Damages of Nuclear War: The Rongelap Report* (Walnut Creek, CA: Left Coast Press, 1998); Holly M. Barker, *Bravo for the Marshallese: Regaining Control in a Post-Nuclear, Post-Colonial World* (Belmont, MA: Cengage Learning, 2012).
32. Emory Jerry Jessee, "Radiation Ecologies: Bombs, Bodies, and Environment during the Atmospheric Nuclear Weapons Testing Period, 1942–1965" (PhD diss., Montana State University, 2013), 161.
33. There were many military objectives of the scientific surveys, as enumerated in "Oceanographic Program at Bikini Atoll," box 3, folder 7, series 3, W. T. Edmondson Papers, Collection No. 2024, University of Washington Libraries Special Collections, Seattle, WA.
34. My account of tests Able and Baker draws from Hines, *Proving Ground,* and trip logbooks, laboratory correspondence, staff meeting minutes, and AEC reports in LRBR and LRER.

35. Box 11, folder 28, LRDP; also "Report on Able and Baker Effects," box 6, folder 4, LRBR.
36. Hines, *Proving Ground,* chapters 2 and 3; "Appendix XIV," box 6, folder "Bikini 1946–1947," LRER.
37. Hines, *Proving Ground.* Press releases can be found in box 2, folder 7, Neal O. Hines Papers, University of Washington Special Collections, Seattle, WA (hereafter NOHP). The Resurvey scrapbook can be found in box 21, LRDP.
38. Hines, *Proving Ground,* 61.
39. The previous summer, AFL member Richard Foster had collected samples of fish from the Columbia River and found them to have concentrations of radioactivity thousands of times higher than their environment. The AFL's findings at the Pacific Proving Grounds resonated with this finding, and with Donaldson's earlier work on fish nutrition. See L.R. Donaldson and R.F. Foster, "Effects of Radiation on Aquatic Organisms," in *The Effects of Atomic Radiation on Oceanography and Fisheries* (Washington, DC: National Academy of Sciences, 1957), 96–102.
40. Neal Hines, "Bikini Atoll and the Scientific Resurveys," box 6, folder 6, LRER.
41. On Marshall Islanders' continuing struggle for self-determination, see M. X. Mitchell, "Offshoring American Environmental Law: Land, Culture, and Marshall Islanders' Struggles for Self-Determination during the 1970s," *Environmental History* 22 (2017): 209–234.
42. Atomic Energy Commission, *Second Semiannual Report* (Washington, DC: Government Printing Office, July 1947), 7; Roger Gale, *The Americanization of Micronesia: A Study on the Consolidation of U.S. Rule in the Pacific* (Washington, DC: University Press of America, 1979).
43. Jessee, "Radiation Ecologies," 94–96.
44. Andrew W. Rogers, *Techniques of Autoradiography,* 3rd ed. (North Holland: Elsevier, 1969).
45. Lauren Donaldson, "Speech" delivered at meeting of the Atomic Energy Project, University of California at Los Angeles, August 11, 1948, folder 2, box 3, LRDP.
46. Lauren R. Donaldson, "Biological Cycles of Fission Products in Aquatic Systems as Studied at the Pacific Atolls of Bikini and Eniwetok," U.S. Atomic Energy Commission Report AECU–3412, box 3, folder 7, LRDP. Letters to Donaldson from the Biology Branch of the Division of Biology and Medicine, and reports on site visits to the AFL, can be found in box 88, entry no. 98, RG 326, National Archives II, College Park, MD (hereafter NARA II).
47. Fred E. Locke to Lauren R. Donaldson, October 5, 1953; Lauren R. Donaldson to Fred E. Locke, October 7, 1953; E. L. Bartlett to Lauren R. Donaldson, August 5, 1954; Lauren R. Donaldson to E. L. Bartlett, August 11, 1954, box 1, folder 1, LRDP. For the role of trace elements in salmon physiology, see Lauren R. Donaldson, "The Inorganic Elements," box 3, folder 3, LRDP.
48. For an overview of the Fern Lake project, see Lauren R. Donaldson, Paul R. Olson, and John R. Donaldson, "The Fern Lake Trace Mineral Metabolism Program," *Transactions of the American Fisheries Society* 88 (1959), 1–5.

49. Quotation from "A Farewell to Doc," box 1, folder 1, LRDP.
50. Lauren R. Donaldson et al., *The Fern Lake Studies* (Seattle: University of Washington Press, 1971). Matt Klingle uses the Fern Lake project to explore the connection between ecology and natural resources management in "Plying Atomic Waters: Lauren Donaldson and the 'Fern Lake Concept' of Fisheries Management," *Journal of the History of Biology* 31 (1998): 1–32, quote on 31.
51. Clipping of Jerry Grosso, "History-Making Tests of Peaceful Atom Uses," in box 89, entry no. 98, RG 326, NARA II.
52. Klingle, "Plying Atomic Waters."
53. W. R. Boss to Dr. Gordon M. Dunning, May 4, 1956, folder 21, box 1, LRBR.
54. "American Fisheries Society Committee on International Relations: Preliminary Report, September 1954," folder 8, box 1, LRDP; R. F. Palumbo, "Radionuclides in Foods from the Central Pacific, 1962," *Nature* 209 (1966): 1190–1192. Correspondence regarding the 1954 accident can also be found in folder 38, box 1, LRBR. On Rongelap, see transcript of telephone conversation, April 3, 1956, W. R. Boss to Dr. Lauren R. Donaldson, box 2, folder 9, LRBR.
55. Robert Conard to Charles Dunham, April 1, 1959, box 89, entry no. 98, RG 326, NARA II.
56. Giff Johnson, "Micronesia: America's 'Strategic' Trust," *Bulletin of the Atomic Scientists* 35 (1979): 10–15.
57. Correspondence between General Mills executive offices and Lauren R. Donaldson can be found in box 1, folders 35–40, LRDP. Additional material can be found in Donaldson's correspondence with W. J. Mullahey of Pan Am in box 6, folder 13, LRDP, and with General E. W. Rawlings in box 6, folder 16, LRDP.
58. A. Jonsgard to Lauren R. Donaldson, November 12, 1963, box 1, folder 35, LRDP.
59. General E. W. Rawlings to Lauren R. Donaldson, "General Order Number 3," box 1, folder 36, LRDP.
60. General E. W. Rawlings to W. J. Mullahey, June 18, 1965, box 1, folder 37, LRDP; W. J. Mullahey to Clarence Hall, May 12, 1965, box 1, folder 37, LRDP.
61. Mullahey to Hall, May 12, 1965.
62. "Notes May 20 1965," box 1, folder 37, LRDP.
63. Ray H. Anderson to Lauren R. Donaldson, February 3, 1967, box 1, folder 38, LRDP.
64. "Domesticating the Sea—Prospects and Problems," delivered to a meeting of the Hawaiian Sugar Technologists, box 3, folder 56, LRDP. On "ocean ranching," see also box 14, folder 38, LRDP.
65. "Sportsmen Benefitting from Atomic Energy," reprint in folder 40, box 3, LRDP.
66. Lauren R. Donaldson to General E. W. Rawlings, September 15, 1967, box 1, folder 39, LRDP; "State of Minnesota Department of Natural Resources," box 2, folder 28, LRDP. Correspondence with the Michigan Department of Conservation can be found in box 2, folder 27, LRDP.

67. See Lauren R. Donaldson to Mr. Milo Moore, November 8, 1966, box 2, folder 30, LRDP; Lauren R. Donaldson to Richard A. Barkley, March 11, 1969, box 3, folder 56, LRDP.
68. Stephen S. Crawford and Andrew Muir, "Global Introductions of Salmon and Trout in the genus Oncorhynchus: 1870–2007," *Reviews in Fish Biology and Fisheries* 18 (2008): 313–344.
69. Estimates range from 29 percent to 88 percent; for overview see Boris Worm, "Averting a Global Fisheries Disaster," *PNAS* 113 (2016): 4895–4897. See also Boris Worm et al., "Rebuilding Global Fisheries," *Science* 325 (2009): 578–585; Christopher Costello et al., "Global Fishery Prospects under Contrasting Management Regimes," *PNAS* 113 (2016): 5125–5129. On freshwater fisheries, see Peter B. McIntyre, Catherine Reidy Liermann, and Carmen Revenga, "Linking Freshwater Fishery Management to Global Food Security and Biodiversity Conservation," *PNAS* 113 (2016): 12880–12885.
70. Food and Agriculture Organization, *The State of World Fisheries and Aquaculture 2020* (Rome: FAO), http://www.fao.org/documents/card/en/c/ca9229en.
71. Dane Klinger and Rosamond Naylor, "Searching for Solutions in Aquaculture: Charting a Sustainable Course," *Annual Review of Environment and Resources* 37 (2012): 247–276.
72. Junji Yuan et al., "Rapid Growth in Greenhouse Gas Emissions from the Adoption of Industrial-Scale Aquaculture," *Nature Climate Change* 9 (2019): 318–322.
73. https://www.dec.ny.gov/outdoor/62477.html, accessed September 30, 2020. Fish stocking is practiced around the world. See R. E. Gozlan et al., "Current Knowledge on Non-Native Freshwater Fish Introductions," *Journal of Fish Biology* 76 (2010): 751–786; Anders Halverson, *An Entirely Synthetic Fish: How Rainbow Trout Beguiled America and Overran the World* (New Haven, CT: Yale University Press, 2011).
74. For examples, see Amy Harig and Mark Bain, "Defining and Restoring Biological Integrity in Wilderness Lakes," *Ecological Applications* 8 (1998): 71–87; Frank J. Rahel, "Homogenization of Fish Faunas across the United States," *Science* 288 (2000): 854–856; K. D. Fausch, "Introduction, Establishment and Effects of Non-Native Salmonids: Considering the Risk of Rainbow Trout Invasion in the United Kingdom," *Journal of Fish Biology* 71 (2007): 1–32; Alexander Alexiades, Alexander Flecker, and Clifford Kraft, "Nonnative Fish Stocking Alters Stream Ecosystem Nutrient Dynamics," *Ecological Applications* 27 (2017): 956–965.
75. "The Philosophers' Stone," *Time*, August 15, 1955, p. 48. See also Carolyn Kopp, "The Origins of the American Scientific Debate over Fallout Hazards," *Social Studies of Science* 9 (1979): 403–422; Creager, *Life Atomic* (2013).
76. "Reports of Standing Committees," *Transactions of the American Fisheries Society* 84 (1955): 330–371.

5. The Specter of Irreversible Change

1. Robert H. Wurtz, "War and the Living Environment," *Nuclear Information* 5 (1963): 1–21.

2. Fallout studies played a central role in the rise of the idea of an interconnected biosphere. See Laura Bruno, "The Bequest of the Nuclear Battlefield: Science, Nature, and the Atom during the First Decade of the Cold War," *Historical Studies in the Physical and Biological Sciences* 33 (2003): 237–260; Joseph Masco, "Bad Weather: On Planetary Crisis," *Social Studies of Science* 40 (2009): 7–40; Emory Jerry Jessee, "Radiation Ecologies: Bombs, Bodies, and Environment during the Atmospheric Nuclear Weapons Testing Period, 1942–1965" (PhD diss., Montana State University, 2013). On the U.S. test ban movement, see Robert Divine, *Blowing on the Wind: The Nuclear Test Ban Debate, 1954–1960* (Oxford: Oxford University Press, 1978).
3. Robert J. Watson, *History of the Office of the Secretary of Defense IV: Into the Missile Age, 1956–1960* (Washington, DC: Office of the Secretary of Defense, 1997), 457, Table 6.
4. Here I engage work on the connection between the concept of "total war" and the rise of the environmental sciences. See Jacob Darwin Hamblin, *Arming Mother Nature: The Birth of Catastrophic Environmentalism* (Oxford: Oxford University Press, 2013).
5. George Perkins Marsh, *Man and Nature, or, Physical Geography as Modified by Human Action* (Cambridge, MA: Belknap Press, 1965, originally published 1864), 41.
6. See, for example, Hugh M. Raup, "Old Field Forests of Southeastern New England," *Journal of the Arnold Arboretum* 21 (1940): 266–273; Juanda Bonck and W. T. Penfound, "Plant Succession on Abandoned Farm Land in the Vicinity of New Orleans," *American Midland Naturalist* 33 (1945): 520–552.
7. William L. Thomas, ed., "Part 2," in *Man's Role in Changing the Face of the Earth* (Chicago: University of Chicago Press, 1956), 677–804. Quotes from Edward Graham, "The Re-creative Power of Plant Communities," in *Man's Role in Changing the Face of the Earth,* ed. William Thomas, 677–691.
8. George Mazuzan and J. Samuel Walker, "Chapter 1," in *Controlling the Atom: The Beginnings of Nuclear Regulation, 1946–1962* (Berkeley: University of California Press, 1984); David Reichle and Stanley Auerbach, *U.S. Radioecology Research Programs of the Atomic Energy Commission in the 1950s* (Oak Ridge, TN: U.S. Department of Energy, 2003).
9. Stanley I. Auerbach et al., "Ecological Research," in *Health Physics Division Annual Progress Report for Period Ending July 31, 1958* (Oak Ridge, TN: Oak Ridge National Laboratory, 1958), 27–41. Auerbach's research is discussed in Stephen Bocking, "Ecosystems, Ecologists, and the Atom: Environmental Research at Oak Ridge National Laboratory," *Journal of the History of Biology* 28 (1995): 1–47.
10. Stanley I. Auerbach, *A History of the Environmental Sciences Division of Oak Ridge National Laboratory* (Oak Ridge, TN: Oak Ridge National Laboratory, 1972). See also Chunglin Kwa, "Radiation Ecology, Systems Ecology and the Management of the Environment," in *Science and Nature: Essays in the History of the Environmental Sciences,* ed. Michael Shortland (Oxford: British Society for the History of Science, 1993); Stephen Bocking,

"Ecosystems, Ecologists, and the Atom"; Sharon Kingsland, "Chapter 7," in *The Evolution of American Ecology, 1890–2000* (Baltimore, MD: Johns Hopkins University Press, 2005).

11. Donald Worster argued, for example, that the ecosystem concept "owed nothing to any of its forebears in the history of science. [. . .] It was born of entirely different parentage: that is, modern thermodynamic physics, not biology." Donald Worster, *Nature's Economy: A History of Ecological Ideas* (New York: Cambridge University Press, 1977), 304. Less restrictively, Sharon Kingsland describes the rise of ecosystem ecology as an effort to convert the "soft science" of ecology into a "hard science" like physics, and thus "show that the subject could command intellectual respect" in *Evolution of American Ecology,* 179. See also Peter J. Taylor, "Technocratic Optimism, H. T. Odum, and the Partial Transformation of Ecological Metaphor after World War II," *Journal of the History of Biology* 21 (1988): 213–244; Paolo Palladino, "Defining Ecology: Ecological Theories, Mathematical Models, and Applied Biology in the 1960s and 1970s," *Journal of the History of Biology* 24 (1991): 223–243; Joel Hagen, *An Entangled Bank: The Origins of Ecosystem Ecology* (New Brunswick, NJ: Rutgers University Press, 1992); Frank Golley, *A History of the Ecosystem Concept in Ecology: More Than the Sum of the Parts* (New Haven, CT: Yale University Press, 1993).
12. Letters between Tom Odum and Eugene Odum, 1940–1957, can be found in carton 3, series 3:1, Eugene P. Odum Papers (MS 3257), Hargrett Rare Book & Manuscript Library, University of Georgia, Athens, GA (hereafter EPOP). Eugene P. Odum, "Variations in the Heart Rate of Birds: A Study in Physiological Ecology," *Ecological Monographs* 11 (1941): 299–326.
13. Tom to Eugene, December 1947, folder 25, carton 3, series 3, EPOP; Tom to Eugene, Summer 1948, folder 25, carton 3, series 3, EPOP.
14. A copy of Tom's comprehensive exams can be found in folder 25, carton 3, series 3, EPOP. Howard T. Odum, "The Stability of the World Strontium Cycle," *Science* 114 (1951): 407–411.
15. Joel B. Hagen, "Eugene Odum and the Homeostatic Ecosystem: The Resilience of an Idea," in *Traditions of Systems Theory: Major Figures and Contemporary Developments,* ed. Darrell P. Arnold (New York: Routledge, 2014), 179–193. On the Cybernetic Conferences, see G. Evelyn Hutchinson to W. T. Edmondson, March 18, 1946, folder 230, box 13, series I, G. Evelyn Hutchinson Papers, Beinecke Rare book and Manuscript Library, Yale University, New Haven, CT; G. Evelyn Hutchinson, "Circular Causal Systems in Ecology," *Annals of the New York Academy of Science* 40 (1948): 221–246.
16. Walter B. Cannon, *The Wisdom of the Body* (New York: Norton, 1932).
17. Peter Galison, "The Ontology of the Enemy: Norbert Wiener and the Cybernetic Vision," *Critical Inquiry* 21 (1994): 228–266. On the history of cybernetic theory, see also Steve J. Heims, *The Cybernetics Group* (Cambridge, MA: MIT Press, 1991); Geof Bowker, "How to Be Universal: Some Cybernetic Strategies, 1943–70," *Social Studies of Science* 23 (1993): 107–127; Paul Edwards, *The Closed World: Computers and the Politics of Discourse in Cold War America* (Cambridge, MA: MIT Press, 1996); Andrew Pickering,

The Cybernetic Brain: Sketches of Another Future (Chicago: University of Chicago Press, 2009); Ronald Kline, *The Cybernetics Moment: Or Why We Call Our Age the Information Age* (Baltimore, MD: Johns Hopkins University Press, 2015).

18. Eugene to Pop [Eugene P. Odum], December 12, 1937, folder 14, box 3, series 3, EPOP; Tom to Eugene, Fall 1949, folder 25, carton 3, series 3, EPOP.
19. H. T. [Odum] to Eugene and Martha, Summer 1951, folder 25, carton 3, series III, EPOP; H. T. to Eugene, c. Spring 1953, box 60, EPOP. Tom edited the book and wrote much of chapters 4–7. Eugene thanked Tom in the preface, but, to his chagrin, not on the title page.
20. Arthur Tansley, "The Use and Abuse of Vegetational Concepts and Terms," *Ecology* 16 (1935): 284–307. Tom Odum, for one, did not read Tansley's paper until years after its publication. Eugene Odum likely suggested it to him while drafting *Fundamentals of Ecology*. In 1950, Tom wrote to Eugene: "Interestingly enough, on reading Tansley (1935), [. . .] in a way it is disappointing to keep finding parts of what one considered original in the works of others." Tom to Eugene, Marther, and Will, c. 1950, folder 22, carton 3, series 3, EPOP.
21. Betty Jean Craige, *Eugene Odum: Ecosystem Ecologist and Environmentalist* (Athens: University of Georgia Press, 2001), 50.
22. Eugene Odum to W. Boss, April 27, 1954; "A Proposal for Studies on the Productivity of Coral Reef Atolls"; and Howard Odum to Sidney Galler, August 14, 1953, in box 1, folder 8, series 1, Eugene Odum Research Files: Eniwetok Atoll (UA06-032), Hargrett Rare Books and Manuscripts Library, University of Georgia, Athens, GA.
23. I analyze the Odums's fieldwork in more detail in Laura J. Martin, "Proving Grounds: Ecological Fieldwork in the Pacific and the Materialization of Ecosystems," *Environmental History* 23 (2018): 567–592.
24. Howard T. Odum and Eugene Odum, "Trophic Structure and Productivity of a Windward Coral Reef Community on Eniwetok Atoll," *Ecological Monographs* 25 (1955): 291–320.
25. Neal O. Hines, "Bikini Atoll and the Scientific Resurveys," box 6, folder 6, Laboratory of Radiation Ecology Records, 1948–1984, University of Washington Special Collections, Seattle, WA.
26. Folder 46, box 12, Laboratory of Radiation Biology Records, 1944–1970, University of Washington Special Collections, Seattle, WA.
27. Robert Jackson, *Guide to U.S. Atmospheric Nuclear Weapons Effects Data* (Alexandria, VA: Defense Nuclear Agency, 1993).
28. Jane Dibblin, *Day of Two Suns: US Nuclear Testing and the Pacific Islanders* (London: Virago, 1988); M. X. Mitchell, "Offshoring American Law: Land, Culture, and Marshall Islanders' Struggles for Self-Determination during the 1970s," *Environmental History* 22 (2017): 209–234.
29. Valerie Kuletz, *The Tainted Desert: Environmental and Social Ruin in the American West* (New York: Routledge, 1998); Joe Masco, *The Nuclear Borderlands: The Manhattan Project in Post-Cold War New Mexico* (Princeton, NJ: Princeton University Press, 2006), 311–315.

30. Eugene Odum's folder on the Geneva conference can be found in carton 22, series 1, EPOP.
31. Eugene Odum to family, September 27, 1957, folder 9, carton 3, EPOP.
32. On model organisms, see Robert E. Kohler, *Lords of the Fly: Drosophila Genetics and the Experimental Life* (Chicago: University of Chicago Press, 1994); *The Life of a Virus: Tobacco Mosaic Virus as an Experimental Model, 1930–1965* (Chicago: University of Chicago Press, 2002), 320–321; Rachel A. Ankeny, "The Conqueror Worm: An Historical and Philosophical Examination of the Use of the Nematode *C. Elegans* as a Model Organism" (PhD diss., University of Pittsburgh, 1997); Karen Rader, *Making Mice: Standardizing Animals for American Biomedical Research, 1900–1955* (Princeton, NJ: Princeton University Press, 2004).
33. Eugene P. Odum, *Fundamentals of Ecology,* 2nd ed. (Philadelphia: Saunders Publishing, 1959), 469.
34. Odum, *Fundamentals of Ecology,* 2nd ed., 452–486. Drafts of the textbook can be found in file 4, box 13, series 1, EPOP.
35. Odum, *Fundamentals of Ecology,* 2nd ed., v.
36. Angela Creager, *Life Atomic: A History of Radioisotopes in Science and Medicine* (Chicago: University of Chicago Press, 2013), 5.
37. Stanley Auerbach, Jerry Olson, and M. Waller, "Landscape Investigations Using Caesium-137," *Nature* 201 (1964): 761; Robert C. Pendleton and A. W. Grundmann, "Use of P-32 in Tracing Some Insect-Plant Relationships of the Thistle, *Cirsium undulatum,*" *Ecology* 35 (1954): 187–191; Eugene Odum and Edward Kuenzler, "Experimental Isolation of Food Chains in an Old-Field Ecosystem with the Use of Phosphorus-32," *Radioecology* 113 (1963): 120. The field notes for the "hot quadrat" studies are in folder 12, carton 87, series I, EPOP.
38. Frank Golley, in Gary W. Barrett and Terry L. Barrett, *Holistic Science: The Evolution of the Georgia Institute of Ecology, 1940–2000* (New York: Taylor & Francis, 2001), 50.
39. Craige, *Eugene Odum,* 46–47.
40. H. H. Mitchell, *Ecological Problems and Post-War Recuperation: A Preliminary Survey from the Civil Defense Viewpoint,* RM-2801 (Santa Monica, CA: The RAND Corporation, 1961). This followed one early report on the ecological effects of nuclear war: John N. Wolfe, *Long-Time Ecological Effects of Nuclear War,* TI-5561 (Washington, DC: USAEC, 1959).
41. For an example, see Lauren Donaldson, draft of "Biological Effect of Atomic Warfare," folder 20, box 17, Lauren R. Donaldson Papers, University of Washington Special Collections, Seattle, WA (hereafter LRDP).
42. G. M. Woodwell, "Design of the Brookhaven Experiment on the Effects of Ionizing Radiation on a Terrestrial Ecosystem," *Radiation Botany* 3 (1963): 125–133; George M. Woodwell and J. K. Oosting, "Effects of Chronic Gamma Irradiation on the Development of Old Field Plant Communities," *Radiation Biology* 5 (1965): 205–222. For another set of early plant irradiation studies, see J. Frank McCormick and R. B. Platt, "Effects of Ionizing Radiation on a Natural Plant Community," *Radiation Botany* 2 (1962): 161–188.

43. George M. Woodwell and A. L. Rebuck, "Effects of Chronic Gamma Radiation on the Structure and Diversity of an Oak-Pine Forest," *Ecological Monographs* 37 (1967): 53–69.
44. G. M. Woodwell, *Effects of Ionizing Radiation on Ecosystems* (Upton, NY: Brookhaven National Laboratory, January 1963), 26.
45. Woodwell and Rebuck, "Effects of Chronic Gamma Radiation on the Structure and Diversity of an Oak-Pine Forest."
46. Woodwell, *Effects of Ionizing Radiation on Ecosystems,* 26.
47. George M. Woodwell, "Effects of Ionizing Radiation on Terrestrial Ecosystems," *Science* 138 (1962): 572–577.
48. See Lester Machta, Robert List, and Lester Hubert "World-Wide Travel of Atomic Debris," *Science* 124 (1956): 474–477; Divine, *Blowing on the Wind,* 129; Jessee, "Radiation Ecologies."
49. Howard T. Odum and R. F. Pigeon, eds., *A Tropical Rain Forest: A Study of Irradiation and Ecology at El Verde, Puerto Rico* (Springfield, VA: National Technical Information Service, 1970).
50. Megan Raby, "'Slash-and-Burn Ecology': Field Science as Land Use," *History of Science* 57 (2019): 441–468. See also Megan Raby, *American Tropics: The Caribbean Roots of Biodiversity Science* (Chapel Hill: University of North Carolina Press, 2017), chapter 4; Ariel E. Lugo, "H. T. Odum and the Luquillo Experimental Forest," *Ecological Monitoring* 178 (2004): 65–74.
51. H.T. Odum, Chapter I-10, in Odum and Pigeon, *A Tropical Rain Forest,* I-191—I-281.
52. Anne Powell, "ORNL Ecologists Study Radiation Effects on Wild, Native Mammals," *Oak Ridge National Laboratory News,* August 15, 1969, reprinted in Auerbach, *A History of the Environmental Sciences Division;* Eugene Odum, *Fundamentals of Ecology,* 3rd ed. (Philadelphia: W.B. Saunders Company, 1971), 457–459.
53. Proceedings are in George M. Woodwell, ed., *Ecological Effects of Nuclear War* (Upton, NY: Brookhaven National Laboratory, 1965).
54. Robert Platt, "Ionizing Radiation and Homeostasis of Ecosystems," in Woodwell, *Ecological Effects of Nuclear War,* 39–60. See also J. F. McCormick and R. B. Platt, "Effects of Ionizing Radiation on a Natural Plant Community," *Radiation Biology* 2 (1962): 161–188.
55. Woodwell, *Ecological Effects of Nuclear War,* iii.
56. Edward O. Wilson and Daniel Simberloff, "Experimental Zoogeography of Islands: Defaunation and Monitoring Techniques," *Ecology* 50 (1969): 267–278; Daniel Simberloff and Edward O. Wilson, "Experimental Zoogeography of Islands: A Two-Year Record of Colonization," *Ecology* 51 (1970): 934–937.
57. Wilson and Simberloff, "Experimental Zoogeography of Islands."
58. Jim Hornbeck, "Events Leading to Establishment of the Hubbard Brook Experimental Forest," May 2001, https://hubbardbrook.org/sites/default/files/includefiles/misc/HBEF_history_Hornbeck.pdf; F. Herbert Bormann and Gene E. Likens, "Nutrient Cycling," *Science* 155 (1967): 424–429; F. Herbert Bormann et al., "Nutrient Loss Accelerated by Clear-Cutting of a

Forest Ecosystem," *Science* 159 (1968): 882–884. For another example of an ecosystem destruction experiment, see D. W. Schindler et al., "Eutrophication of Lake 227, Experimental Lakes Area, Northwestern Ontario, by Addition of Phosphate and Nitrate," *Journal of the Fisheries Research Board of Canada* 28 (1971): https://doi.org/10.1139/f71-261.

59. Gene E. Likens et al., "Effects of Forest Cutting and Herbicide Treatment on Nutrient Budgets in the Hubbard Brook Watershed-Ecosystem," *Ecological Monographs* 40 (1970): 23–47.
60. Frederic E. Clements, "Experimental Ecology in the Public Service," *Ecology* 16 (1935): 342–363.
61. George Woodwell and Arnold Sparrow, "Effects of Ionizing Radiation on Ecological Systems," in Woodwell, *Ecological Effects of Nuclear War.*
62. Robert Ayres, "Environmental Effects of Nuclear Weapons," Prepared under Contract No. OCD-OS-62-218, Department of Defense, Office of Civil Defense, December 1, 1965, HI-518-RR.
63. Wurtz, "War and the Living Environment."
64. *The Effects of Nuclear War* (Washington, DC: Office of Technology Assessment, May 1979).
65. David Takacs, *The Idea of Biodiversity: Philosophies of Paradise* (Baltimore, MD: Johns Hopkins University Press, 1996).
66. Robert MacArthur, "Fluctuations of Animal Populations, and a Measure of Community Stability," *Ecology* 36 (1955): 533–536.
67. Charles Elton, *Ecology of Invasions by Animals and Plants* (London: Chapman & Hall, 1958). Other important texts on the diversity-stability hypothesis include MacArthur, "Fluctuations of Animal Populations, and a Measure of Community Stability"; Robert May, *Stability and Complexity in Model Ecosystems* (Princeton, NJ: Princeton University Press, 1973). For a recent review, see Kevin McCann, "The Diversity-Stability Debate," *Nature* 405 (2000): 228–233.
68. Platt, "Ionizing Radiation and Homeostasis of Ecosystems."
69. H.T. Odum, Chapter A-2, in Odum and Pigeon, *A Tropical Rain Forest,* A-10.
70. George M. Woodwell, "Radiation and the Patterns of Nature," *Science* 156 (1967): 461–470.
71. Robert E. Park, "Succession, an Ecological Concept," *American Sociological Review* 1 (1936): 171–179.
72. Howard T. Odum, "Energy, Ecology, and Economics," *Ambio* 2 (1973): 220–227.
73. Eugene Odum, "Energy, Ecosystem Development and Environmental Risk," *Journal of Risk and Insurance* 43 (1976): 1–16.
74. Odum, "Energy, Ecology, and Economics," 222.
75. G. Woodwell, "Effects of Pollution on the Structure and Physiology of Ecosystems," *Science* 168 (1970): 429–433. See also G. Woodwell, "Toxic Substances and Ecological Cycles," *Scientific American* 216 (1967): 24–31.
76. Donald Worster, "Chapter 16," in *Nature's Economy: A History of Ecological Ideas* (Cambridge, UK: Cambridge University Press, 1977).

77. Robert Jenkins and W. Brian Bedford, "The Use of Natural Areas to Establish Environmental Baselines," *Biological Conservation* 5 (1973): 168–174.
78. Lyndon B. Johnson, "Special Message to the Congress on Conservation and Restoration of Natural Beauty," February 8, 1965, *Public Papers of the Presidents of the United States: Lyndon B. Johnson, 1965,* volume I, entry 54 (Washington, DC: Government Printing Office, 1966): 155–165.
79. Chunglin Kwa, "Ecology and Science Policy: The Case of the International Biological Programme," *Social Studies of Science* 17 (1987): 413–442.
80. Dillon Ripley, as quoted in Chunglin Kwa, "Representations of Nature Mediating between Ecology and Science Policy: The Case of the IBP," *Social Studies of Science* 17 (1987): 424.
81. "Man's Survival in a Changing World," folder 28, box 14, LRDP.

6. Extinct Is Forever

1. "This is the Bottle for the Age of Ecology," Coca Cola advertisement no. 70-C-254, *Sterling Daily Gazette,* November 10, 1970.
2. Barry Commoner, "Can We Survive," 91st Cong., 1st sess., *Congressional Record* 115 (December 17, 1969): 39741.
3. On postwar environmentalism, see Robert Gottlieb, *Forcing the Spring: The Transformation of the American Environmental Movement* (Washington, DC: Island Press, 1993); Kirkpatrick Sale, *The Green Revolution: The American Environmental Movement, 1962–1992* (New York: Hill and Wang, 1993); Sheila Jasanoff, "Image and Imagination: The Formation of Global Environmental Consciousness," in *Changing the Atmosphere: Expert Knowledge and Environmental Governance,* ed. Clark A. Miller and Paul N. Edwards (Cambridge, MA: MIT Press, 1996), 309–336; Adam Rome, *The Bulldozer in the Countryside: Suburban Sprawl and the Rise of American Environmentalism* (New York: Cambridge University Press, 2001); Andrew J. Kirk, *Counterculture Green: The Whole Earth Catalog and American Environmentalism* (Lawrence: University of Kansas Press, 2007); Adam Rome, *The Genius of Earth Day: How a 1970 Teach-In Unexpectedly Made the First Green Generation* (New York: Hill and Wang, 2014).
4. "The Real Thing," *Time,* December 14, 1970.
5. E. P. Odum, "Questions and Answers CBS News, New York for Earth Day Tape, April 22, 1970," box 104, folder 27, series 1, Eugene P. Odum Papers (MS 3257), Hargrett Rare Book & Manuscript Library, University of Georgia Libraries, Athens, GA.
6. John G. Mitchell and Constance L. Stallings, eds., *Ecotactics: The Sierra Club Handbook for Environment Activists* (New York: Simon and Schuster, 1970).
7. U.S. Congress, House of Representatives, Committee on Merchant Marine and Fisheries, *Report to Accompany H.R. 37,* 93rd Cong., 1st sess., Report No. 93-412, 143, 145.
8. "Predator and Rodent Control Policy of the U.S. Fish and Wildlife Service," April 3, 1956, box 105, Folder: Policy + Philosophy W.S., Entry P230, Records

of the U.S. Fish and Wildlife Service (RG 22), National Archives Record Administration II, College Park, MD (hereafter NARA II).

9. Draft 7/6/64, "Branch of Wildlife Management," box 2, folder: Branch Programs & Policy, Series P246, RG 22, NARA II.
10. Fish and Wildlife Service, "Reserving Wetlands for Wildlife," reprinted from the Proceedings of the MAR Conference, November 12/16, 1962, box 31, entry no. 234, RG 22, NARA II.
11. Robert Wilson, "Directing the Flow: Migratory Waterfowl, Scale, and Mobility in Western North America," *Environmental History* 7 (2002): 247–266.
12. Folder: Wildlife Refuges, box 43, series P253: Office Files of Dr. Frederick C Lincoln 1917–1960, RG 22, NARA II.
13. A 1940 reorganization plan in the Department of the Interior consolidated the Bureau of Fisheries and the Bureau of Biological Survey into one agency to be known as the Fish and Wildlife Service. The Bureau of Sport Fisheries and Wildlife was created as a part of the U.S. Fish and Wildlife Service by the Fish and Wildlife Act of 1956. That act was amended on July 1, 1974, by Public Law 93-271, abolishing the position of Commissioner of Fish and Wildlife and designating the Bureau as the U.S. Fish and Wildlife Service. This chapter spans the periods of the Bureau of Sport Fisheries and Wildlife and the U.S. Fish and Wildlife Service. For simplicity's sake, in this chapter I refer to both as the Fish and Wildlife Service (FWS).
14. On the elk controversy, see James Pritchard, *Preserving Yellowstone's Natural Conditions: Science and the Perception of Nature* (Lincoln: University of Nebraska Press, 1999); Wendy Zirngibl, "Elk in the Greater Yellowstone Ecosystem: Conflicts over Management and Conservation Prior to Natural Regulation" (MA thesis, Montana State University, 2006).
15. A. Starker Leopold et al., "Wildlife Management in the National Parks," in *Transactions of the Twenty-eighth North American Wildlife and Natural Resources Conference,* ed. James B. Trefethen (Washington, DC: Wildlife Management Institute, 1963).
16. Advisory Board on Wildlife Management to Stewart Udall, Secretary of the Interior, March 9, 1964, Folder: Starker Leopold Report, box 1, series P230, RG 22, NARA II.
17. Eric A. Peacock, "Sodium Monofluoroacetate (Compound 1080), c. 1964, unpublished manuscript, Folder: Compound 1080, box 103, series P230, RG 22, NARA II.
18. Box 32, series: P177—Branch of Wild Research, Research Reports 1912 1951, RG 22, NARA II. See also Folder: Wildlife Research, box 43, series P253, RG 22, NARA II.
19. Thomas Dunlap, *Saving America's Wildlife: Ecology and the American Mind, 1850–1990* (Princeton, NJ: Princeton University Press, 1991), chapter 8. Dunlap argues that science "brought us from poisoning 'varmints' to reintroducing wolves" in the twentieth-century United States (p. x). This chapter seeks to understand the role of scientific ecologists in shifting wildlife management while not upholding ecology as a source of prescient or static knowledge.

20. Dunlap, *Saving America's Wildlife,* 117.
21. Charles Cadieux to Howard Matley, January 16, 1961, as quoted in Dunlap, *Saving America's Wildlife,* 118.
22. Advisory Board on Wildlife Management to Stewart Udall, Secretary of the Interior, March 9, 1964, box 1, Folder: Starker Leopold Report, series P230, RG 22, NARA II.
23. Leleand C. Bacus to Regional Supervisor, PARC, September 11, 1964, box 1, Folder: Accidents—Dog Poisoning, series P230—General Correspondence Relating to Wildlife Services 1890–1972, RG 22, NARA II.
24. The Federal Animal Control Program of the US Department of the Interior, Washington, DC, February 1964, series P230, RG 22, NARA II.
25. Peacock, "Sodium Monoflouroacetate," memo 121, October 24, 1945, as quoted in Dunlap, *Saving America's Wildlife,* 115.
26. Craig L. Thomas to John S. Cottschalk, December 30, 1966, box 104, Folder: Policy—Corres. Re Control Policy, series P230, RG 22, NARA II.
27. Charles Callison to Jack Berryman, February 3, 1967, box 104, Folder: Policy—Corres. Re Control Policy, series P230, RG 22, NARA II.
28. An Animal Damage Control Policy, Review Draft, November 9, 1966, box 104, Folder: Policy—Corres. Re Control Policy, series P230, RG 22, NARA II.
29. "Birth Control for Predators," *Rapid City Journal,* February 21, 1965, box 4, Folder: Animals, Birth Control File, series P230, RG 22, NARA II.
30. Fish and Wildlife Service, "Birth Control Is Latest Weapon against Coyote," Press release, February 22, 1965, box 4, Folder: Animals, Birth Control File, Series P230, RG 22, NARA II.
31. Section of Animal Depredations Control Studies, South Lincoln, Massachusetts, Progress Report January 1–December 31, 1965, series P230, RG 22, NARA II.
32. "A Statement of Philosophy and Policy for Animal Damage Control," box 104, Folder: Policy Brochure ADC-Policy, series P230, RG 22, NARA II.
33. Guy Connolly, "Development and Use of Compound 1080 in Coyote Control, 1944–1972," *Proceedings of the 21st Vertebrate Pest Conference* (Davis: University of California Press, 2004), 221–239.
34. Dunlap, *Saving America's Wildlife,* 135.
35. Faith McNulty, "The Prairie Dog and the Black-Footed Ferret," *New Yorker,* June 6, 1970.
36. Advisory Committee on Predator Control, *Predator Control—1971* (Ann Arbor, MI: Institute for Environmental Quality, 1971); Executive Order 11643 of February 8, 1972, Environmental Safeguards on Activities for Animal Damage Control on Federal Lands, Federal Register 37: 2875.
37. Mark V. Barrow Jr., *Nature's Ghosts: Confronting Extinction from the Age of Jefferson* (Chicago: University of Chicago Press, 2009).
38. Mentioned in William Palmer, "Endangered Species Protection: A History of Congressional Action," *Boston College Environmental Affairs Law Review* 255 (1975): 255–293.
39. D. A. Janzen, "Administrative Manual, Subject: Committee on Rare and Endangered Wildlife Species," January 30, 1964, as cited in Johnny Winston,

"Science, Practice, and Policy: The Committee on Rare and Endangered Wildlife Species and the Development of U.S. Federal Endangered Species Policy, 1956–1973" (PhD diss., Arizona State University, 2011).

40. As quoted in Thomas R. Dunlap, "Organization and Wildlife Preservation: The Case of the Whooping Crane in North America," *Social Studies of Science* 21 (1991): 197–221, 200.
41. John B. French Jr., Sarah J. Converse, and Jane E. Austin, "Whooping Cranes Past and Present," in *Whooping Cranes: Biology and Conservation* (London: Academic Press, 2019): 3–16, 25–48.
42. Quoted in Faith McNulty, *The Whooping Crane: The Bird That Defies Extinction* (New York: E. P. Dutton, 1966), 136.
43. Whooping Crane Conference, Minutes of Meeting—October 29, 1956, Secretary's Conference Room, Department of the Interior, Washington, DC, box 18, Folder: Conferences—Whooping crane, entry P246, RG 22, NARA II.
44. Box 18, Folder: Correspondence (Informational) Whooping Crane, entry P246, RG 22, NARA II.
45. Richard Griffith to Assistant Director for Wildlife, Subject: Service policy on proposed whooping crane propagation, August 22, 1956, box 18, Folder: Correspondence (Informational) Whooping Crane, entry P246, RG 22, NARA II.
46. Paul H. Baldwin, "The Hawaiian Goose—Its Distribution and Reduction in Numbers," *The Condor* 47 (1945): 27–37; J. Donald Smith, "The Hawaiian Goose (Nene) Restoration Program," *Journal of Wildlife Management* 16 (1952): 1–9. See also box 18, series P254, RG 22, NARA II.
47. William H. Elder, "Ne-ne in Hawaii: Preliminary Report of the Ne-ne in Hawaii," *Wildfowl Trust Ninth Annual Report, 1956–1957* (Gloucestershire, UK: Wildfowl & Wetlands Trust, 1957), 112–117; S. Dillon Ripley, "Saving the Nene, World's Rarest Goose," *National Geographic,* November 1965.
48. Barrow, *Nature's Ghosts,* 301.
49. Barrow, *Nature's Ghosts,* chapter 10.
50. Richard Griffith to Assistant Director for Wildlife, Subject: Service policy on proposed whooping crane propagation, August 22, 1956, box 18, Folder: Correspondence (Informational) Whooping Crane, entry P246, RG 22, NARA II.
51. Ray C. Erickson, "A Federal Research Program for Endangered Wildlife," *Transactions of the North American Wildlife and Natural Resources Conference* 33 (1968): 418–433.
52. "Department of the Interior and Related Agencies Appropriation Bill, 1967—Conference Report," 89th Cong., 2nd sess., *Congressional Record* 112, pt. 9 (June 6, 1966): 11054.
53. Barrow, *Nature's Ghosts,* chapter 10.
54. A number of species on the draft list, including bighorn sheep, Utah prairie dogs, and Pacific right whales, would not make it onto the first official endangered species list, approved by Secretary Udall in 1967. Secretary of the Interior, "Native Fish and Wildlife: Endangered Species," *Federal Register* 32, no. 48 (March 11, 1967): 4001.

55. "Wildlife: The Vanishing Americans," *Washington Post,* October 3, 1965.
56. Letter from Stewart Udall to John McCormack, June 5, 1965, reprinted in U.S. House Committee on Merchant Marine and Fisheries, *Protection of Endangered Species of Fish and Wildlife,* Report No. 1168, 89th Cong., 1st sess. (Washington, DC: Government Printing Office, 1965): 12–14.
57. *An Act to provide for the conservation, protection, and propagation of native species* [. . .], Pubic Law 89-669, *U.S. Statues at Large* 80 (1966): 926. On the history of the 1966 law, see Shannon Petersen, "Congress and Charismatic Megafauna: A Legislative History of the Endangered Species Act," *Environmental Law* 29 (1999): 463–491; Shannon Petersen, *Acting for Endangered Species: The Statutory Ark;* Steven Lewis Yaffee, *Prohibitive Policy: Implementing the Federal Endangered Species Act* (Cambridge, MA: MIT Press, 1982).
58. Statement of Stanley Cain, "Some Thoughts on the Ecological Basis of Administrative Goals and Policies in a Natural Resource Program," January 12, 1966, box 3, entry no. 246, RG 22, NARA II.
59. *An Act to prevent the importation of endangered species of fish or wildlife into the United States* [. . .], Pubic Law 91-135, *U.S. Statues at Large* 83 (1969): 275.
60. Alston Chase, *In a Dark Wood: The Fight over Forests and the Rising Tyranny of Ecology* (New York: Houghton Mifflin Co., 1995), 90.
61. *Endangered and Threatened Species Conservation Act of 1973,* on September 18, 1973, 93rd Cong., 1st sess., *Congressional Record* 119, pt. 23: 30157. The final version defined "conservation" to encompass restoration and captive breeding, as well as including propagation, transplantation, and even regulated taking. See "Conference Report on S. 1983," on September 18, 1973, 93rd Cong., 1st sess., *Congressional Record* 119, pt. 33: 42627.
62. Richard Nixon, Special Message to the Congress Outlining the 1972 Environmental Program Online by Gerhard Peters and John T. Woolley, The American Presidency Project https://www.presidency.ucsb.edu/node/255047; Executive Order 11643 of February 8, 1972, Environmental Safeguards on Activities for Animal Damage Control on Federal Lands, Federal Register 37: 2875.
63. HR 37 (introduced 1/03/1973), HR 4758 (introduced 2/27/1973), S 1983 (introduced 6/12/1973). Interestingly, these versions did not mention ecosystems. Language about ecosystem protection and restoration had reentered drafts by September 18, 1973.
64. U.S. Congress, House of Representatives, *Hearings before the Subcommittee on Fisheries and Wildlife Conservation and the Environment of the Committee on Merchant Marine and Fisheries,* 93rd Cong., 1st sess., March 15, 26, 27, 1973, Serial No. 93-5 (Washington, DC: U.S. Government Printing Office, 1973), 192.
65. Shannon Petersen, "Congress and Charismatic Megafauna: A Legislative History of the Endangered Species Act," *Environmental Law* 29 (1999): 463–491.
66. As Peter Alagona details in chapter 4 of *After the Grizzly: Endangered Species and the Politics of Place in California* (Berkeley: University of California

Press, 2013), in the decades after passage of the Endangered Species Act, habitat protection provisions expanded greatly through legislative amendments, administrative rules, and court decisions. In *Federal Ecosystem Management: Its Rise, Fall, and Afterlife* (Lawrence: University Press of Kansas, 2015), James Skillen argues that the ESA's "strict regulatory provisions and political durability" led to the emergence of the field of ecosystem management in the late 1980s and 1990s.

67. Box 3, series P320: Endangered and Threatened Species Files, RG 22, NARA II.
68. Office of Endangered Species and International Activities, *Threatened Wildlife of the United States,* 1973 edition, Resource Publication 114 (Washington, DC: Bureau of Sport Fisheries and Wildlife, March 1973).
69. *Science News* 108 (August 9, 1975): 95, as quoted in Yaffee, *Prohibitive Policy,* 71.
70. U.S. Congress, Senate, *Amending the Endangered Species Act of 1973: Hearings before the Subcommittee on Resource Protection of the Committee on Environment and Public Works,* 95th Cong., 2nd sess., 1978, 375.
71. Yaffee, *Prohibitive Policy,* 97–103.
72. The statutory basis for mitigation is truly byzantine. For an early overview, see Leo Krulitz, "Federal Legal Background for Mitigation," in *The Mitigation Symposium: A National Workshop on Mitigating Losses of Fish and Wildlife Habitats,* July 16–20, 1979, General Technical Report RM-65 (Fort Collins: Colorado State University), 19–26.
73. Director to Regional Director—Region 4, January 6, 1978, box 1, Folder: FWS 4-78-C-001d, entry P322, RG 22, NARA II.
74. U.S. Government Accountability Office, *Improved Federal Efforts Needed to Equally Consider Wildlife Conservation with Other Features of Water Resource Developments,* Report B-118370 (Washington, DC: USGAO, March 8, 1974).
75. Allyn J. Sapa, "Restoration of Wildlife Habitat to Offset Project Losses, Garrison Diversion Unit, North Dakota," in *Mitigation Symposium,* 318.
76. W. N. Lindall et al., "Estuarine Habitat Mitigation Planning the Southeast," in *Mitigation Symposium,* 129.
77. Laurence R. Jahn, "Summary of the Symposium," in *Mitigation Symposium,* 6.
78. Nebraska alleged that the project had an inadequate environmental impact statement under NEPA. Nebraska also filed a second suit on the grounds that the Army Corps had issued its 404 permit (described in chapter 8) inappropriately. Other lawsuits filed by environmental organizations alleged inadequate environmental impact statement as well as violation of the Endangered Species Act.
79. The Grayrocks Dam case, mentioned in Yaffee, *Prohibitive Policy,* 101, is treated in more depth in Julia Wondolleck, "*Bargaining for the Environment: Compensation and Negotiation in Energy Facility Siting*" (Master in City Planning thesis, Massachusetts Institute of Technology, 1979).
80. Seth King, "A Study Finds Birds and Dam Able to Coexist," *New York Times,* December 9, 1978; U.S. Department of the Interior, "Proceedings of the Endangered Species Committee," January 23, 1979, https://lawdigitalcommons

.bc.edu/darter_materials/2; Department of the Interior News Release, "Endangered Species Committee Completes Report on Grayrocks and Tellico," February 8, 1979.

81. John Aronson and Scott Ellis, "Monitoring, Maintenance, Rehabilitation and Enhancement of Critical Whooping Crane Habitat, Platte River, Nebraska," in *Mitigation Symposium;* G. R. Lingle, "Control of Woody Vegetation in Sandhill Crane Habitat on Riverine Islands," *Restoration & Management Notes* 1 (1981): 28–29; Kenneth J. Strom, "Protecting Critical Whooping Crane Habitat on the Platte River, Nebraska," *Natural Areas Journal* 5 (1985): 3–13.
82. "The Lousewort and the Law," *Washington Post,* April 4, 1977.
83. The committee declined to exempt the Tellico Dam project, however, and Congress ultimately stepped in to exempt the project from the Endangered Species Act on its own direct authority.
84. *Endangered Species Act Amendments,* Pubic Law 95-632, *U.S. Statues at Large* 92 (1978): 3766.
85. Statement of Lynn Greenwalt, *Endangered Species: Hearings before the Subcommittee on Fisheries and Wildlife Conservation and the Environment of the Committee on Merchant Marine and Fisheries, House of Representatives,* 96th Cong. 1st sess., 1979, 231.
86. Preface, *Recovery Plan for the Eastern Timberwolf* (Washington, DC: U.S. Fish and Wildlife Service, May 1978).
87. Warren B. King et al., "Report of the American Ornithologists' Union Committee on Conservation, 1976–77," *The Auk* 94 (1977): 3DD–19DD; Clayton White, "Strategies for the Preservation of Rare Animals," *Great Basin Naturalist Memoirs* 3 (1979): 101–111.
88. The FWS and National Marine Fisheries Service provided that species can be delisted as recovered in 1980, and formally defined the term "recovery" in 1986: "Rules for Listing Endangered and Threatened Species, Designating Critical Habitat, and Maintaining the Lists," *Federal Register* 45, no. 40 (February 27,1980): 13010; "Interagency Cooperation–Endangered Species Act of 1973, as Amended," *Federal Register* 51 (June 3, 1986): 19957. See Dale D. Goble, "The Endangered Species Act: What We Talk About When We Talk About Recovery," *Natural Resources Journal* 49 (2009): 1–44.
89. "Factors for Listing, Delisting, or Reclassifying Species," *Code of Federal Regulations* 50 (1984) Section 424.11(d)(1)-(3).
90. A useful database of listed and delisted species can be found at https://ecos.fws.gov/ecp/; "Proposal to Remove the Brown Pelican in Southeastern United States," *Federal Register* 48, no. 219 (November 10, 1983): 51736; "Removal of the Brown Pelican in the Southeastern United States from the List of Endangered and Threatened Wildlife," *Federal Register* 50, no. 23 (February 4, 1985): 4938; "Reclassification of the American Alligator to Threatened Due to Similarity of Appearance throughout the Remainder of its Range," *Federal Register* 52, no. 107 (June 4, 1987): 21059.
91. Michael D. Lemonick, "Coming Back from the Brink: Alligators and Leopards Are No Longer Seen as Endangered," *Time,* July 20, 1987.

92. "Final Rule to Remove the Aleutian Canada Goose from the Federal List of Endangered and Threatened Wildlife," *Federal Register* 66, no. 54 (March 20, 2001): 15643; "Determination of *Potentilla robbinsiana* To Be an Endangered Species, with Critical Habitat," *Federal Register* 45 (September 17, 1980): 61944; "Removal of *Potentilla robbinsiana* (Robbins's cinquefoil) from the Federal List of Endangered and Threatened Plants," *Federal Register* 67 (August 27, 2002): 54,968.
93. David R. Zemmerman, "Death Comes to the Peregrine Falcon," *New York Times,* August 9, 1970; Harold Faber, "Peregrine Falcons Gain, Thanks to Lab Breeding," *New York Times,* September 2, 1973; Nelson Bryant, "Wood, Field and Stream, Cornell Plans a Project to Release Peregrine Falcons," *New York Times,* February 16, 1975.
94. "Falcons Taken for Jet Ride in Survival Bid," *New York Times,* Saturday, May 14, 1977.
95. Tom Cade to Harold Olson, June 25, 1979, as cited in Nick Fox, *The Use of Exotic and Hybrid Raptors in Falconry* (Carmathen, Wales: International Wildlife Consultants Ltd., 1999).
96. Nicholas Wade, "Bird Lovers and Bureaucrats at Loggerheads over Peregrine Falcon," *Science* 199 (1978): 1053–1055; Department of the Interior, "Endangered Peregrine's Flight Honors Rachel Carson, 17 Years After 'Silent Spring,'" News Release, July 11, 1979; Tom J. Cade and William Burnham, *Return of the Peregrine* (Boise, ID: The Peregrine Fund, 2003).
97. CEQ Draft, "The President's 1977 Environmental Message to the Congress of the United States," March 31, 1977, container 20, folder 5/12/77, Records of the Office of the Staff Secretary (Accession no. 80-1), Jimmy Carter Presidential Library, Atlanta, GA (hereafter JCPL).
98. On the Carter administration's approach to environmental regulation, see Paul Sabin, "'Everything Has a Price': Jimmy Carter and the Struggle for Balance in Federal Regulatory Policy," *Journal of Policy History* 28 (2016): 1–47.
99. CEQ Draft, "The President's 1977 Environmental Message"; "Appendix A: The President's Message on the Environment and Executive Orders," in *Environmental Quality—1977* (Washington, DC: Executive Office of the President, 1977); "Highlights of Activities, Council on Environmental Quality, 1977–1979," container 143, folder 12/21/79, JCPL.
100. Department of the Interior, "Peregrine Falcons for the Nation's Capital," News Release, June 20, 1979.
101. Department of the Interior, Endangered Species Technical Bulletin, vol. 5, no. 8, August 1980; Darryl McGrath, *Flight Paths: A Field Journal of Hope, Heartbreak, and Miracles with New York's Bird People* (Albany: State University of New York Press, 2016).
102. Harrison Tordoff and Patrick Redig, "Role of Genetic Background in the Success of Reintroduced Peregrine Falcons," *Conservation Biology* 15 (2001): 528–532.
103. Holly Doremus, "Restoring Endangered Species: The Importance of Being Wild," *Harvard Environmental Law Review* 23 (1999): 1–92. On section 10(j) reintroductions through the end of 1995, see Mimi S. Wolok, "Experimenting with Experimental Populations," in *The Endangered Species Act: Law, Policy,*

and Perspectives, ed. William R. Irvin and Donald C. Baur (Chicago: American Bar Association, 2001). See "Natural Heritage Data Center Network and The Nature Conservancy," in *Perspectives on Species Imperilment* (Arlington: The Nature Conservancy, 1993).

104. U.S. Fish and Wildlife Service, "Endangered and Threatened Wildlife and Plants," *Federal Register* 49, no. 167 (August 27, 1984): 33885–33894; Doremus, "Restoring Endangered Species"; Wolok, "Experimenting with Experimental Populations." As of July 2019, fifty-four nonessential experimental populations had been listed under the Endangered Species Act, including populations of Chinook salmon, whooping crane, gray wolf, red wolf, and black-footed ferret.
105. Doremus, "Restoring Endangered Species."
106. Timothy Tear et al., "Status and Prospects for Success of the Endangered Species Act: A Look at Recovery Plans," *Science* 262 (1993): 976–977.
107. Ian McTaggart Cowan, "Conservation and Man's Environment," *Nature* 208 (1965): 1145–1151.
108. Harry Goodwin and Eley Denson, Office of Endangered Species, "Status of Endangered Species Program," Department of the Interior and Related Agencies Appropriations for 1972, *Hearings before the U.S. House of Representatives Committee on Appropriations,* 92nd Cong. (Washington, DC: Government Printing Office, 1971).
109. Remarks of Dr. Stanley A. Cain at the Wildlife Interpretation and Recreation Planning Workshop of the Bureau of Sports Fisheries and Wildlife, October 2, 1967, box 3, entry P246, RG 22, NARA II.
110. Ray C. Erickson, "Propagation Studies of Endangered Wildlife at the Patuxent Center," *International Zoo Yearbook* 20 (1980): 40–47.
111. Adolph Murie, *Fauna of the National Parks of the United States: Ecology of the Coyote in the Yellowstone National Park* (Washington, DC: National Park Service, 1940); Department of the Interior, *Northern Rocky Mountain Wolf Recovery Plan* (Denver, CO: U.S. Fish and Wildlife Service, 1987); Paul Schullery, *Searching for Yellowstone: Ecology and Wonder in the Last Wilderness* (Boston: Houghton Mifflin Co., 1997), 125.
112. Richard Sellars, *Preserving Nature in the National Parks: A History* (New Haven, CT: Yale University Press, 1997), 258.

7. The Mood of Wild America

1. Elisabeth Bullimer, "Bush Promotes Wetlands Plan to Counter Kerry's Attack," *New York Times,* April 24, 2004.
2. John Rather, "Green Invaders Spread Their Tentacles," *New York Times,* June 15, 2003; Christopher West Davis, "An Invasion of Hungrier, Bigger Worms," *New York Times,* July 20, 2003; David M. Lodge, "Biological Hazards Ahead," *New York Times,* June 19, 2003.
3. Peter Coates, *American Perceptions of Immigrant and Invasive Species: Strangers on the Land* (Berkeley: University of California Press, 2006); Philip J. Pauly, chapter 5, in *Fruits and Plains: The Horticultural Transformation of America* (Cambridge, MA: Harvard University Press, 2007).

4. Mark Davis, *Invasion Biology* (New York: Oxford University Press, 2009).
5. On the xenophobic rhetoric of invasion biology, see Jonah Peretti, "Nativism and Nature: Rethinking Biological Invasion," *Environmental Values* 7 (1998): 183–192; Banu Subramaniam, "The Aliens Have Landed! Reflections on the Rhetoric of Biological Invasions," *Meridians* 2 (2001): 26–40. Responses include David Simberloff, "Confronting Introduced Species: A Form of Xenophobia?," *Biological Invasions* 5 (2003): 179–192. The most thorough response to Peretti's linking of concern about invasive species and xenophobia can be found in Coates, *American Perceptions of Immigrant and Invasive Species.*
6. On the public trust doctrine, see Eric T. Freyfogle and Dale D. Goble, *Wildlife Law: A Primer* (Washington, DC: Island Press, 2009).
7. Sherry Morgan and Jim Wilson, "Plant Recovery Activities at the State Level," *Natural Areas Journal* 2 (1982): 7–10.
8. William R. Jordan III and George M. Lubick, *Making Nature Whole: A History of Ecological Restoration* (Washington, DC: Island Press, 2011).
9. https://www.ser.org/page/IndividualMembership, accessed July 2019.
10. Scholars have offered a number of reasons for the cleaving of the Committee on the Preservation of Natural Conditions from the ESA and the eventual formation of The Nature Conservancy. Sara Tjossem emphasizes the role of one individual, Robert Griggs, in eliminating the preservation committee. Abby Kinchy argues that the ESA ultimately rejected the preservation committee in an effort to negotiate a rigidifying postwar boundary between science and politics. Zoe Nyssa cites changes to U.S. tax law as an important motivator for the split. See Sara Tjossem, "Preservation of Nature and Academic Respectability: Tensions in the Ecological Society of America, 1915–1955" (PhD diss., Cornell University, 1994); Abby J. Kinchy, "On the Borders of Post-War Ecology: Struggles over the Ecological Society of America's Preservation Committee, 1917–1946," *Science as Culture* 15 (2012): 23–44; Zoe Nyssa, "Why Scientists Succeed yet Their Organizations Splinter: Historical and Social Network Analyses of Policy Advocacy in Conservation," *Environmental Science and Policy* 98 (2019): 88–94.
11. Record in folder 8, box 10, The Nature Conservancy Records, 1931–2016, Denver Public Library, Denver, CO (hereafter TNCR).
12. Records of the Ecologists Union can be found in box 10, folders 8–15, and box 13, folders 1–27, TNCR. See also Victor Shelford, "Two Open Letters," *Ecological Society of America Bulletin* 25, no. 2 (1944): 12–15; "Referendum," *ESA Bulletin* 26, no. 3/4 (1945); *Ecological Society of America Bulletin* 27, no. 4 (1946): 58; Patrick F. Noonan, *The Gorge: A History of the Nature Conservancy,* unpublished manuscript, in folder 23, box 10, TNCR.
13. Noel Grove, *The Nature Conservancy: Preserving Eden* (New York: Abrams, 1992).
14. "The Ecologists Union Newsletter," May 23, 1950, folder 226, box 13, series 1, G. Evelyn Hutchinson Papers, Manuscripts and Archives, Yale Sterling Memorial Library, New Haven, CT (hereafter GEHP); "Ecologists Union Circular Number 6: Minutes of the New York Meeting," April 1950, folder

225, box 13, series 1, GEHP; "Living Museums of Primeval America: A Need and an Opportunity," May 1950, folder 226, box 13, series 1, GEHP. There are additional Ecologists Union materials in folder 226, box 13, series 1, GEHP.

15. "A System of Nature Reserves," Spring 1952, folder 16, box 10, TNCR; "The Need for Natural Areas," c. 1953, folder 16, box 10, TNCR; Daniel Smiley, "Interpretation of Nature Conservancy Objectives," February 1958, folder 16, box 10, TNCR.
16. "The Need for Natural Areas," TNCR. "Fragments of Wild America," *Nature Conservation News,* February 18, 1957, as quoted in Walter P. Cottam, "Our Social Responsibilities," box 1, folder 21, TNCR.
17. Report to Board, July 18, 1955, folder 6, box 2, TNCR.
18. "Nature Conservancy President's Annual Report," August 28, 1958, box 1, folder 7, TNCR.
19. On the organizational history of TNC, see Ralph W. Dexter, "History of the Ecologists' Union," *Bulletin of the Ecological Society of America* 59 (1978): 146–147; Noel Grove, *The Nature Conservancy;* Bill Birchard, *Nature's Keepers: The Remarkable Story of How The Nature Conservancy Became the Largest Environmental Organization in the World* (San Francisco: Jossey-Bass, 2005).
20. Minutes of Nature Conservancy Board of Governors Meeting, December 5, 1954, folder 6, box 2, TNCR.
21. 1953 pamphlet, folder 19, box 13, TNCR.
22. Birchard, *Nature's Keepers.*
23. Patrick F. Noonan, *The Gorge: A History of the Nature Conservancy,* folder 23, box 10, TNCR; executive committee meeting, Washington, DC, October 25, 1969, folder 1, box 8, TNCR; Richard H. Goodwin, "The Scientific Role of The Nature Conservancy," 1969, folder 45, box 14, TNCR; "Report of the Scientific Focus Committee of The Nature Conservancy," 1969, folder 45, box 14, TNCR. On Jenkin's hiring see box 1, folder 34, series 1, TNCR.
24. Robert Jenkins, "Ecology of Three Species of Saltators in Costa Rica with Special Reference to Their Frugivorous Diet" (PhD diss., Harvard University, 1970), in folder 9, box 113, TNCR.
25. Robert Jenkins, undated notes, c. 1972, folder 20, box 114, TNCR.
26. Robert Jenkins, "Research. Development, and Application—A Proposal," c. 1971, folder 9, box 6, TNCR.
27. Truman Temple, "The Marsh Maker of St. Michaels," *Amicus Journal* (1983), folder 19, box 110, series 3, TNCR. John and Mildred Teal, *Life and Death of the Salt Marsh* (Boston: Little, Brown, 1969).
28. Robert Jenkins, "Ecosystem Restoration," Third Midwest Prairie Conference Proceedings, Kansas State University, Manhattan, September 22–23, 1972, clipping in folder 24, box 11, George R. Cooley Papers, SC18858, New York State Library, Albany, NY; Edgar W. Garbisch, "Hambleton Island Restoration: Environmental Concern's First Wetland Creation Project," *Ecological Engineering* 24 (2005): 289–307.
29. Robert H. Boyle, "The Man Who Makes Marshes," *Sports Illustrated,* October 20, 1975.

30. Jenkins, "Research. Development, and Application."
31. Jenkins, "Ecosystem Restoration," 25.
32. "Home," Environmental Concern, http://wetland.org.
33. William R. Jordan III, "Hint of Green," *Restoration & Management Notes* 1 (1983): 4–10.
34. Daniel Smiley, "Interpretation of Nature Conservancy Objectives," February 1958, folder 16, box 10, TNCR.
35. *The Nature Conservancy Preserve Management Manual,* August 1972, folder 32, box 13, TNCR.
36. Daniel Smiley, "Interpretation of Nature Conservancy Objectives," February 1958, folder 16, box 10, TNCR.
37. Jenkins, "Research. Development, and Application."
38. Folder 26, box 10, TNCR; Ray Culter to Dorothy Behlen, "Book Revision: Stewardship," 2/17/81, folder 30, box 124, TNCR.
39. Jenkins, "Ecosystem Restoration," 24.
40. "Statement of Request," 1971, folder 9, box 6, TNCR.
41. Jenkins, "Ecosystem Restoration," 24.
42. TNC, "Skills Training: Ecological Restoration on Preserves," August 10, 1990, folder 23, box 126, TNCR.
43. "A System of Nature Reserves," Spring 1952, folder 16, box 10, TNCR; "The Need for Natural Areas," c. 1953, folder 16, box 10, TNCR; Daniel Smiley, "Interpretation of Nature Conservancy Objectives," February 1958, folder 16, box 10, TNCR.
44. A. Starker Leopold et al., "Wildlife Management in the National Parks," in *Transactions of the Twenty-Eighth North American Wildlife and Natural Resources Conference,* ed. James B. Trefethen (Washington, DC: Wildlife Management Institute, 1963).
45. A. Starker Leopold et al., "Wildlife Management in the National Parks."
46. Richard Sellars, *Preserving Nature in the National Parks: A History* (New Haven, CT: Yale University Press, 1997), chapter 6; Victor Cahalane, "The Evolution of Predator Control Policy in National Parks," *Journal of Wildlife Management* 4 (1939): 229–237.
47. Howard Zahniser, "Guardians Not Gardeners," *Living Wilderness* 83 (Spring 1963): 2.
48. As quoted in Jordan Smith, *Engineering Eden: The True Story of a Violent Death, A Trial, and the Fight over Controlling Nature* (New York: Crown, 2016), 129.
49. Stephen Pyne, "Vignettes of Primitive America: The Leopold Report and Fire Policy," *Forest History Today,* Spring 2017, 12–18.
50. Smith, *Engineering Eden,* 113, 124.
51. Stephen Pyne, "Vignettes of Primitive America: The Leopold Report and Fire Policy," 17.
52. Richard Sellars, *Preserving Nature in the National Parks: A History* (New Haven, CT: Yale University Press, 1997), 258.
53. David M. Graber, "Rationalizing Management of Natural Areas in National Parks," *George Wright Forum* 3 (1983): 48–56; David M. Graber, "Man-

aging for Uncertainty: National Parks as Ecological Reserves," *George Wright Forum* 4 (1985): 4–7.

54. Reed F. Noss, "On Characterizing Presettlement Vegetation: How and Why," *Natural Areas Journal* 5 (1985): 5–19.
55. See G. M. Day, "The Indian as an Ecological Factor in the Northeastern Forest," *Ecology* 34 (1953): 329–346; H. J. Lutz, "Aboriginal Man and White Man as Causes of Fires in the Boreal Forest," *Yale University School of Forestry Bulletin* 65 (1959); J. G. Ogden III, "Pleistocene Pollen Records from Eastern North America," *Botanical Review* 31 (1964): 481–504; Emily W. B. Russell, "Indian-Set Fires in the Forests of the Northeastern United States," *Ecology* 64 (1983): 78–88.
56. Thomas Bonnicksen and Edward Stone, "Reconstruction of a Presettlement Giant Sequoia-Mixed Conifer Forest Community Using the Aggregation Approach," *Ecology* 63 (1982): 1134–1148.
57. As quoted in Smith, *Engineering Eden,* 301.
58. Carol Holleufer, Sierra Club Oral History Project, as quoted in Smith, *Engineering Eden,* 303.
59. Rezneat Darnell and Robert Burgess, section 1, and Robert Jenkins, "Voices of the Vanishing Wilderness," in "Ecological Reserves in Natural Resource Management," 1976, box 15, folders 5–7, Ecological Society of America Records, UA97-061, Hargrett Rare Book & Manuscript Library, University of Georgia Libraries, Athens, GA.
60. Reed F. Noss, "On Characterizing Presettlement Vegetation," 13.
61. M. Kat Anderson and Michael Barbour, "Simulated Indigenous Management: A New Model for Ecological Restoration in National Parks," *Ecological Restoration* 21 (2003): 269–277.
62. It is worth noting that a small group of paleoecologists were interested in a Pleistocene baseline. See, for example, Daniel Janzen and Paul S. Martin, "Neotropical Anachronisms: The Fruits the Gomphotheres Ate," *Science* 215 (1982): 19–27. "Pleistocene Re-wilding" would later become an international conversation with Josh Donlan et al., "Re-wilding North America," *Nature* 436 (2005): 913–914.
63. U.S. National Park Service, *Compilation of the Administrative Policies for the National Parks and National Monuments of Scientific Significance* (Washington, DC: Department of the Interior, 1968); U.S. National Park Service, *Management Policies* (Washington, DC: Department of the Interior, 1978).
64. On paleoecology and historical baselines for restoration, see N. L. Christensen et al., "The Report of the Ecological Society of America Committee on the Scientific Basis for Ecosystem Management," *Ecological Applications* 6 (1996): 665–691; Thomas W. Swetnam, Craig D. Allen, and Julio L. Betancourt, "Applied Historical Ecology: Using the Past to Manage for the Future," *Ecological Applications* 9 (1999): 1189–1206; D. Egan and E. Howell, *The Historical Ecology Handbook: A Restorationist's Guide to Reference Ecosystems* (Washington, DC: Island Press, 2001).
65. Charles H. Lamoureux, "Restoration of Native Ecosystems," in Charles P. Stone and J. Michael Scott, eds., *Hawai'i's Terrestrial Ecosystems: Preservation*

and Management, Proceedings of a symposium held June 5–6 at Hawai'i Volcanoes National Park (Honolulu: University of Hawai'i, 1985).

66. J. K. Baker and D. W. Reeser, "Goat Management Problems in Hawaii Volcanoes National Park: A History, Analysis, and Management Plan," Natural Resources Report No. 2 (Washington, DC: National Park Service, 1972); Kenneth Brower, "The Pig War," *The Atlantic,* August 1985; W. Edwin Bonsey, "Goats in Hawai'i Volcanoes National Park: A Story to Be Remembered," available at https://www.nps.gov/havo/learn/nature/upload/Goats-4-26-11_508.pdf.
67. Rezneat Darnell and Robert Burgess, section 3, "Ecological Reserves in Natural Resource Management," 1976, box 15, folders 5–7, Ecological Society of America Records, UA97-061, Hargrett Rare Book & Manuscript Library, University of Georgia Libraries, Athens, GA.
68. David M. Graber, "Managing for Uncertainty: National Parks as Ecological Reserves," *George Wright Forum* 4 (1985): 4–7.
69. The Nature Conservancy, *Stewardship* 1 (1974), folder 1, box 126, TNCR; Stewardship Guide for Preserve Committees, 1978, box 125, folder 25, TNCR.
70. David R. Stoddart, "Catastrophic Human Interference with Coral Atoll Ecosystems," *Geography* 53 (1968): 25–40.
71. On the history of invasion biology, see Sarah Hayden Reichard and Peter White, "Invasion Biology: An Emerging Field of Study," *Annals of the Missouri Botanical Garden* 90 (2003): 64–66; Mark Davis, "Invasion Biology 1958–2005: The Pursuit of Science and Conservation," in *Conceptual Ecology and Invasion Biology: Reciprocal Approaches to Nature,* ed. Marc Cadotte, Sean McMahon, and Tadashi Fukami (Dordrecht: Springer, 2006); Matthew Chew and A. H. Hamilton, "The Rise and Fall of Biotic Nativeness: A Historical Perspective," in *Fifty Years of Invasion Ecology: The Legacy of Charles Elton,* ed. David M. Richardson (Oxford: Wiley-Blackwell), 35–47.
72. Peter M. Vitousek, "Biological Invasions and Ecosystem Processes: Towards an Integration of Population Biology and Ecosystem Studies," *Oikos* 57 (1990): 7–13.
73. Paul Gobster, "Invasive Species as Ecological Threat: Is Restoration an Alternative to Fear-Based Resource Management?" *Ecological Restoration* 23 (2005): 261–270.
74. IUCN, *IUCN Guidelines for the Preservation of Biodiversity Loss Caused by Alien Invasive Species* (Gland: IUCN Council, 2000).
75. "Lythrum salicaria," Fire Effects Information System, Database, https://www.fs.fed.us/database/feis/plants/forb/lytsal/all.html.
76. "Lymantria dispar," The Virtual Nature Trail at Penn State New Kensington, Species Pages, https://www.psu.edu/dept/nkbiology/naturetrail/speciespages/gypsymoth.htm.
77. Richard N. Mack, "Plant Naturalizations and Invasions in the Eastern United States: 1634–1860," *Annals of the Missouri Botanical Garden* 90 (2003): 77–90; Christopher J. Costello and Andrew R. Solow, "On the Pattern of Discovery of Introduced Species," *Proceedings of the National Academy of*

Sciences of the United States of America 100 (2003): 3321–3323; Claude Lavoie et al., "Naturalization of Exotic Plant Species in North-Eastern North America: Trends and Detection Capacity," *Diversity and Distributions* 18 (2012): 180–190.

78. Frank E. Egler. "Indigene vs. Alien in the Development of the Arid Hawaiian Vegetation," *Ecology* 23 (1942): 14–23.
79. Richard Southwood and J. R. Clarke, "Charles Sutherland Elton: 29 March 1900—1 May 1991," *Biographical Memoirs of Fellows of the Royal Society* 45 (1999): 130–146.
80. Peter Crowcroft, *Elton's Ecologists: A History of the Bureau of Animal Population* (Chicago: University of Chicago Press, 1991); Matthew Chew, "Ending with Elton: Preludes to Invasion Biology" (PhD diss., Arizona State University, 2006).
81. Charles S. Elton, *The Ecology of Invasions by Animals and Plants* (Chicago: University of Chicago Press, [1958] 2000), 116, 117, 155.
82. Garrett Hardin, "The Competitive Exclusion Principle," *Science* 131 (1960): 1292–1297.
83. Walter Courtenay and C. Richard Robins, "Exotic Organisms: An Unsolved, Complex Problem," *Bioscience* 25 (1975): 306–313.
84. *The Nature Conservancy News* 29, no. 1 (Spring 1970); *The Nature Conservancy News* 30, no. 1 (January/February 1980).
85. John Schewgman, "Letter from the President," *Journal of the Natural Areas Association* 1 (1981): 1. Renamed *Natural Areas Journal* after publication of the first volume.
86. "Natural Area Notes," *Journal of the Natural Areas Association* 1 (1981): 12.
87. William R. Jordan III, "Restoration and Management Notes: A Beginning," *Restoration & Management Notes* 1 (1981): 1.
88. "Front Matter," *Natural Areas Journal* 4 (1984).
89. I. A. MacDonald, F. J. Kruger, and A. A. Ferrar, eds., *The Ecology and Management of Biological Invasions in Southern Africa* (Cape Town, SA: Oxford University Press, 1986); Harold A. Mooney and James A. Drake, eds., *Ecology of Biological Invasions of North America and Hawaii* (New York: Springer-Verlag, 1986); Laura Huenneke et al., "SCOPE Program on Biological Invasions: A Status Report," *Conservation Biology* 2 (1988): 8–10.
90. John P. Rieger and Bobbie A. Steele, eds., *Proceedings of the Native Plant Revegetation Symposium,* November 15, 1984, San Diego, CA (San Diego: California Native Plant Society, 1985).
91. *Weekly Stewardship News,* July 20, 1987, folder 5, box 136, TNCR.
92. Pamphlet, *The Nature Conservancy Stewardship Program,* c. 1987, folder 16, box 126, TNCR.
93. Florida Administrative Code, Ch. 16D-2, as mentioned in Reed F. Noss, "On Characterizing Presettlement Vegetation," 5–19.
94. "Extinct Ecosystem to Be Restored," press release, November 13, 1989, folder 24, box 154, series 4, TNCR; James Shaw and Bryan Coppedge, "The Initial Effects of Bison Reintroduction on the Landscape of the Tallgrass Prairie Preserve," research proposal submitted December 14, 1992, folder 38, box 158, series 4, TNCR.

95. Draft material, no author, 1992, folder 33, box 125, TNCR.
96. Weekly Stewardship News Report, August 15, 1990, box 126, TNCR.
97. G. Tanner Girard, Brian D. Anderson, and Taylor De Laney, "Managing Conflicts with Animal Activists: White-tailed Deer and Illinois Nature Preserves," *Natural Areas Journal* 13 (1993): 10–17; Stewardship News Report, March 26, 1993, folder 11, box 126, TNCR. On the conflict between the conservation and animal welfare movements, see also Ursula Heise, *Imagining Extinction: The Cultural Meanings of Endangered Species* (Chicago: University of Chicago Press, 2016), chapter 4.
98. Holmes Rolston III, *Conserving Natural Values* (New York: Columbia University Press, 1994).
99. William R. Jordan III and George M. Lubick, *Making Nature Whole: A History of Ecological Restoration* (Washington, DC: Island Press, 2011).
100. John Cairns Jr., ed., *The Recovery and Restoration of Damaged Ecosystems* (Charlottesville: University Press of Virginia, 1975); John Cairns Jr., ed., *The Recovery Process in Damaged Ecosystems* (Ann Arbor, MI: Ann Arbor Science Publishers, 1980).
101. John J. Berger, *Environmental Restoration: Science and Strategies for Restoring the Earth* (Washington, DC: Island Press, 1990).
102. "Making Nature Whole Again," *Newsweek,* January 18, 1988, folder 19, box 110, series 3, TNCR; Margo Freistadt, "Ecology Conference Highlights Value of Environmental Repair Work," *Christian Science Monitor,* February 2, 1988; Berger, *Environmental Restoration.*
103. William R. Jordan III, "A New Society," *Restoration & Management Notes* 6, no. 1 (1988): 2–3; *Newsletter of the Society for Ecological Restoration* 27, no. 2 (2013).
104. William R. Jordan III, "Restoration and Management Notes: A Beginning," *Restoration & Management Notes* 1 (1981): 2.
105. "Earthkeeping: A Program of Environmental Healing and Learning," undated, c. 1989, folder 23, box 126, TNCR.
106. Peter White and Susan Bratton, "After Preservation: Philosophical and Practical Problems of Change," *Biological Conservation* 18 (1980): 241–255.
107. John Cairns Jr., "Restoration and Management: An Ecologist's Perspective," *Restoration & Management Notes* 1 (1981): 6–8.
108. Michel E. Gilpin, "Restoration Ecology: A Note on the Theory and Practice," *Restoration & Management Notes* 1 (1983): 11–13.
109. A. D. Bradshaw, "Restoration Ecology as a Science," *Restoration Ecology* 1 (1993): 71–73.
110. Michael Soulé, "History of the Society for Conservation Biology: How and Why We Got Here," *Conservation Biology* 1 (1987): 4–5.
111. Truman Young, "Restoration Ecology and Conservation Biology," *Biological Conservation* 92 (2000): 73–83.
112. Draft, "Abating the Threat to Biodiversity from Invasive Alien Species: A Business Plan," 3/27/01, folder 12, box 116, TNCR; white paper, "TNC's Strategies for Success against Invasive Alien Species," 1/03/01 draft, folder 12, box 116, TNCR.

113. Christopher Rudolf, "Globalization, Sovereignty, and Migration: A Conceptual Framework," *UCLA Journal of International Law and Foreign Affairs* 3 (1998): 325–356; Wendy Brown, *Walled States, Waning Sovereignty* (Cambridge, MA: MIT Press, 2010); I. M. Destler, "America's Uneasy History with Free Trade," *Harvard Business Review,* April 28, 2016.
114. Executive Order 13112 of February 3, 1999, Invasive Species, *Federal Register* 64 (February 8, 1999): 6183–6186.
115. James T. Carlton, "A Journal of Biological Invasions," *Biological Invasions* 1 (1999): 1.
116. See, for example, Jonathan Levine and Carla M. D'Antonio, "Forecasting Biological Invasions with Increasing International Trade," *Conservation Biology* 17 (2003): 322–326; Charles Perrings et al., "How to Manage Biological Invasions under Globalization," *Trends in Evolution and Ecology* 20 (2005): 212–215; David Lodge et al., "Biological Invasions: Recommendations for U.S. Policy and Management," *Ecological Applications* 16 (2006): 2035–2054; Laura Meyerson and Harold Mooney, "Invasive Alien Species in an Era of Globalization," *Frontiers in Ecology and the Environment* 5 (2007): 199–208.
117. David Wilcove et al., "Quantifying Threats to Imperiled Species in the United States: Assessing the Relative Importance of Habitat Destruction, Alien Species, Pollution, Overexploitation, and Disease," *BioScience* 48 (1998): 607–615.
118. The Nature Conservancy, *Stewardship* 7 (1980), folder 2, box 126, TNCR; "Addressing the Challenge of Invasive Species," draft 9/5/00, folder 28, box 187, TNCR. See also Elizabeth A. Chornesky and John Randall, "The Threat of Invasive Alien Species to Biological Diversity: Setting a Future Course," *Annals of the Missouri Botanical Garden* 90 (2003): 67–76.
119. Richard Pough, "An Inventory of Threatened and Vanishing Species," *Second North American Wildlife Conference* (1937): 599–604.
120. The Nature Conservancy, *Tools for Intelligent Tinkering: A Steward's Handbook* (Arlington County, VA: TNC, 1995), folder 34, box 125, TNCR. Emphasis original.
121. See Jessica Gurevitch and Dianna Padilla, "Are Invasive Species a Major Cause of Extinctions?" *Trends in Ecology and Evolution* 19 (2004): 470–474; Dov Sax and Steven Gaines, "Species Invasions and Extinction: The Future of Native Biodiversity on Islands," *Proceedings of the Natural Academy of Sciences of the United States of America* 105 (2008): 11480–11497; Mark Davis et al., "Don't Judge Species on Their Origins," *Nature* 474 (2011): 153–154; Céline Bellard, Phillip Cassey, and Tim Blackburn, "Alien Species as a Driver of Recent Extinctions," *Biology Letters* 12 (2016).
122. Laura J. Martin and Bernd Blossey, "The Runaway Weed: Costs and Failures of *Phragmites australis* Management in the USA," *Estuaries and Coasts* 36 (2013): 626–632.
123. Peter Vitousek et al., "Biological Invasions as Global Environmental Change," *American Scientist* 84 (1996): 218–228; Will Steffen et al., "Planetary Boundaries: Guiding Human Development on a Changing Planet," *Science* 347 (2015): 736.

124. "Wildlife Invasive Species Program: Weeds on the Web," 1/18/01, folder 12, box 116, TNCR.
125. The Nature Conservancy, *Florida Keys GreenSweep: A Volunteer-Based Habitat Restoration Initiative,* pamphlet c. 2002, folder 37, box 161, TNCR.
126. https://www.ser.org/page/IndividualMembership#, accessed May 16, 2021.

8. An Ecological Tomorrowland

1. The Nature Conservancy, "The Disney Wilderness Preserve Story," white paper, 2014, http://njurbanforest.files.wordpress.com/2014/03/dwp-story-for-emailing.pdf.
2. "We Say: 'Mystery' Industry Is Disney," *Orlando Sentinel,* October 23, 1965; "Osceola Officials Voice Varied View," *Orlando Sentinel,* November 11, 1965; "Disney Tells of $100 Million Project," *Orlando Sentinel,* November 16, 1965; Chad Denver Emerson, *Project Future: The Inside Story behind the Creation of Disney World* (New York: Ayefour Publishing, 2010).
3. Thomas E. Dahl and Gregory J. Alford, *Technical Aspects of Wetlands, History of Wetlands in the Conterminous United States* (Washington, DC: U.S. Geological Survey, 1996); Ann Vileisis, *Discovering the Unknown: A History of America's Wetlands* (Washington, DC: Island Press, 1999); William Lewis, *Wetlands Explained: Wetland Science, Policy, and Politics in America* (New York: Oxford University Press, 2001).
4. "Mr. Alligator Cleans Sewers," *The Independent,* September 18, 1920, p. 347. On shifting perceptions of the Everglades, see Archie Carr, "Alligators: Dragons in Distress," *National Geographic,* October 1967, pp. 133–148; Christopher Meindl, "Past Perceptions of the Great American Wetland: Florida's Everglades during the Earth Twentieth Century," *Environmental History* 5 (2000): 378–395.
5. S. P. Shaw and C. G. Fredine, *Wetlands of the United States: Their Extent and Their Value to Waterfowl and Other Wildlife,* USFWS Cir 39, 1971, https://www.fws.gov/wetlands/documents/Wetlands-of-the-United-States-Their-Extent-and-Their-Value-to-Waterfowl-and-Other-Wildlife.pdf.
6. W. E. Frayer et al., *Status and Trends of Wetlands and Deepwater Habitats in the Conterminous United States, 1950's to 1970's* (Washington, DC: Fish and Wildlife Service, 1983); Thomas Dahl and Craig Johnson, *Wetlands Status and Trends in the Conterminous United States Mid-1970's to Mid-1980's* (Washington, DC: Fish and Wildlife Service, 1990).
7. David McCally, *The Everglades: An Environmental History* (Gainesville: University Press of Florida, 1999); Laura A. Ogden, *Swamplife: People, Gators, and Mangroves Entangled in the Everglades* (Minneapolis: University of Minnesota Press, 2011).
8. U.S. Army Corps of Engineers and South Florida Water Management District, *Rescuing an Endangered Ecosystem: The Plan to Restore America's Everglades* (West Palm Beach: South Florida Management District, 1999).
9. Martin Kessler and Larry Teply, "Jetport: Planning and Politics in the Big Cypress Swamp," *University of Miami Law Review* 25 (1971): 713–748.

10. B. F. McPherson et al., *The Environment of South Florida: A Summary Report* (Washington, DC: U.S. Government Printing Office, 1976).
11. U.S. Department of the Interior, *Environmental Impact of the Big Cypress Swamp Jetport* (Washington, DC: U.S. Department of the Interior, 1969).
12. Marti Mueller, "Everglades Jetport: Academy Prepares a Model," *Science* 166 (1969): 202–203; Robert Gilmour and John McCauley, "Environmental Preservation and Politics: The Significance of the 'Everglades Jetport,'" *Political Science Quarterly* 90 (1976): 719–739.
13. *An Act to Amend the Federal Water Pollution Control Act,* Public Law 92-500, *U.S. Statues at Large* 86 (1972): 816.
14. Edward Schiappa, "When Are Definitions Political?" in *Defining Reality: Definitions and Political Meaning* (Carbondale: Southern Illinois University, 2003), 75.
15. Jennifer Neal, "Paving the Road to Wetlands Mitigation Banking," *Boston College Environmental Affairs Law Review* 27 (1999): 161–192.
16. "U.S. Fish and Wildlife Service Mitigation Policy; Notice of Final Policy," *Federal Register* 46 (January 23, 1981): 7644–7663. The 1981 policy did not apply to species listed under the Endangered Species Act; rather, Endangered Species Act Section 7 describes the requirements for compensation for unavoidable (residual) impacts to listed species.
17. 40 CFR §1508.20(a-e).
18. Palmer Hough and Morgan Robertson, "Mitigation under Section 404 of the Clean Water Act: Where It Comes From, What It Means," *Wetlands Ecology and Management* 17 (2009): 15–33.
19. Margaret Race and Mark Fonseca, "Fixing Compensatory Mitigation: What Will It Take?" *Ecological Applications* 6 (1996): 94–101.
20. Oliver Houck, "Hard Choices: The Analysis of Alternatives under Section 404 of the Clean Water Act and Similar Environmental Laws," *University of Colorado Law Review* 60 (1989): 773–840.
21. Stream restoration has followed a similar trajectory, as analyzed by Rebecca Lave, Martin Doyle, and Morgan Robertson, "Privatizing Stream Restoration in the US," *Social Studies of Science* 40 (2010): 677–703; Rebecca Lave, *Fields and Streams: Stream Restoration, Neoliberalism, and the Future of Environmental Science* (Athens: University of Georgia Press, 2012).
22. "Restoration Key to Assessing Environmental Damages Liability: Interior Seeks Aid," *Restoration & Management Notes* 1 (1983): 14–15.
23. "Restoration Key to Assessing Environmental Damages Liability."
24. In 1980, the §404(b)(1) Guidelines were adopted as final regulations. On the legal basis of mitigation, see chapter 3 of Jessica B. Wilkinson et al., "The Next Generation of Mitigation: Linking Current Plans and Future Mitigation Programs with State Wildlife Action Plans and Other State and Regional Plans," white paper, Environmental Law Institute and The Nature Conservancy, 2008, https://www.eli.org/sites/default/files/eli-pubs/d19_08.pdf.
25. Ross A. Dobberteen, "Scientific Analysis and Policy Evaluation of Wetland Replication in Massachusetts" (PhD diss., Tufts University, 1989).

26. Ralph Tiner, *Wetlands of the United States: Current Status and Recent Trends* (Washington, DC: Fish and Wildlife Service, 1984).
27. U.S. General Accounting Office, "Assessments Needed to Determine Effectiveness of In-Lieu-Fee Mitigation," Report GAO-01-325 (Washington, DC: General Accounting Office, 2001).
28. Hough and Robertson, "Mitigation under Section 404 of the Clean Water Act," 15–33.
29. Memorandum of Agreement between the Department of the Army and the Environmental Protection Agency: The Determination of Mitigation under the Clean Water Act Section 404(b)(1) Guidelines. Signed February 6, 1990, Washington, DC.
30. Morgan M. Robertson, "Emerging Ecosystem Service Markets: Trends in a Decade of Entrepreneurial Wetland Banking," *Frontiers in Ecology and the Environment* 4 (2005): 297–302; Thompson J. Wilkinson, *2005 Status Report on Compensatory Mitigation in the United States* (Washington DC: Environmental Law Institute, 2006).
31. Hough and Robertson, "Mitigation under Section 404 of the Clean Water Act." The first commercial sale of Section 404 compensation credits occurred at the LaTerre Bank in southern Louisiana in February 1986. Robertson, "Emerging Ecosystem Service Markets," 297–302; Wilkinson, *2005 Status Report on Compensatory Mitigation.*
32. Morgan M. Robertson, "The Neoliberalization of Ecosystem Services: Wetland Mitigation Banking and Problems in Environmental Governance," *Geoforum* 35 (2004): 361–373.
33. Robertson, "Emerging Ecosystem Service Markets," 297–302; "Federal Guidance for the Establishment, Use and Operation of Mitigation Banks," *Federal Register* 60 (1995): 58605–58614.
34. Wilkinson, *2005 Status Report on Compensatory Mitigation.*
35. "Disney Deal Sets Healthy Precedent," *Tampa Tribune,* May 14, 1993, clipping in folder 22, box 160, CONS 245, The Nature Conservancy Records, Denver Public Library, Denver, CO (hereafter TNCR); The Nature Conservancy, "The Disney Wilderness Preserve Story," white paper, 2014, http://njurbanforest.files.wordpress.com/2014/03/dwp-story-for-emailing.pdf. On the 1989 plans, see Richard E. Foglesong, *Married to the Mouse: Walt Disney World and Orlando* (New Haven, CT: Yale University Press, 2001), chapter 9.
36. Records in folder 13, box 161, TNCR.
37. The Nature Conservancy, "Fact Sheet," folder 22, box 160, TNCR.
38. Robert Jenkins, "Statement of Request," 1971, folder 9, box 6, TNCR.
39. Charles Lee, "The Disney Wilderness Preserve—An Ecological Tomorrowland," *Florida Naturalist,* Summer 1993, clipping in folder 22, box 160, TNCR.
40. Charles Adams, as interviewed in Andrew Ross, *The Celebration Chronicles: Life, Liberty, and the Pursuit of Property Value in Disney's New Town* (New York: Ballantine, 1999), 278.
41. Mireya Navarro, "Disney Announces Plans for a Wildlife Theme Park," *New York Times,* June 21, 1995; Christine Shenot, "Animal Kingdom Coming to Life at Disney World," *Orlando Sentinel,* July 8, 1996.

42. The Staff of the Disney Wilderness Preserve, *Fifth Annual Management Report,* May 1998, folder 22, box 160, TNCR.
43. The Nature Conservancy, *Management and Restoration Plan for the Greater Orlando Aviation Authority Mitigation Lands* (TNC: January 1994), folder 35, box 160, TNCR.
44. Lee, "The Disney Wilderness Preserve."
45. Beverly Keneagy, "Some Disney Preserve Hogs Will Become Food for Needy," *South Florida Sun-Sentinel,* July 15, 1993.
46. Andrea Povinelli, "Brining Back a Central Florida Wilderness," *Chapter News,* Fall 1995, clipping in folder 22, box 160, TNCR.
47. The Staff of the Disney Wilderness Preserve, *Fifth Annual Management Report.*
48. The Nature Conservancy, *Wilderness Times,* Winter 1999, folder 22, box 160, TNCR.
49. The Staff of The Disney Wilderness Preserve, *Fifth Annual Management Report.*
50. "Nature Preserve Lauded," *The Ledger,* April 24, 1993, clipping in folder 22, box 160, TNCR; "Disney Deal Sets Healthy Precedent," *Tampa Tribune,* May 14, 1993, clipping in folder 22, box 160, TNCR.
51. The Nature Conservancy, "Fact Sheet," folder 22, box 160, TNCR.
52. John Flicker, "The Disney Wilderness Preserve," folder 22, box 160, TNCR.
53. The Disney Wilderness Preserve Fact Sheet, folder 22, box 160, TNCR; TNC, Pamphlet, "Picture Yourself Here . . . The Disney Wilderness Preserve," folder 22, box 160, TNCR; The Nature Conservancy, "Walt Disney World Co., The Nature Conservancy Create Disney Wilderness Preserve," Press Release, April 1993, folder 22, box 160, TNCR.
54. No author, untitled speech in Arkansas, no date, c. 1998, folder 22, box 160, TNCR.
55. Leslie Roberts, "Wetlands Trading Is a Loser's Game, Say Ecologists," *Science* 260 (1993): 1890–1892.
56. Handout, "The Disney Wilderness Preserve Gateway Center," folder 22, box 160, TNCR.
57. Natural Resources Conservation Service, *Emergence Watershed Protection Program: Draft Programmatic Environmental Impact Statement* (December 1999), https://www.nrcs.usda.gov/Internet/FSE_DOCUMENTS/stelprdb1044020.pdf, chapter 4.
58. Gina Keating, "Disney to Give $7 Million to Reforestation Projects," *Reuters,* November 3, 2009; Environmental News Service, "Disney Spends $7 Million to Conserve Forests in Peru, Congo, USA," January 12, 2013.
59. Alcoa, *2017 Alcoa Sustainability Report,* https://www.alcoa.com/sustainability/en/pdf/2017-Sustainability-Report.pdf.
60. Keith Rizzardi, "Alligators and Litigators: A Recent History of Everglades Regulation and Litigation," *Florida Bar Journal* 75 (2001): 18–34.
61. Alice Clarke and George Dalrymple, "$7.8 Billion for Everglades Restoration: Why Do Environmentalists Look So Worried?" *Population and Environment* 24 (2003): 541–569.

62. Tony Reichhardt, "Everglades Plan Flawed, Claim Ecologists," *Nature* 397 (1999): 462. On the history of CERP, see Michael Grunwald, part 3, *The Swamp: The Everglades, Florida, and the Politics of Paradise* (New York: Simon & Schuster, 2006).
63. C. Zaneski, "Big Ecological Guns Fault Plan for Everglades," *Miami Herald,* January 30, 1999, pp. 1A, 11A; Robert W. Blythe, *Wilderness on the Edge: A History of Everglades National Park,* chapter 28, https://evergladeswildernessontheedge.com/.
64. Keith Kloor, "Everglades Restoration Plan Hits Rough Waters," *Science* 288 (2000): 1166–1167.
65. Laura Ogden, "The Everglades Ecosystem and the Politics of Nature," *American Anthropologist* 110 (2008): 21–32.
66. Grunwald, *The Swamp.*
67. U.S. Army Corps of Engineers, *Central and Southern Florida Project Comprehensive Review Study: Final Integrated Feasibility Report and Programmatic Environmental Impact Statement* (Jacksonville, FL: U.S. Army Corps of Engineers, 1999).
68. Nicole T. Carter, *South Florida Ecosystem Restoration and the Comprehensive Everglades Restoration Plan* (Washington, DC: Congressional Research Service Report for Congress, 2003).
69. Pervaze A. Sheikh and Nicole T. Carter, *Everglades Restoration: The Federal Role in Funding* (Washington, DC: Congressional Research Service Report for Congress, 2006).
70. Michael Voss, "The Central and Southern Florida Project Comprehensive Review Study: Restoring the Everglades," *Ecology Law Quarterly* 27 (2000): 751–770.
71. Council on Environmental Quality, "Chapter 6: Ecosystem Approach to Management and Biodiversity" in *Twenty-Fourth Annual Report* (Washington, DC: Executive Office of the President, 1993); James R. Skillen, *Federal Ecosystem Management: Its Rise, Fall, and Afterlife* (Lawrence: University Press of Kansas, 2015), chapter 4.
72. Interagency Ecosystem Management Task Force, *The Ecosystem Approach: Healthy Ecosystems* and *Sustainable Economies* (Washington, DC: National Technical Information Service, 1995).
73. Skillen, *Federal Ecosystem Management.* See also C. S. Holling, *Adaptive Environmental Assessment and Management* (New York: John Wiley & Sons, 1978); C. J. Walters, *Adaptive Management of Renewable Resources* (New York: Macmillan Publishers, 1986); Steven Davis and John Ogden, eds., *Everglades: The Ecosystem and Its Restoration* (Delray Beach, FL: St. Lucie Press, 1994); R. Edward Grumbine, "What Is Ecosystem Management?" *Conservation Biology* 8 (1994): 27–38; John Freemuth, "The Emergence of Ecosystem Management: Reinterpreting the Gospel?" *Society and Natural Resources* 9 (1996): 411–417; Hanna J. Cortner and Margaret Ann Moote, *The Politics of Ecosystem Management* (Washington, DC: Island Press, 1999).
74. Don Peterson, "The Relationship of Fish and Wildlife Service's Ecosystem Approach to the National Environmental Policy Act," 11/24/95, box 1,

entry no. 328, Entry P230, Records of the U.S. Fish and Wildlife Service (RG 22), National Archives Record Administration II, College Park, MD.

75. Robert Elliot, "Faking Nature," *Inquiry* 25 (1982): 81–93 (87); Robert Elliot, *Faking Nature: The Ethics of Environmental Restoration* (London: Routledge, 1997). See also Eric Katz, "Restoration and Redesign: The Ethical Significance of Human Intervention in Nature," *Restoration & Management Notes* 9 (1991): 90–96.
76. William Jordan III, "Restoration as Realization," *Restoration & Management Notes* 7, no. 1 (1989): 2–3. Emphasis original.
77. Constance I. Millar and William J. Libby, "Disneyland or Native Ecosystem: Genetics and the Restorationist," *Restoration & Management Notes* 7 (1989): 18–24.
78. "Making Nature Whole Again," *Newsweek,* January 18, 1988, pp. 78–79, clipping in folder 19, box 110, series 3, TNCR.
79. Susan M. Galatowitsch and Arnold G. Van der Valk, "The Vegetation of Restored and Natural Prairie Wetlands," *Ecological Applications* 6 (1996): 102–112.
80. Joy B. Zedler, "Ecological Issues in Wetland Mitigation: An Introduction to the Forum," *Ecological Applications* 6 (1996): 33–37.
81. Zedler, "Ecological Issues in Wetland Mitigation"; Margaret S. Race and Donna Christie, "Coastal Zone Development: Mitigation, Marsh Creation, and Decision Making," *Environmental Management* 6 (1982): 317–328.
82. Stephen Brown and Peter Veneman, "Effectiveness of Compensatory Wetland Mitigation in Massachusetts, USA," *Wetlands* 21 (2001): 508–518.
83. Kevin Erwin, *An Evaluation of Wetland Mitigation in the South Florida Water Management District,* vol. 1 (West Palm Beach, FL: SFWMD, 1991).
84. See, for example, Margaret S. Race, "Critique of Present Wetlands Mitigation Policies in the United States Based on an Analysis of Past Restoration Projects in San Francisco Bay," *Environmental Management* 9 (1985): 71–82; John A. Kusler and Mary E. Kentula, eds., *Wetland Creation and Restoration: The Status of the Science* (Corvallis: Environmental Research Laboratory, US EPA, 1989); Zedler, "Ecological Issues in Wetland Mitigation: An Introduction to the Forum"; Margaret S. Race and Mark S. Fonseca, "Fixing Compensatory Mitigation: What Will It Take?" *Ecological Applications* 6 (1996): 94–101; Barbara Bedford, "The Need to Define Hydrologic Equivalence at the Landscape Scale for Freshwater Wetland Mitigation," *Ecological Applications* 6 (1996): 57–68; National Research Council, *Compensating for Wetland Losses under the Clean Water Act* (Washington, DC: National Academy Press, 2001); Deborah A. Campbell, Charles Andrew Cole, and Robert P. Brooks, "A Comparison of Created and Natural Wetlands in Pennsylvania, USA," *Wetlands Ecology and Management* 10 (2002): 41–49.
85. National Research Council, *Compensating for Wetland Losses under the Clean Water Act.*
86. U.S. Army Corps of Engineers, "Draft Environmental Assessment, Finding of No Significant Impact, and Regulatory Analysis for Proposed Compensatory Mitigation Regulation" (Washington, DC: U.S. Army Corps of Engineers, 2006).
87. Wilkinson, *2005 Status Report on Compensatory Mitigation.*

88. Followed by 12.5 percent by compensatory mitigation under the Endangered Species Act. Environmental Law Institute, *Mitigation of Impacts to Fish and Wildlife Habitat: Estimating Costs and Identifying Opportunities* (Washington, DC: Environmental Law Institute, 2007).
89. As reported in Wilkinson et al., "The Next Generation of Mitigation."
90. Freeman Dyson, "Can We Control the Carbon Dioxide in the Air?" *Energy* 2 (1977): 287–291.
91. Tia Nelson and Sonal Pandya, "Carbon Sequestration Paper for the BOG—DRAFT," April 28, 1998, folder 9, box 187, TNCR.
92. Nelson and Pandya, "Carbon Sequestration Paper." See also The Nature Conservancy, "Approaches and Strategies for Forest Conservation and Climate Change Mitigation," April 1998, folder 9, box 187, TNCR.
93. Nelson and Pandya, "Carbon Sequestration Paper."
94. "Summary of Results: Buenos Aires Climate Change Meeting," folder 12, box 187, TNCR; International Alliance of NGOs for the Inclusion of Forest-Based Joint Implementation in the Kyoto Protocol, signatures as of 12/11/97, folder 21, box 198, TNCR.
95. Records in box 10, folder 25, TNCR.
96. Nelson and Pandya, "Carbon Sequestration Paper."
97. "Nature Conservancy Carbon Sequestration Projects under Development," c. 2001, folder 35, box 53, series 2, TNCR; "The Nature Conservancy's Climate Change Program," c. 2001, folder 35, box 53, series 2, TNCR.
98. "Nature Conservancy Carbon Sequestration Projects under Development."
99. "Carbon Sequestration in the Forests of Belize," white paper, c. 1998, folder 9, box 187, TNCR.
100. "Carbon Sequestration Building Capacity—The Next Steps for the Conservancy," draft of February 13, 1998, folder 9, box 187, TNCR.
101. Global Canopy Programme, *The Little REDD+ Book* (Oxford: Global Canopy Foundation, 2009).
102. Stephen Donofrio et al., "Voluntary Carbon and the Post-Pandemic Recovery," *Ecosystem Marketplace,* September 2020, https://app.hubspot.com/documents/3298623/view/88656172
103. Oscar Venter et al., "Harnessing Carbon Payments to Protect Biodiversity," *Science* 326 (2009): 1368; Susan M. Galatowitsch, "Carbon Offsets as Ecological Restorations," *Restoration Ecology* 17 (2009): 563–570; Celia A. Harvey, Barney Dickson, and Cyril Kormos, "Opportunities for Achieving Biodiversity Conservation through REDD," *Conservation Letters* 3 (2010): 53–61; Bernardo Strassburg et al., "Global Congruence of Carbon Storage and Biodiversity in Terrestrial Ecosystems," *Conservation Letters* 3 (2010): 98–105; Jean-Francois Bastin et al., "The Global Tree Restoration Potential," *Science* 365 (2019): 76–79.
104. Bronson Griscom et al., "Natural Climate Solutions," *Proceedings of the National Academy of Sciences* 114 (2017): 1645–1650; Bastin et al., "The Global Tree Restoration Potential."
105. Karen Holl and Pedro Brancalion, "Tree Planting Is Not a Simple Solution," *Science* 368 (2020): 580–581.

106. F. E. Putz and Kent Redford, "Dangers of Carbon-Based Conservation," *Global Environmental Change* 19 (2009): 400–401; David B. Lindenmayer et al., "Avoiding Bio-Perversity from Carbon Sequestration Solutions," *Conservation Letters* 5 (2012): 28–36.
107. https://www.forest-trends.org/publications/.
108. David Diaz, Katherine Hamilton, and Evan Johnson, *State of the Forest Carbon Markets 2011: From Canopy to Currency* (Forest Trends, 2011), https://www.forest-trends.org/wp-content/uploads/imported/state-of-forest-carbon-markets_9292011_web-pdf.pdf.
109. Matt Grainger and Kate Geary, "The New Forests Company and Its Uganda Plantations," Oxfam Case Study, September 2011, https://www.oxfam.org/en/research/new-forests-company-and-its-uganda-plantations-oxfam-case-study.
110. Kristen Lyons and Peter Westoby, "Carbon Colonialism and the New Land Grab: Plantation Forestry in Uganda and Its livelihood Impacts," *Journal of Rural Studies* 36 (2014): 13–21.
111. See, for example, Emily Boyd, "Governing the Clean Development Mechanism: Global Rhetoric versus Local Realities in Carbon Sequestration Projects," *Environment and Planning A* 41 (2009): 2380–2395; Hannah Wittman and Cynthia Caron, "Carbon Offsets and Inequality: Social Costs and Co-Benefits in Guatemala and Sri Lanka," *Society and Natural Resources* 22 (2009): 710–726; Anne M. Larson, "Forest Tenure Reform in the Age of Climate Change: Lessons for REDD," *Global Environmental Change* 21 (2011): 540–549; Lyons and Westoby, "Carbon Colonialism and the New Land Grab"; Connor Cavanagh and Tor A. Benjaminsen, "Virtual Nature, Violent Accumulation: The 'Spectacular Failure' of Carbon Offsetting at a Ugandan National Park," *Geoforum* 56 (2014): 55–65.
112. Associated Press, "Disney Co. Spending $7M on Conservation Projects," *The Oklahoman,* November 3, 2009.
113. The Nature Conservancy, "Cause Marketing: The Walt Disney Company," https://www.nature.org/en-us/about-us/who-we-are/how-we-work/working-with-companies/cause-marketing/disney-oceans/.
114. Todd BenDor et al., "Estimating the Size and Impact of the Ecological Restoration Economy," *PLOS One,* https://doi.org/10.1371/journal.pone.0128339. See also Robertson, "The Neoliberalization of Ecosystem Services."
115. Karen D. Holl and Richard B. Howarth, "Paying for Restoration," *Restoration Ecology* 8 (2000): 260–267.
116. https://www.nature.org/en-us/what-we-do/our-priorities/protect-water-and-land/land-and-water-stories/invasive-plant-species-invasive-species-education-1/, accessed October 2020.

Epilogue

1. For review, see Nicole E. Heller and Erika Zavaleta, "Biodiversity Management in the Face of Climate Change: A Review of 22 Years of Recommendations," *Biological Conservation* 2009: 14–32; Emma Marris, *Rambunctious Garden: Saving Nature in a Post-Wild World* (New York: Bloomsbury, 2011);

Richard Hobbs et al., "Intervention Ecology: Applying Ecological Science in the Twenty-First Century," *BioScience* 61 (2011): 442–450; Laura J. Martin et al., "Conservation Opportunities across the World's Anthromes," *Diversity and Distributions* 20 (2014): 745–755.

2. Camille Parmesan and Gary Yohe, "A Globally Coherent Fingerprint of Climate Change Impacts across Natural Systems," *Nature* 421 (2003): 37–42; Allison Perry et al., "Climate Change and Distribution Shifts in Marine Fishes," *Science* 24 (2005): 1912–1915; Rachael Hickling et al., "The Distributions of a Wide Range of Taxonomic Groups Are Expanding Polewards," *Global Change Biology* 12 (2006): 450–455; Michael Burrows et al., "Geographical Limits to Species-Range Shifts Are Suggested by Climate Velocity," *Nature* 507 (2014): 492–495.
3. Thomas Martin, "Climate Correlates of 20 Years of Trophic Changes in a High-Elevation Riparian System," *Ecology* 88 (2007): 367–380.
4. O. Hoegh-Guldberg et al., "Assisted Colonization and Rapid Climate Change," *Science* 321 (2008): 345–346; D. M. Richardson et al., "Multidimensional Evaluation of Managed Relocation," *Proceedings of the National Academy of Sciences USA* 106 (2009): 9721–9724; Pati Vitt et al., "Assisted Migration of Plants: Changes in Latitudes, Changes in Attitudes," *Biological Conservation* 143 (2010): 18–27; N. Hewitt et al., "Taking Stock of the Assisted Migration Debate," *Biological Conservation* 144 (2011): 2560–2572; Mary Williams and R. Kasten Dumroese, "Preparing for Climate Change: Forestry and Assisted Migration," *Journal of Forestry* 111 (2013): 287–297.
5. Torreya Guardians, http://www.torreyaguardians.org/, accessed January 5, 2019; Government of Canada, "Assisted Migration," https://www.nrcan.gc.ca/climate-change/impacts-adaptations/climate-change-impacts-forests/adaptation/assisted-migration/13121, accessed April 10, 2021.
6. Intergovernmental Panel on Climate Change, *Global Warming of 1.5°C. An IPCC Special Report on the Impacts of Global Warming of 1.5°C above Pre-industrial Levels and Related Global Greenhouse Gas Emission Pathways* (IPCC, 2018).
7. See Madeleine J. H. van Oppen et al., "Building Coral Reef Resilience through Assisted Evolution," *PNAS* 112 (2015): 2307–2313; Ken Anthony et al., "New Interventions Are Needed to Save Coral Reefs," *Nature Ecology & Evolution* 1 (2017): 1420–1422; Phillip Cleves et al., "CRISPR/Cas9-mediated Genome Editing in a Reef-Building Coral," *PNAS* 115 (2018): 5235–5240.
8. For overview, see Linda Laikre et al., "Compromising Genetic Diversity in the Wild: Unmonitored Large-Scale Release of Plants and Animals," *Trends in Ecology and Evolution* 25 (2010): 520–529; Iris Braverman, *Coral Whisperers: Scientists on the Brink* (Oakland: University of California Press, 2018); Karen Filbee-Dexter and Anna Smajdor, "Ethics of Assisted Evolution in Marine Conservation," *Frontiers in Marine Science*, January 30, 2019, https://doi.org/10.3389/fmars.2019.00020; Christopher J. Preston, *The Synthetic Age: Outdesigning Evolution, Resurrecting Species, and Reengineering Our World* (Cambridge, MA: MIT Press, 2018).

9. On the debate, see Jason S. McLachlan, Jessica Hellmann, and Mark Schwartz, "A Framework for Debate of Assisted Migration in an Era of Climate Change," *Conservation Biology* 21 (2007): 297–302; Anthony Ricciardi and Daniel Simberloff, "Assisted Colonization Is Not a Viable Conservation Strategy," *Trends in Ecology and Evolution* 24 (2009): 248–254; Martin Schlaepfer et al., "Assisted Colonization: Evaluating Contrasting Management Actions (and Values) in the Face of Uncertainty," *Trends in Ecology and Evolution* 24 (2009): 471–472; Dov Sax, Katherine Smith, and Andrew Thompson, "Managed Relocation: A Nuanced Evaluation Is Needed," *Trends in Ecology and Evolution* 24 (2009): 472–473; Philip J. Seddon et al., "The Risks of Assisted Colonization," *Conservation Biology* 23 (2009): 788–789; Ben Minteer and James Collins, "Move It or Lose It? The Ecological Ethics of Relocating Species under Climate Change," *Ecological Applications* 20 (2010): 1801–1804.
10. For examples, see S. G. Willis et al., "Assisted Colonization in a Changing Climate: A Test-Study Using Two U.K. Butterflies," *Conservation Letters* 2 (2009): 45–51; Aglaen Carbajal-Navarro et al., "Ecological Restoration of *Abies religiosa* Forests Using Nurse Plants and Assisted Migration in the Monarch Butterfly Biosphere Reserve, Mexico," *Frontiers in Ecology and Evolution* 7 (2019): 421, doi:10.3389/fevo.2019.00421.
11. For overview, see Young Choi, "Restoration Ecology to the Future: A Call for New Paradigm," *Restoration Ecology* 15 (2007): 351–353. Richard Hobbs, Eric Higgs, and James Harris, "Novel Ecosystems: Implications for Conservation and Restoration," *Trends in Ecology and Evolution* 24 (2009): 599–605; Richard Hobbs et al., "Intervention Ecology: Applying Ecological Science in the Twenty-First Century," *BioScience* 61 (2011): 442–450.
12. On recent definitions of ecological restoration, see Eric Higgs, *Nature by Design: People, Natural Process, and Ecological Restoration* (Cambridge, MA: MIT Press, 2003); Stuart K. Allison, *Ecological Restoration and Environmental Change: Renewing Damaged Ecosystems* (New York: Routledge, 2012).
13. Choi, "Restoration Ecology to the Future," 351–353; Hobbs, Higgs, and Harris, "Novel Ecosystems," 599–605; Carolina Murcia et al., "A Critique of the 'Novel Ecosystem' Concept," *Trends in Ecology and Evolution* 29 (2014): 548–553; Eric Higgs et al., "The Changing Role of History in Restoration Ecology," *Frontiers in Ecology and the Environment* 12 (2014): 499–506.
14. Example from Hobbs, Higgs, and Harris, "Novel Ecosystems."
15. Edward O. Wilson, *Half Earth: Our Planet's Fight for Life* (New York: Liveright Publishing Corporation, 2016); E. Dinerstein et al., "A Global Deal for Nature: Guiding Principles, Milestones, and Targets," *Science Advances* 5 (2019), doi:10.1126/sciadv.aaw2869.
16. Judith Schleicher et al., "Protecting Half of the Planet Could Directly Affect Over One Billion People," *Nature Sustainability* 2 (2019): 1094–1096.
17. Charles Geisler and Ragendra de Sousa, "From Refuge to Refugee: The African Case," *Public Administration and Development* 21 (2001):159–170; Paige West, James Igoe, and Dan Brockington, "Parks and Peoples: The Social

Impact of Protected Areas," *Annual Review of Anthropology* 35 (2006): 251–277; Mark Dowie, *Conservation Refugees: The Hundred-Year Conflict between Global Conservation and Native Peoples* (Cambridge, MA: MIT Press, 2009).

18. William Cronon, "The Trouble with Wilderness; or, Getting Back to the Wrong Nature," in William Cronon, ed., *Uncommon Ground: Rethinking the Human Place in Nature* (New York, NY: W. W. Norton & Co., 1995), 69–90.
19. Katharine Suding et al., "Committing to Ecological Restoration," *Science* 348 (2015): 638–640.
20. UN Environment, "New UN Decade on Ecosystem Restoration," press release, March 1, 2019, https://www.unenvironment.org/news-and-stories/press-release/new-un-decade-ecosystem-restoration-offers-unparalleled-opportunity. See also https://www.decadeonrestoration.org/about-un-decade.
21. R. E. Turner, A. Redmond, and J. Zedler, "Count It by Acre or Function—Mitigation Adds Up to Net Loss of Wetlands," *National Wetlands Newsletter* 23 (2001): 5–6; T. BenDor, J. Sholtes, and M. W. Doyle, "Landscape Characteristics of a Stream and Wetland Mitigation Banking Program," *Ecological Applications* 19 (2009): 2078–2092; A. Moilanen, A. van Teeffelen, Y. Ben-Haim et al., "How Much Compensation Is Enough?" *Restoration Ecology* 17 (2009): 470–478.
22. Workshop Planning Team, "Brown Trout below Glen Canyon Dam: A Preliminary Analysis of Risks and Options," September 21, 2017, https://www.usbr.gov/uc/progact/amp/amwg/2017-09-20-amwg-meeting/BT03.pdf.
23. In 2018 the Arizona Game and Fish Department began stocking rainbow trout again for the first time since 1998; see https://www.azgfd.com/azgfd-to-stock-rainbow-trout-into-lees-ferry/.
24. Kurt Dongoske, "Dissenting Report on the Technical Work Groups Recommendation Concerning the FY 2010 & 2011 Work Plan and Budget for the Glen Canyon Dam Adaptive Management Program," July 10, 2009, https://www.usbr.gov/uc/progact/amp/amwg/2009-08-12-amwg-meeting/Attach_08e.pdf.
25. Bureau of Reclamation, "Environmental Assessment: Non-native Fish Control Downstream from Glen Canyon Dam," December 30, 2011, https://www.usbr.gov/uc/envdocs/ea/gc/nnfc/NNFC-EA.pdf; Department of the Interior, press release, "Salazar Announced Improvements to Glen Canyon Dam Operations to Restore High Flows and Native Fish in Grand Canyon," May 23, 2012.
26. Kurt E. Dongoske, Theresa Pasqual, and Thomas King, "The National Environmental Policy Act and the Silencing of Native American Worldviews," *Environmental Practice* 17 (2015): 36–45. See also Governor Val Panteah to Brent Rhees, February 6, 2017, http://gcdamp.com/images_gcdamp_com/3/33/LtrBrentRheesBrownTroutProblemLeesFerry_ZuniGovernorSigned_06February2017.pdf.
27. A recent ecology textbook states, for example, "Ecologists understand in a deep and fundamental way the factors that regulate populations, and we use

this knowledge to manage our natural resources effectively (or at least we try to—politics often gets in the way)." Gary Mittelbach, *Community Ecology* (New York: Oxford University Press, 2012), 340.

28. Aldo Leopold, "The Upshot," in *A Sand County Alamac: And Sketches Here and There* (New York: Oxford University Press, 2020 [1949]), 192.
29. For an introduction to Environmental Justice, see Robert Bullard, *Dumping in Dixie: Race, Class, and Environmental Quality* (Nashville, TN: Westview Press, 1990); Dorceta Taylor, *Toxic Communities: Environmental Racism, Industrial Pollution, and Residential Mobility* (New York: New York University Press, 2014); David Pellow, *What Is Critical Environmental Justice?* (Cambridge, UK: Polity Press, 2018).
30. Secretary of the Interior, Order No. 3390, "Transfer of Functions and Property Related to the National Bison Range," January 15, 2021, https://www.doi.gov/sites/doi.gov/files/elips/documents/est-17934-so-national-bison-range-transfer.pdf; Anna V. Smith, Reclaiming the National Bison Range," *High Country News,* January 26, 2021.
31. "Iinnii Buffalo Spirit Center," https://blackfeetnation.com/iinnii-buffalo-spirit-center/; Michelle Nijhuis, "The Bison and the Blackfeet," *Sierra,* June 14, 2021.
32. Victoria Reyes-García et al., "The Contributions of Indigenous Peoples and Local Communities to Ecological Restoration," *Restoration Ecology* 27 (2019): 3–8. For a critical look at the project of integrating traditional knowledge and science, see Paul Nadasdy, "The Politics of TEK: Power and the 'Integration' of Knowledge," *Arctic Anthropology* 36 (1999): 1–18.
33. Steve Jackson, "Rethinking Repair," in Tarleton Gillespie, Pablo Boczkowski, and Kirsten Foot, eds., *Media Technologies: Essays on Communication, Materiality and Society* (Cambridge, MA: MIT Press, 2014); Andrew Russell and Lee Vinsel, "After Innovation, Turn to Maintenance," *Technology and Culture* 59 (2018): 1–25; Shannon Mattern, "Maintenance and Care," *Places Journal,* November 2018.
34. Donna Haraway, *When Species Meet* (Minneapolis: University of Minnesota Press, 2007); Maria Puig de la Bellacasa, "Matters of Care in Technoscience: Assembling Neglected Things," *Social Studies of Science* 41 (2011): 85–106; Michelle Murphy, "Unsettling Care: Troubling Transnational Itineraries of Care in Feminist Health Practices," *Social Studies of Science* 45 (2015): 717–737.
35. Aryn Martin, Natasha Myers, and Ana Viseu, "The Politics of Care in Technoscience," *Social Studies of Science* 45 (2015): 625–641.

ACKNOWLEDGMENTS

This book is the result of countless conversations and sustained institutional support. I would like to thank Sara Pritchard, whose mentorship and whose foundational work at the intersection of Science and Technology Studies and Environmental History has deeply shaped this project; Aaron Sachs, whose care for writing and history as crafts is my constant inspiration; Clifford Kraft and Sue Stein, for modeling intellectual curiosity and generosity; Peter Galison and Janet Browne, for shaping my scholarship and for hosting me at Harvard; and Joyce Chaplin and Harriet Ritvo, for welcoming me to Cambridge and for their influential work.

For their crucial feedback on this manuscript, special thanks to Megan Black, Stephen Bocking, Angela Creager, Ezra Feldman, Nick Howe, Karen Merrill, Daegan Miller, Sara Pritchard, Megan Raby, Aaron Sachs, David Singerman, Paul Sutter, and the book's reviewers.

I am lucky to be indebted to a few incredible interdisciplinary groups. To begin, I thank Historians are Writers (HAW!) at Cornell University, my first intellectual home, and especially Daegan Miller for his brilliance and friendship. Thank you, also, to the members of the Cornell Roundtable on Environmental Studies (CREST), which Aaron Sachs, Amy Kohout, and I founded in 2010. I am grateful for the continued community of the Social Science Research Council Ecological History Group, convened by Peter Perdue and Stevan Harrell. In 2015, Joyce Chaplin and I established the Harvard Environmental History Group: thank you for the camaraderie, inspiration, and fruitful feedback. I completed this project while a fellow at the Stanford Humanities Center, and I thank them for the time and space to write.

Portions of this work were presented at meetings of the American Society for Environmental History, the History of Science Society, and the Ecological Society of America, as well as the Yale New Perspectives in Environmental History conference and the University of Pittsburgh Life Sciences after WWII conference. The project has also benefited from audiences at the Franke Program in Science and the Humanities at Yale University, Brown University, the Center for Humanistic Inquiry at Amherst College, the University of Virginia Environmental Humanities Colloquium, the University of Cambridge Department of History and Philosophy of Science, the Harvard STS Circle, the Boston Environmental History Seminar, the University of Michigan STS colloquium, the MIT Seminar on Environmental and Agricultural History, the Stanford Animal Studies Working Group, the Stanford

Program in History & Philosophy of Science, and the Williams College History Department colloquium.

I am grateful to the staff of the Cornell Rare and Manuscript Collections; the Yale Sterling Memorial Library; the University of Washington Special Collections, especially John Bolcer; the University of Georgia Archives, especially Gilbert Head and Margie Compton; the National Archives at College Park, especially Joe Schwartz; the New York State Library; and Martin Leuthauser and Abby Hoverstock at the Denver Public Library.

While writing this book I was supported by fellowships from the National Science Foundation, the Doris Duke Conservation Leadership program, Cornell University, the Harvard University Center for the Environment, and the Stanford Humanities Center. Research funding for this project was provided by Williams College, Harvard University, the Cornell Institute for Social Sciences, the Cornell Society for the Humanities, a Social Science Research Council Dissertation Proposal Development Grant, and a National Science Foundation Dissertation Improvement Grant (Award No. 1329750) from the Program in Science, Technology, and Society.

I thank my colleagues at Williams College, including the Environmental Studies program, the Oakley Center for the Humanities, the History Department, and the Science & Technology Studies program; Alex Bevilacqua, Ralph Bradburd, Christine DeLucia, Pia Kohler, Sarah Jacobson, James Manigault-Bryant, Gage McWeeny, Brittany Meché, Christina Simko, Yana Skorobogatov, Jason Josephson Storm, and my research assistants, Sofia Barandiaran and Amber Lee.

My gratitude to others who have helped to shape this project: Anurag Agrawal, Peter Alagona, Paolo Bocci, Dennis Bogusz, Angelo Caglioti, Zachary Caple, Gene Cittadino, Susan Cook-Patton, Alan Covich, Bill Cronon, Raf De Bont, Ron Doel, Samuel Dolbee, Nathan Ela, Erle Ellis, David Fedman, Simon Feldman, the late Mark Finlay, Jenny Goldstein, Harry Greene, Nils Güttler, Marcus Hall, Donna Haraway, Evan Hepler-Smith, Richard Hobbs, Sheila Jasanoff, Timothy Johnson, Dolly Jørgensen, Sharon Kingsland, Brandon Kraft, Nancy Langston, Erika Milam, Gregg Mitman, Paul Nadasdy, Lynn Nyhart, Joanna Radin, Lukas Rieppel, Margaret Ronsheim, Helen Rozwadowski, Paul Sabin, Caterina Scaramelli, Daniel Schrag, Sujit Sivasundaram, Sverker Sörlin, Abby Spinak, Dalena Storm, Jim Tantillo, Maria Taylor, Greg Thaler, and Jay Turner.

Thank you to my editors at Harvard University Press, Jeff Dean, who believed in this book in its early stages, and Janice Audet, who shepherded this book into being, and to Emeralde Jensen-Roberts and Stephanie Vyce for all their help.

Ezra Feldman, I will always be grateful for, and wonderstruck by, our shared writing life and our shared life.

INDEX

Note: Page numbers in italics refer to illustrations.

Adams, Charles C., 45, 48
Adirondack League Club, 26
Advisory Board on Wildlife Management, 142
AES/Barbers Point Co., 217–218
Age of Ecology, 15, 139–140, 147
Alcoa, 210–211
Allee, Warder Clyde, 46
Allen, Robert Porter, 149
alligators, 156, 160, 197, 199, 200
American Bison Society: disbanding of, 41; founding of, 23, 25–30; funding from, 82, 150; FWS and, 84; initial goal of, 26–27; precedent for assisted evolution in, 225; work of, 13, 30–41, 165
American Breeders' Association, 28–29, 165
American Committee for International Wild Life Protection, 148
American Electric Power, 218
American Farm Bureau Federation, 146
American Fisheries Society, 157
American Game Association, 83–84
American Game Protection and Propagation Society, 83
American Plants for American Gardens (Roberts and Rehmann), 75–76, 181
American Society for the Prevention of Cruelty to Animals, 21
Anderson, Mary Perle, 24–25, 63–64, 71
Andrus, Cecil, 163
Animal Kingdom Theme Park, 16, 197, 207
Anthropocene, 12
Applied Fisheries Laboratory, 93–94, 96–99, 110, 271n2, 274n39
aquaculture: alterations from, 111; Donaldson's advocacy of, 14, 94, 109–112; pollution from, 110–111
Aransas Migratory Waterfowl Refuge, 149
Archibald, George, 1–2
Army Corps of Engineers. *See* U.S. Army Corps of Engineers
Arnold Arboretum (Boston), 71
assisted evolution, 224, 224–227
assisted migration, 224–227
Atomic Age, 14, 94
atomic bomb: detonation of first (The Gadget), 90, 100; Doomsday simulation, 14, 114–115, 123–135; Hiroshima and Nagasaki, 99–100; testing at Pacific Proving Ground, 14, 98–106, *101, 102*, 118–119; U.S. detonations (1945–1992), 94
Atomic Energy Act of 1946, 271n2
Atomic Energy Act of 1954, 119, 121
Atomic Energy Commission: dismissal of contamination concerns, 107; Donaldson's relationship with, 94, 99; Doomsday simulation, 14, 114–115, 123–135; funding for ecology, 94, 105–106, 112, 116, 121–122, 133–134; investment in breeding applications, 97–98; Odum (Eugene) and, 118–120; waste handling/disposal, 116
atomic (nuclear) technology: Donaldson's defense of, 99; Doomsday simulation, 14, 114–115, 123–135; fear of irreversible change from, 113–115; and fish nutrition, 14, 105–106; and human nutrition, 106–110; mutation breeding of fish, 14, 97–98; radioisotopes for biological studies, 14, 94–95, 103–106
Atoms for Peace campaign, 97
Audubon Society, 39, 82, 146, 158; and Disney Wilderness Preserve, 206–207; and Everglades plan/restoration, 212, 213; and funding, 82, 166; and Grayrocks dam controversy, 158; and peregrine falcon,

Audubon Society (*continued*) 161–162; on predator control, 146; and volunteer efforts, 39; and whooping crane, 149–150; and wildflower preservation, 24, 63
Auerbach, Stanley, 116, 121, *127*
authenticity, 214–217
automobiles, as threat to nature, 65, 87–88
autonomy, 5–6, 8, 214, 226, 227
Ayres, Robert, 130

Bacon Bits, 108, 110
Badlands National Monument, 60
Bankhead-Jones Farm Tenant Act, 60
Barnum, P. T., 19
Baynes, Ernest Harold, 26–28, *27*
Baysinger, Earl, 154
bears, 77–78, 221
Beck, Thomas H., 81, 84
Berger, John, 192
Bergh, Henry, 21
Bessey, Charles E., 45, 65, 68
Bikini Atoll, 98–107; contamination from thermonuclear weapon, 106–107; control specimens for, 100; test Able, 100–101; test Baker, 101, *101*
Bikini Radiobiological Survey, 100–106, 273n33
Bikini Scientific Resurvey, 101–103
Biltmore Forest School, 50
bioaccumulation, 99, 119, 123–124
biocide, 128–129
biodiversity, 131–132, 133, 194, 219–220
biological control, 146, *147*
Biological Invasions (journal), 187, 194
Biological Survey. *See* Bureau of Biological Survey
"bio-perversity," 219–220
"Biotic View of Land, A" (Leopold), 88
bird restoration: peregrine falcon, 160–164, *161*, 188; President's Committee on Wildlife Restoration and, 81–85; whooping crane, 1–5, 148–153; wildflowers (plants) and, 89
bird sanctuaries, establishment of, 32, 39
birth control, for predators, 146, *147*
bison: campaign to eradicate, 21–22; campaign to popularize, 26–27, 248n28; confinement vs. wildness, 30, 250n45; domestication/economic potential of, 26–29; founding of American Bison Society, 23, 25–30; game reservations for, 30–41, *40*; hunting and killing of, *20*, 20–21; museum acquisitions (taxidermy) of, 19–22; Native American management of, 233; in private game parks, 26; rebound/surplus of, 39–40; shipping and transport of, *34*, 34–35, *36*; survey of (1902), 25–26; work of American Bison Society, 13, 30–41; in zoological parks, 22–23, *23*, 30
Blackfoot Confederacy, 233
bobcats, 77–78, 143, 156
Boone and Crockett Club, 23–24, 32, 37, 82
Bormann, F. Herbert, 129
botanical gardens, 66–76; aesthetics of wildness and, 68–71; cosmopolitan vs. naturalistic, 69–70; Vassar College Ecological Laboratory, 73–76, *74–75*, 88, 264n41; Wild Botanic Garden (Minneapolis), 69–72, *72*, 88, 263n36; Wisconsin Arboretum, 66–67, 84–90, *87*, 269n91
botany, gender and study of, 65–67
Bradshaw, Anthony D., 193
Branch of Predator and Rodent Control, 143, 145, 148, 154
Branch of Wildlife Refuges, 150, 151
Brattleboro Rifle Club, 83–84
Braun, Emma Lucy: background of, 47; on check-area (experimental control), 54; criticism of CCC, 51; descriptive studies of, 55; efforts to create nature reservations, 47–49, 171; justification for nature reservations, 59; membership in WFPS, 65
Breton Islands, Louisiana, 32
Britton, Elizabeth, 24, 63–69, 89
Broadhurst, Jean, 64–65
Brookhaven National Laboratories, 98, 124–128, *125*, *126*
Browner, Carol, 209
brown pelicans, 160
Buffalo Treaty, 233
Bureau of Biological Survey: Darling's leadership of, 83; formation of, 252n73; and game reservations, 37, 39; on game restoration, 84; incorporation into FWS, 84, 252n73, 284n13; and predator control, 77–79, *79*, 82, *144*; and President's Committee on Wildlife Restoration and, 81–82; Stoddard's work for, 80
Bureau of Fisheries, 84, 252n73, 268n81, 284n13
Bureau of Indian Affairs, 31, 82
Bureau of Land Management, 13, 38–39
Bureau of Reclamation, 82, 158, 231–232
Bush, George W., 168–170, *169*, 205

Butler, Eloise, 68–72, *72*; background of, 68; botanical garden established by, 69–72; essays of, 262n26; legacy of, 89; pedagogical aim of, 69; propagation techniques of, 70–71; regional approach/collecting by, 71

Cabo Blanco Strict Nature Reserve (Costa Rica), 7
Cade, Tom, *161*, 161–163, 188
Cain, Stanley, 142, 154, 166
Cairns, John, 193
California condors, 145
Camp Fire Club of America, 83–84
Camp Madison, 86–87
Canadian Wildlife Service, 3
Cannon, Walter, 117
Capitalocene, 12
captive breeding: endangered species legislation and, 140–141, 162–164; FWS involvement in, 150–151; new meanings of wildness and, 164–167; peregrine falcon, *161*, 162–164; whooping crane, 1–5, 149–153, *152*
carbon colonialism, 221
carbon offsetting, 16, 198, 210–211, 217–221; business/financial aspects of, 218, 221–222; impact on local inhabitants, 219–220, 229
carbon sequestration, 217–222
care: for companion species, 5; defining, 246n48; ecological restoration as form of, 12–13, 233–234; for wild species, 5, 24
Carnegie Institution, 28, 45, 56
Carson, Rachel, 163
Carter, Jimmy, 162–163, 167, 203
Celebration, Florida, 207
Center for Applied Research in Environmental Sciences (CARES), 175–177, *176*, 179
Central & Southern Florida Project (C&SF), 200–201, 208, 211–212, *213*
Chamberlain, Josephine, 30
check-area, 44, 54–58, *57*, 60, 83, 173
Chiles, Lawton, 212
Chinook salmon, 96, 98
Cincinnati Acclimatization Society, 170
Cinergy, 219
Civilian Conservation Corps (CCC), 49–54, *50*; ecologists' critique of, 51, 61; Luquillo Experimental Forest, 126; tree planting by, 52–54; work at Wisconsin Arboretum, 86–87, *87*
Clean Water Act: and Disney Wilderness Preserve, 16, 197–198, 207; and Environmental Concern, Inc., 176; environmental opportunities under, 207; mitigation mechanisms under, 204–208; Section 404 permitting, 8, 202–204, 216
clear-cutting, 128–130
Clements, Edith, 45–46, 60–61, 88
Clements, Frederic, 45–46; background of, 45; on importance of experimental studies, 56–58, 60; justification for nature reservations, 59; on natural recovery, 61; *Plant Ecology*, 56–57, 117; promotion of ecology, 51–52; on restoration and resettlement, 60–61; Santa Barbara gardens, 88; on succession theory, 130
climate change: carbon offsetting and, 16, 198, 210–211, 217–221, 229; ecological restoration as solution to, 222; half-earth and, 227–228; identification of, 9; international agreements on, 6, 218, 228; irreversibility of damage from, 11; new ways of caring for species necessitated by, 223–224
climate stabilization areas, 227
climax community, 44–45, 132
Clinton, Bill, 194, 212
Clute, Willard, 66
Coastal Protection and Restoration Program, 8
Coca-Cola, 139–140, 147
Colorado River, 230–232
Columbia River, 93–94, 99, 113–114, 116, 274n39
Committee on Breeding Wild Mammals of the American Breeders' Association, 28–29, 165
Committee on Rare and Endangered Wildlife Species, 148, 149, 151–154, 199
Committee on Restoration of the Greater Everglades Ecosystem, 212
Committee on the Preservation of Natural Conditions for Ecological Study (ESA), 48, 171–172, 256n23, 292n10
Commoner, Barry, 139
compensatory mitigation, 15, 140, 157–159; business/financial aspects of, 218, 221–222; carbon offsetting, 16, 198, 210–211, 217–221, 229; definition of, 203; Disney Wildlife Preserve, 16, 197–211; federal laws enabling rise of, 202; first formal policy on (FWS), 203–204; mechanisms of, 204–208; natural vs. restored ecosystem, 214–217, 229; "No Net Loss," 205; Section 404 permitting and, 202–204; wetlands mitigation banks, 205–206, 216–217
competitive exclusion principle, 188

Compound 1080, 78, 143–148, 151
Comprehensive Everglades Restoration Plan, 211–214, 229–230
Confederated Salish and Kootenai Tribes, 233
conservation: assumption about human behavior in, 8; Boone and Crockett Club, 23–24; Endangered Species Act definition of, 287n61; game restoration and, 13, 25; gender and interests in, 24; goal of, 7; management ethos of, 7; preservation vs., 6–7, 228; racism in early efforts, 23–24; relationship with ecological restoration, 7–8, 15–16, 228. *See also specific programs and efforts*
conservation biology, 193
Conservation Biology (journal), 193
conservation corridors, 227
"Conservation Ethic, The" (Leopold), 89
Conservation Fund, 221
Conservation International, 218, 221
conservation refugees, 13, 227
controlled burn, 182–185, *211*
"controlled wild culture," 89–90
Cooperative Fish and Wildlife Research Units, 83
COP 3, 218
coral: assisted evolution of (super coral), *224*, 224–225, 230; atomic testing and, 103, 118–119; destruction of, 223–224; restoration in Florida, 5–6
Coral Restoration Foundation, 5–6
Corbin Park, 26, 30
Cornell College of Forestry, 50
cosmopolitanism, 69–70
Council on Environmental Quality, 162–163, 203
Cowan, Ian, 165
Cowles, Henry Chandler, 44–46, 65, 67, 72, 73
coyotes, 77–78, 143, 146–147, *147*
CRISPR-Cas9, 14, 225
Crutzen, Paul, 12
Curtis, John, 88–89
Curtis Prairie, 67. *See also* Wisconsin Arboretum
cybernetic theory, 117–118

Dade County Port Authority, 201–202
Darling, Jay "Ding," 81–83
Darwin, Charles, 44
Davenport, Charles B., 28
Davis, Opal, 73
Dawes Act of 1887, 32, 232–233
DDT, 133, 160, 161, 163
de-extinction, 14
defaunation, 128–129, *129*
Defenders of Wildlife, 148
Defense Advanced Research Projects Agency, 128–129, *129*, 134
Delano, Columbus, 21
delisting, of endangered species, 160, 289n88, 289n90
Design of Experiments, The (Fisher), 55
Detroit Edison, 219
Deutschlands Pflanzengeographie (Bessey), 45
Diamond, Jared, 193
Dingell, John, 154–155
Disney, Walt, 198
Disney Wilderness Preserve: environmental and regulatory background for, 198–207; expansion of, 209–210; multiparty agreement for, 16, 197–198, 206–207; nonnative species in, 208–209; precedents set by, 211; restoration of, 208–211, *211*; Section 404 of Clean Water Act and, 16, 197–198, 207; as win-win, 210
Disney World: off-site mitigation for, 16, 197–211; original development, 198–199, *200*
distanced design, 172
Diston, Hamilton, 208
distributive justice, 232
disturbed community, 44–45
diversity-stability hypothesis, 131–132
Division of Economic Ornithology and Mammalogy, 77, 252n73, 265n52
Division of Wildlife Services, 145–146, 148, 154
Dock, Mira, 65
Donaldson, Lauren, 14, 93–112, *97*, 230; advocacy of ocean ranching (aquaculture), 14, 94, 109–112; biological fieldwork at Pacific Proving Grounds, 14, 98–106, 273n33; dismissal of contamination concerns, 106–107; early studies of radiation effects on fish, 93–97, 113–114, 270n1; fish nutrition studies of, 105–106; human nutrition studies of, 106–110; mutation breeding of fish, 14, 97–98; radioisotopes as tool for study, 103–106; relationship with AEC, 94, 99; selective breeding of fish, 95–96, 99; use of off-site mitigation, 112; vision of ecological restoration, 109; work with General Mills, 94, 107–110
Donaldson trout, 110, 154, 230
Dongoske, Kurt, 231

Doomsday simulation, 14, 114–115, 123–135; distinction between ecosystem structure and ecosystem function in, 131, 132; disturbance beyond repair in, 130–133; diversity-stability hypothesis in, 131–132; ecosystem destruction studies in, 123–130, 134; homeostasis in, 114–115, 117, 128, 132; interrelated principles in, 132
Doomsday/World War III, 14, 113–115
Dresden Botanic Garden (Germany), 69
Drude, Carl Georg Oscar, 45, 69
Drury, Newton, 51
Duck Stamp Act of 1934, 82
Ducks Unlimited, 81, 141, 150
Duff, Joe, 4–5
Dukakis, Michael, 205
Dunlap, Thomas, 240n18, 284n19
Dust Bowl, 13, 42–43, 50–62; and apocalypse planning, 114; disciplinary conflict (ecology vs. forestry) and, 50–54; experimental studies in, 56–58; Leopold's plans for Wisconsin Arboretum and, 86; recovery site (grassland reservation) sought in, 58–62
Dyson, Freeman, 217

eagles, 147, 166
Earth (film), 221
Earth Day, 139–140
Earth Repair (Hall), 240n18
ecocentric restoration, 242n24
Ecological Effects of Nuclear War (Woodwell), 128
ecological restoration, 5–16; active nature of, 10, 193–196; aesthetics vs. survival, 139–141; archives of, 8–9; assumption about human behavior in, 8; care in, 12–13, 233–234; definitions of, 5, 8, 10, 12, 225–226, 228; diverse geographies included in, 8; Donaldson's vision of, 109; early promoters of, 45; economy/business of, 6, 221–222; ecosystem functions in, 15, 131, 132, 226–227; emergence of discipline, 192–193; ethics of, 232–234; future of, 223–234; hazards of, 229; historical baselines for, 10–11, 172, 179–187, 191, 295n62; historical fidelity in, 38, 172, 189, 192–193, 197–198, 225–226; history of, 6–7, 241n19; homeostasis and, 114–115; hope found in, 9–10, 228, 230, 233–234; in ideal form, 16; interventions encompassed in, 5–6; irreversible change vs., 11, 113–115, 133–135; Jenkins' view of, 174–175; Leopold (Aldo) as inventor of, 6, 67; natural vs. restored ecosystem, 214–217, 229; professionalization of, 171, 187–192, 204; radiation as tool for, 14, 94–95, 110–112; scale of projects, 229–230; science and technology studies in, 9, 243nn28–29, 273n28; Wisconsin Arboretum as first major site of, 67–68. *See also specific programs and efforts*
Ecological Society of America: advocacy of check-areas (experimental control), 58; archives of, 8; disciplinary conflict (ecology vs. forestry) and, 50–54; efforts to create nature reservations, 42–43, 46–49; formation of, 46; leaders of, 45; Leopold (Aldo) and, 86, 88; lobbying restrictions on, 172–173; recovery site (grassland reservation) sought by, 58–62; roots of The Nature Conservancy in, 48, 62, 170–172, 292n10
Ecologists Union, The, 173
ecology (scientific discipline): Age of, 15, 139–140, 147; check-area or experimental control in, 44, 54–58, *57*, 60, 83, 173; disciplinary conflict with forestry, 50–54, 61; Dust Bowl/New Deal and, 13, 42–43, 50–62; environmental movement and, 133–135; ideal study site for, 48; invasion biology as subdiscipline of, 187; origin of term, 44; research infrastructure of, 47; rise of, 42, 44–46; textbooks on, 56–57, 117–120
Ecology of Invasions by Plants and Animals, The (Elton), 188
ecosystem(s): emergence of paradigm, 116, 278n11; IBP definition of, 134–135; idea enshrined in federal regulation, 213–214; natural vs. restored, 214–217, 229; novel, 12, 226–227; Odum's definition of, 118, 124–125; organismal physiology and, 116–117; origin of term, 118; radioisotope visualization of, 14, 94–95, 103–106; restoration of, 173–179; science and theory, 115–118
ecosystem destruction studies, 14, 123–130, 134; Brookhaven (Woodwell), 124–126, *125*, *126*, 127–128; DARPA defaunation, 128–129; Hubbard Brook, 129–130, 187; Luquillo Experimental Forest, 125–127; Oak Ridge National Laboratory, 127–128
ecosystem functions: ecosystem structure vs., 131, 132; restoration of, 15, 131, 226–227

ecosystem management, 213–214
ecosystem structure, 131, 132
efficient community hypothesis, 215
Egler, Frank, 188
Ehrenfeld, David, 193
E.I. du Pont de Nemours Powder Company, 83
Eisenhower, Dwight, 97, 106
electrofishing, 231–232
elk, 37, 142, 180
Elliott, Robert, 214–215
Elrod, M. J., 36
Elton, Charles, 117, 131, 188
Embody, George Charles, 95–96
endangered species: delisting of, 160, 289n88, 289n90; draft list of, 151–153, 286n54; experimental populations of, 164, 291n104; managing for, 148–156; Red Book or Red List of, 151–153; whooping crane as symbol of, 148. *See also* bison; whooping crane
Endangered Species Act of 1973, 15, 83, 140–141, 148–167; amendment (1982), 164; amendments (1978), 157, 159; and compensatory mitigation, 202; conservation defined in, 287n61; habitat protection under, 155, 287n66; implementation of, 156; inventory of natural lands under, 187; new meanings of wildness and, 164–167; nonnative species control under, 170; origins and passage of, 148–156; predator reintroduction under, 160–167, *166*; recovery plans under, 159–160; restoration under, 156–159
Endangered Species Preservation Act of 1966, 153–154
Endangered Species Research Program, 1
Enewetak Atoll, 103–106, 118–119
Environmental Concern, Inc., 176, *177*, 192
environmental justice, 16, 232–232
environmental management: conservation vs. preservation in, 6–7; ecological restoration in, 7–8
Environmental Protection Agency: and Compound 1080, 148; and Disney Wilderness Preserve, 206, 209; and mitigation policy, 204, 205–206; and Section 404 permitting, 202–204
Eradication Methods Laboratory, 77–78, 143
Erickson, Ray, 151, 154
Errington, Paul, 80
ethics, 89, 232–234
ethnobotany, 65
eugenics, 29–30
Everglades: Comprehensive Everglades Restoration Plan, 211–214, 229–230; Disney Wilderness Preserve, 16, 197–211; drainage and hydrology of, 198–202, *201*; as endangered ecosystem, 199, 202; endangered species in, 151, 166, 199, 212–213; environmental impact state on jetport, 201–202; federal lawsuit against Florida, 212; federal spending on wetlands restoration in, 6; national park, 58, 211–212; recovery potential of, 212
Everglades Coordinating Council, 213
evolution, assisted, 224, 224–227
Executive Order 11643, 155
Executive Order 11987, 162
Executive Order 13112, 194
experimental control, 44, 54–58, *57*, 60, 83, 173
experimental populations, 164, 291n104
Extermination of the American Bison, The (Hornaday), 22, 25, 30
extinction: breeding programs vs., 29–30; de-extinction of species, 14; irreversibility of, 26; rate of, 11; saving species from, 38, 39. *See also* cndangered species

"Faking Nature" (Elliott), 214–215
Federal Aid in Wildlife Restoration Act of 1937, 82–83, 150
Federal Water Pollution Control Act, 202
Fell, George, 173
Fern Lake, 105–106
Finger Lakes Native Plant Society, 6
fire/burning: CCC and management of, 49–50, *50*; controlled use of, 182–185, *211*; destruction studies, 128; Disney Wilderness Preserve, 209, *211*
fish: decline in wild fisheries, 110–111, 276n69; Donaldson's nutrition studies of, 105–106; farming of (aquaculture), 14, 94, 109–112; mutation breeding of, 14, 97–98; nonnative, vs. humpback chub, 230–232; off-site mitigation for, 112; radiation effects on, 93–106, *102*, *104*, 113–114; selective breeding of, 95–96, 99
Fish and Wildlife Service. *See* U.S. Fish and Wildlife Service
Fisher, R. A., 55
fisheries biology, 95–96, 272n10
fisheries ecology, 272n10
fish hatcheries, 95–96

fish stocking, 109–112, 232–234. *See also* *specific fish*
Florida: bird sanctuaries in, 32; Comprehensive Everglades Restoration Plan, 211–214, 229–230; coral restoration in, 5–6; Disney Wilderness Preserve, 16, 197–211; federal lawsuit against, 212; restoration of whooping crane population, 3; *Torreya taxifolia* in, 223–224; wetlands drainage and hydrology of, 198–202, *201*
Florida Game and Freshwater Fish Commission, 206, 213
food web, 103–106, 111
forestry: disciplinary conflict with ecology, 50–54, 61; Shelter-Belt Program, *52*, 52–54
forests, and carbon sequestration, 217–221
"forever chemicals," 10
Fort Niobrara Game Preserve, 37
Foster, Richard, 274n39
Friends of the Everglades, 212, 213
Fukuryū Maru (fishing boat), 106–107
functional equivalency, 204, 216
Fundamentals of Ecology (Odum), 117–120, 122

game conservationists, 24
"game farms itself" (Leopold), 79
game management, 6–8; Leopold (Aldo) and, 77–81; Leopold Report on FSW (1964), 142–148; Leopold Report on NPS (1963), 142; predator control in, 38, 77–79, *79*, 141–148, *144*, 180. *See also* game restoration
game parks, private, 26
game production, 141
game reservation, 30–41; advocacy by Leopold (Aldo), 77–81; appropriation of Native American land for, 13, 31–37, 232–233; ecological study in, 58; land expanses required for, 30–31; managers of, 37; National Wildlife Refuge System of, 13, 37–41, 84, 148–149, 153, 232–233, 252n74
game restoration: as alternative to conservation, 25; compensatory mitigation and, 157–159; federal expansion into, 83–85; founding of American Bison Society, 23, 25–30; gun manufacturers and, 83–84; Leopold (Aldo) and, 77–81, *80*, 84–87; predator reintroduction, 160–167, *166*; President's Committee on Wildlife Restoration and, 81–85; proposal for Great Plains National Monument, 59–60; recovery plans for, 159–160; Section 7 of ESA and, 156–159; whooping crane, 1–5, 148–153; work of American Bison Society, 13, 30–41
Ganong, William, 56
Garbisch, Edgar, 175–176, *177*, 192
Garden Club of America Conservation Committee, 73
gardening: maintenance, 242n24; reparative, 242n24. *See also* naturalistic gardening
Garretson, Martin, 40
gender: and botany / WFPS, 65–67; and conservation interests, 24; and Wild Flower Preservation Society, 63–68
General Mills, 94, 107–110
genetic modification, 14, 94–95, 224, *224*–227, 230
genome editing systems, 14
Gilpin, Michael, 193
Glacier Bay National Monument, 48–49
Glen Canyon Dam, 230–232
globalization: carbon market and, 217–221; fears of, 15–16; restoration as global practice, 222; roles of North vs. South, 16, 220, 229
Golley, Frank, 121–122
Goode, George, 22
Goodnight, Charles, 34
Gore, Al, 213
Gorongosa National Park (Mozambique), 5
Graber, David, 183, 184, 186
Grand Canyon National Park, 230–232
Grange, Wallace, 51
Grant, Madison, 24, 29
grassland reservation, 58–62
Grayrocks Dam controversy, 158–159, 288n78
Grays Lake National Wildlife Refuge, 3
Great Barrier Reef, 223–224
Great Plains National Monument, 59–60
Greene, Henry, 88–89
Greenpeace, 218
Griggs, Robert, 292n10
Grinnell, George Bird, 23–24
Grinnell, Joseph, 46
gun manufacturers, 83–84

habitat: protection under ESA, 155, 287n66; restoration of, 15
Haeckel, Ernst, 44, 254n8
Half-Earth (Wilson), 227–228
Hambleton Island (Maryland), 175–176, *176*
Hanford Engineer Works, 93–94, 99, 113–114, 116

Hardin, Garrett, 188
Haring, Inez, 73
Harshberger, John, 65, 67, 76
Hawaii: assisted evolution in, 224, *224–225*; nēnē in, 150; restoration baseline in, 185–186
Heath, Fannie Mahood, 71
Hetch Hetchy dam, 6–7
Hines, Neil O., 270n1
historical baselines, 10–11, 172, 179–187, 191, 295n62
historical fidelity, 38, 172, 189, 192–193, 197–198, 225–226
homeostasis: atomic testing and, 118–119; disturbance beyond repair, 130–133; Doomsday simulation and, 114–115, 117, 128, 132; ecosystem destruction studies and, 123–130; explanation of, 114–115; Hutchinson on, 117; problem of achieving, 134
Homestead Act of 1862, 31, 58
Hooper, Franklin, 37, 38
Hoover Dike (Florida), 200
hormonal birth control, 146, *147*
Hornaday, William Temple, *34*; on bird species, 39; bison confinement criticized by, 30; bison donated for game reservations, 33–36; bison survey (1902), 25–26; as curator of Department of Living Animals, 22–23; in founding of American Bison Society, 23, 26; influence on Leopold (Aldo), 77, 265n49; membership in Boone and Crockett Club, 23–24; museum acquisitions (taxidermy) of, 19–22; opposition to predatory animals, 38; orangutans living with, 30; racism/anti-immigrant attitudes of, 24; resignation as ABS president, 37
Hubbard Brook Experimental Forest, 129–130, 187
Hubbs, Carl, 186
human disturbance, 15–16. *See also specific disturbances*
humpback chub, 230–232
humpback whale, 221
hunting: bison, *20*, 20–21; Boone and Crocket Club, 23–24; lobbying by gun manufacturers, 83–84; wildlife management funding from, 82–83
Huntington, Ellsworth, 53
Hutchinson, G. Evelyn, 117–118, 131, 173
hybrid species, 162–164
hydroids, atomic testing and, 102–103
Iinnii Initiative, 233
Illinois Sportsmen's League, 83–84
immigrants: attitudes of white conservationists toward, 24–25, 29; blame for Dust Bowl, 43; racism/eugenics and, 29; Wild Flower Preservation Society and, 64–65
Indian Removal Act, 31
Interagency Ecosystem Management Task Force, 213–214
International Bamboo Development Co., 218
International Biological Programme, 134–135
International Conference on the Peaceful Uses of Atomic Energy, 119
International Crane Foundation, 1–2, 10
International Geophysical Year, 134
International Phytogeographic Excursion, 46
International Principles and Standards for the Practice of Ecological Restoration, 12
International Society for the Preservation of European Bison, 39
International Union for Conservation of Nature, 148, 151–153, 187
invasion biology, 187
invasive species: cosmopolitanism and, 69; as ecosystem disturbers, 15–16, 187; Elton on, 188; executive order on, 194; fundraising potential of, 193–194; killing/eradication of, 15, 183, 190–196, *195*; political targeting of, 168–170, *169*; professionalization and attitudes toward, 187–192; SCOPE on, 189–190; as threat to carbon storage, 222. *See also* nonnative species
island biogeographical theory, 174

Janzen, Dan, 192
Jenkins, Robert, 134, 174–179, 184–187, 207
Johnson, Lyndon B., 134
Johnson, Noye, 129
Jordan, William, III, 189, 192, 214, 241n19, 242n24
Joshua Tree National Park, 11
justice, 13, 16, 229, 232–234

Kachung Forest Reserve (Uganda), 220
Kellogg, Royal S., 53–54
Kendeigh, S. Charles, 116
Kennedy, John F., 151
Kerry, John, 168
King, Clarence, 24
Kings Canyon National Park, 183
Kiowa-Comanche Reservation, 32–33

Kissimmee River, 200–201, 208
Kyoto meeting/protocol, 218

Lacey Act of 1900, 25, 67
Lake Okeechobee, 199–200
Lake Powell, 230
landscape architecture, 75–76. *See also* naturalistic gardening
land trusts, 15, 170–171, 189
Langley, Samuel, 23
Lee Park Wild Flower and Bird Sanctuary (Virginia), 88
Leopold, Aldo: background of, 76–77; death of, 66; ethics of, 89, 232; fire observations of, 182–183; Forest Service work, 76–81; Hornaday's influence on, 77, 265n49; influence on Leopold Report (1963), 181; as inventor of ecological restoration, 6, 67; legacy of, 89; membership in Boone and Crockett Club, 24; preference for private vs. federal role, 84–85; productivity approach to game, 78–81, *80*; *Report on a Game Survey,* 79–80, 81; service on President's Committee on Wildlife Restoration, 81–85; UW hiring of, 81, 85; values of wildness and nativity merged by, 88; Wisconsin Arboretum, 66–67, 85–90, 269n91
Leopold, A. Starker, 15, 142, 180–187
Leopold, Estella, 145
Leopold, Luna, 202
Leopold Report: on Fish and Wildlife Service (1964), 142–148; on National Park Service (1963), 142, 180–187
Libby, William, 214–215
Life and Death of the Salt Marsh (M. Teal and J. Teal), 175
Likens, Gene, 129, 193
Lodge, David, 168
Longenecker, William, 85–86
Loxahatchee National Wildlife Refuge, 212
Lubick, George, 241n19, 242n24
Luquillo Experimental Forest, 125–127
Lynch, John J., 150

MacArthur, Robert, 131
maintenance gardening, 242n24
Malheur Lake, Oregon, 32
managed relocation, 224–227
Man and Nature (Marsh), 115
Manhattan Engineer District (MED), 93, 96, 100, 271n2
Manhattan Project (atomic bomb), 90, 100
Marsh, George Perkins, 115
Marshall Islands, 14, 98–107, *101, 102,* 118–119, 120
Mbaracayú Forest Nature Reserve, 218
McCormick, Richard, 21
McNulty, Faith, 147–148
meliorative land management, 242n24
Mendel, Gregor, 28
Merriam, John, 81
Miccosukee tribe, 213
migration: assisted, 224–227; human-guided, of whooping cranes, 3–5, *4*
Migratory Bird Treaty (1916), 149
Millar, Constance, 214–215
mitigation, 15, 140, 157–159; business/financial aspects of, 218, 221–222; carbon offsetting, 16, 198, 210–211, 217–221, 229; definition of, 203; Disney Wildlife Preserve, 16, 197–211; federal laws enabling rise of, 202; first formal policy on (FWS), 203–204; mechanisms of, 204–208; natural vs. restored ecosystem, 214–217, 229; "No Net Loss," 205; Section 404 permitting and, 202–204; wetlands mitigation banks, 205–206, 216–217
Mondo Cane (film), 98–99
Monsanto Chemical Corporation, 143
Montana, game reservation established in, 35–36
Moore, Barrington, 80
More Game Birds in America Foundation, 81, 266n66
mountain lions, 77–80, *79*
Muir, John, 6–7
mule deer, 37
Murie, Adolph, 182
museums, acquisition of bison specimens, 19–22
mutation breeding, 14, 97–98

National Bison Range, 36–37, 233
National Environmental Policy Act, 148, 157, 202, 203, 231
National Marine Fisheries Service, 159, 202–203
National Memorial to the American Indian, 41
National Museum, 19–23, *23*
National Museum Department of Living Animals, 22–23, 36
National Oceanic and Atmospheric Administration (NOAA), 156, 159, 170

National Park Service: CCC projects for, 49–54; critique of Everglades plan, 212; disciplinary conflict and, 51; establishment of, 13; grassland reservation sought from, 58–62; jurisdiction in Everglades, 213; native species management by, 191; nature reservations sought from, 46–49; precolonial baseline for, 179–187; predator control by, 142; purpose of, 180; reorientation to restoration approach, 182; review of wildlife practices (Leopold Report), 142, 180–187; stocking practices vs. humpback chub, 230–232; work of American Bison Society and, 38–39
National Research Council, 212
National Science Foundation, 112, 119, 129, 174
National Wildlife Federation, 158, 162
National Wildlife Refuge System, 13, 37–41, 84, 148–149, 153, 232–233, 252n74
Native Americans: in Anglo-American imagination, 40–41; assimilation vs. segregation of, 32; campaign against bison and, 21–22; erosion of sovereignty, 13, 21, 32, 250n52; forced relocation of, 31; forcible removal for Yellowstone, 7; land lost to game reservations, 13, 31–37, 232–233; objections to killing nonnative trout, 231–232; precolonial baseline and, 179–187; regaining of land and bison control, 233; sale of land taken from, *33*; tribal rights in Everglades, 213
Native Plant Reserve (Minneapolis), 71
natural areas, 15, 47–48. *See also specific types*
Natural Areas Association, 189
Natural Areas Journal, 189–190
natural areas management, 187–192
naturalistic gardening, 14, 62, 66–76; *American Plants for American Gardens* (Roberts and Rehmann), 75–76; cosmopolitanism vs., 69–70; nonnative vs. native species in, 87–88; restoration projects, 87–90; Vassar College Ecological Laboratory, 73–76, *74–75*, 88, 264n41; Wild Botanic Garden (Minneapolis), 69–72, *72*, 88, 263n36; Wisconsin Arboretum, 66–67, 84–90
Natural Resources Conservation Service, 202–203
natural selection, 44
Nature Conservancy, The, 8, 15–16; corporate partnerships of, 210–211; Disney reforestation project, 221; Disney Wilderness Preserve agreement, 16, 206–207; Disney Wilderness Preserve restoration, 208–211; first land acquisition by, 173; hands-off management vs. intervention, 172–179, 186, 193–196; history of, 48, 62, 170–172, 292n10; increase in holdings, 189; influence of ecologists in, 174; interest in Everglades, 213; Jenkins and, 174–179; killing/eradication of nonnative species, 190–196; as major landholder, 62; precolonial baseline for, 179–187; promotion of carbon market, 217–218; revenue and membership of, 173; stewardship focus of, 178–179; and whooping crane, 149–150, 158–159
nature reserve/nature reservation: accelerating nature's recovery in, 58–62; as check-area, 44, 54–58, *57*, 60, 83, 173; ecologists' justifications for, 59; efforts to create, 13–14, 42–43, 46–49; half-earth, 227–228; hands-off management vs. intervention in, 172–179, 186; list of potential sites (1926), 48; nonnative species control in, 170–171; precolonial baseline for, 179–187
Necedah National Wildlife Refuge, 3
nēnē, 150, 153
Newbold, Frederic, 264n41
New Deal, 13, 43, 49–54. *See also specific programs*
New York Botanical Garden, 63
New York Zoological Park, 23, 26, 30, 33–35
Niering, Bill, 215
Nixon, Richard, 78, 148, 155, 202
Noel Kempff Mercado National Park, 218–219
"No Net Loss," 205
nonnative species: controlled burns and, 182–183; in Disney Wilderness Preserve, 208–209; as ecosystem disturbers, 15–16, 187; Elton on, 188; fundraising potential of, 193–194; historical view of, 87–88, 168–170, 187–189, 269n95; IUCN on, 187; killing/eradication of, 15, 183, 190–196, *195*; natural areas management and, 187–192; political targeting of, 168–170, *169*; professionalization and attitudes toward, 187–192; purposefully introduced to U.S., 170, 187; records on, 187–188; SCOPE on, 189–190; as threat to native species, 87. *See also* invasive species
North American Game Breeders Association, 83–84
Noss, Reed, 183, 185
novel ecosystems, 12, 226–227

nuclear colonialism, 110–111, 119
nuclear pollution/waste, 10, 116
nuclear technology: Donaldson's defense of, 99; Doomsday simulation, 14, 114–115, 123–135; fear of irreversible change from, 113–115; and fish nutrition, 14, 105–106; and human nutrition, 106–110; mutation breeding of fish, 14, 97–98; radioisotopes for biological study, 14, 94–95, 103–106
nutrient cycles, 15, 99, 109, 111, 118–119, 131, 227
nutrition: fish, nuclear technology and, 14, 105–106; human, nuclear technology and, 106–110

Oak Ridge National Laboratory, 116, 121, *122, 127,* 127–128
ocean ranching, 109–110. *See also* aquaculture
Odum, Eugene, 116–122, *122,* 187; background of, 116–117; definition of ecosystem, 118, 124–125; on environmental destruction, 140; *Fundamentals of Ecology,* 117–120, 122; NSF fellowship, 119–120; studies at Pacific Proving Ground, 118–119, 120; on succession vs. ecosystem, 133
Odum, Howard "Tom," 116, 117–119, 124–127, 132–133
Office of Endangered Species, 154, 156, 158, 159
Office of Scientific Research and Development, 93, 143
Office of Technology Assessment, 130–131
off-site mitigation: business/financial aspects of, 218, 221–222; carbon offsetting, 16, 198, 210–211, 229; Disney Wilderness Preserve, 16, 197–211; Donaldson's fisheries and, 112; emergence of concept and practice, 198; federal laws enabling rise of, 202; natural vs. restored ecosystems, 214–217, 229; normalization of, 217; Section 404 permitting and, 202–204
on-site mitigation, 204–205
Operation Crossroads, 99–100
Operation Ivy, 119
Operation Migration, 3–5, 16
organismal physiology, 116–117
Orlando International Airport, 209
Osborn, Henry Fairfield, 23
Our Federal Lands: A Romance of American Development (Yard), 38
Our Vanishing Wild Life (Hornaday), 24–25, 77, 265n49

PacifiCorp, 218, 219
Pacific Proving Grounds, 14, 98–106, *101, 102,* 118–119, 120
Pan American World Airways, 108
Pan-Atomic Canal, 126
Park, Robert, 132
Parsons, Dave, 184
Partial Test Ban Treaty, 125
passenger pigeon, 14
Passing of the Great Race, The (Grant), 29
Patuxent Wildlife Research Center, 1–3, 83, 143, 151–153, *152,* 166, 244n31
Pautzke, Clarence, 95–96
Pease, Cora, 68–71
Pelican Island, 32
peregrine falcon, 160–164, *161, 163,* 188
Peregrine Fund, *161,* 162–164, *163,* 225
permittee-responsible mitigation, 204–205
Pinchot, Gifford, 6, 24, 50
Pittman-Robertson Act, 82–83, 150
plant communities, 44–46; climax, 44–45; disturbed, 44–45; relict, 56–58
Plant Ecology (Weaver and Clements), 56–57, 117
Plantesamfund (Plant Community) (Warming), 44
plant restoration: and game management, 77–81; hands-off management vs. intervention, 172–179; nonnative vs. native species in, 87–88, 170–171; Wisconsin Arboretum, 84–87
Plato, 43
Platt, Robert, 128, 132
Platte River Whooping Crane Maintenance Trust, 158–159
Pleistocene rewilding, 12, 295n62
Pough, Richard, 194
prairie: disciplinary conflict (ecology vs. forestry) and, 50–54; Dust Bowl, 13, 42–43, 50–56; recovery of (grassland reservation), 58–62; relict sites, 56–58; Shelter-Belt Program, *52,* 52–54; University of Illinois project, 88; Wisconsin Arboretum restoration, 84–87; Wisconsin Prison for Women project, 88
prairie dogs, 147–148
Prairie States Forestry Project, 52–54
precolonial baseline, 179–187, 191
predator control, 38, 77–79, *79,* 82, 141–148, *144,* 155, 180
predator reintroduction, 160–167, *166*
Prescott, Elizabeth, 88

preservation: assumption about human behavior in, 8; conservation vs., 6–7, 228; efforts to create nature reservations, 13–14, 42–43, 46–49; half-earth, 227–228; hands-off approach of, 7; relationship with ecological restoration, 7–8, 16, 228
Preservation Committee (ESA), 48, 171–172, 256n23, 292n10
President's Committee on Wild Life Restoration, 81–85
"primitive scene," 180–181
procedural justice, 232
pronghorn antelope, 37
protected areas, 7, 170–171, *171*. *See also specific types*
Pueblo of Zuni, 231–232

racism: Anglo-American imagination and, 40–41; biodiversity protection and, 13; campaign against bison and, 21–22; early conservation efforts and, 24–25; ecological restoration and, 229; eugenics and, 29–30; succession theory and, 132–133. *See also* Native Americans
radiation: Doomsday simulation, 14, 114–115, 123–135; early study of effects on fish, 93–94; ecosystem destruction studies, 123–128, *125*, *126*; fear of irreversible change from, 113–115; mutation breeding of fish, 14, 97–98; nutritional applications of, 106–110; Pacific Proving Grounds, 14, 98–106; as tool for restoration, 14, 94–95, 110–112
Radiation Biology Laboratory, 108–109
radioautograph, 103, *104*, 118–119
radioecology: destruction studies, 123–128; first decade of studies, 121; funding and program development, 116; Odum (Eugene) on, 118, 119–120
radioisotopes: demarcation of study sites with, 120; disposal of, 116; mass production of, 121; as study tool, 14, 94–95, 103–106
radiosensitivity, 123–124, 127
rainbow trout, 96, 106, 110, 111, 230–232
Ramsar Convention on Wetlands, 6
RAND Corporation, 123–124, 127–128
Rawlings, Edwin W., 107–110
recovery plans, 159–160
Red Book or Red List, 151–153
REDD+, 219
Rehmann, Elsa, 68, 74–76, 181, 215
reintroduction, 160–167, *166*
relict sites, 56–58
Remington Repeating Arms Company, 83
reparative gardening, 242n24
reparative naturalizing, 242n24
Report on a Game Survey (Leopold), 79–80, 81
Resettlement Administration, 49, 58–59, 81–83
Restoration and Management Notes, 158–159, 189–190, 192, 204
restoration ecology. *See* ecological restoration
reversibility, 11–12, 14, 113–115, 133–135
Richards, Ellen Swallow, 69
Ricker, Percy, 65–66
Riis, Paul, 87
Riley, Smith, 37–38
Rio Bravo Carbon Sequestration Pilot Project, 218–219
Ripley, S. Dillon, 134, 150
Roberts, Edith, 68, 72–76, 215; background of, 72–73; book on naturalistic gardening, 75–76, 181; influence on Leopold (Aldo), 86; legacy of, 89; membership in WFPS, 65; Vassar College Ecological Laboratory, 73–76
Rongelap Atoll, 106–107
Rookery Bay National Estuarine Research Reserve, 168–170, *169*
Roosevelt, Franklin D.: election and "new deal" of, 49; establishment of first national wildlife experiment station, 83; land management approach of, 13, 43; opposition from Leopold (Aldo), 84; President's Committee on Wild Life Restoration, 81–85; Resettlement Administration under, 58–59, 81–83; Shelter-Belt Program, *52*, 52–54; whooping crane refuge established by, 149
Roosevelt, Theodore: appropriation of Native American land, 32; Baynes and, 26; as conservationist, 6; establishment of bird sanctuaries, 32, 39; establishment of game reservations, 32–37; founding of Boone and Crockett Club, 23–24; interest in bison specimens, 22
Rush, Frank, 34

salmon: farming (aquaculture), 109–112; Fern Lake, 106; mutation breeding of, 14, 97–98; off-site mitigation for, 112; radiation effects on, 14, 93–94, 96–99; selective breeding of, 95–96, 99
Sanborn, Elwin, 35

Sand County Almanac, A (Leopold), 6, 85, 90, 232
sandhill crane, 151
Savannah River Ecology Laboratory, 121–122
Savannah River Plant, 116, 118, 120
Save-the-Redwoods League, 51
Schreiner, Keith, 156–157
Schwartz, Edith. *See* Clements, Edith
science and technology studies (STS), 9, 243nn28–29, 273n28
Scientific Committee on Problems of the Environment (SCOPE), 189–190
Sears, Paul Bigelow, 45, 53
Section 4 of ESA, 155
Section 7 of ESA, 15, 140, 155–159
Section 9 of ESA, 155
Section 404 of the Clean Water Act: and Disney Wilderness Preserve, 16, 197–198, 207; mitigation mechanisms under, 204–208; permitting under, 8, 202–204, 216
selective breeding, of fish, 95–96, 99
Seligmann, Peter, 221
Seminoles, 213
settler colonialism: and game restoration, 41; and graduated record, 57–58; myth of, 11; and precolonial baseline, 179–187, 191; and succession theory, 132
Shantz, Homer, 61
Shelford, Victor Ernest: background of, 47; on check-area (experimental control), 54, 55–56, 83; Clements' mentorship of, 45; as Ecologist Union founder, 173; efforts to create nature reservations, 47–49; as ESA founder and leader, 45, 46, 171; justification for nature reservations, 59; on land use/management, 61; on prairie recovery, 58; prairie restoration project, 88
Shelter-Belt Program, *52*, 52–54
Sherman, William Tecumseh, 21
Shreve, Forrest, 56
Sierra Club, 182–183, 184, 212, 213, 215
Sierra Club Handbook for Environmental Activists, 140
Silent Spring (Carson), 163
Simberloff, Daniel, 128–129, 193
Singleton, W. Ralph, 98
Smithsonian Institution, 19, 22–23, 162
snail darter, 159
snail kites, 151, 212–213
Society for Conservation Biology, 193
Society for Ecological Restoration/Society for Ecological Restoration and Management, 5–6, 171–172, 192–193, 196, 216; archives of, 8; definitions of restoration, 5, 10, 12, 225–226; establishment of, 6, 15; goal of, 15, 171; membership of, 15; on Wisconsin Arboretum, 68
Society for the Preservation of Native Plants, 260n1
Society of American Foresters, 50, 88
Society of American Taxidermists, 19
sodium fluoroacetate (Compound 1080), 78, 143–148, 151
Soil Conservation Service, 203
Soulé, Michael, 193
South Florida Water Management District, 206, 210
Special Advisory Board on Wildlife Management, 142
Sperry, Theodore, 86–87
Statistical Methods for Research Workers (Fisher), 55
Stazione Zoologica Anton Dohrn (Italy), 47
steelhead trout, 96, 106
stewardship: The Nature Conservancy and, 178–179, 186; Shelford on, 59
Stewardship (magazine), 186
stilbestrol, 146, *147*
stocking, fish, 109–112, 232–234. *See also specific fish*
stocking, game. *See* game restoration
Stoddard, Herbert L., 80
Stump Lake, North Dakota, 32
succession: academic research on, 89; Clements (Frederic) on, 130; and Doomsday simulation, 14, 114–115, 128, 132–133; and early ecological studies, 44–46; ecosystem concept vs., 132–133; and experimental studies, 60; and natural recovery, 61; and nature reservations, 13
Sumner, Francis, 47
Suncor, 219
Suter, J. Estes, *40*
Swamp Land Acts, 199–200

Taft, William Howard, 41
Tansley, Arthur, 46, 118, 279n20
taxidermy, 19–20
Taxidermy and Zoological Collecting (Hornaday), 20
Taylor, Walter P., 42, 51, 61
Teal, John, 175
Teal, Mildred, 175
Tellico Dam controversy, 159, 289n83
test Able, 100–101

test Baker, 101, *101*
thallium, 147
thermonuclear bomb, 106–107, 118–119
Torreya Guardians, 224
Torreya taxifolia, 223–224
Torrey Botanical Society, 66
"trashfish," 111–112
tree planting: for carbon sequestration, 217–221; Shelter-Belt Program, *52,* 52–54
trout: Donaldson, 110, 154, 230; farming (aquaculture), 109–112; Fern Lake, 106; mutation breeding of, 97–98; nonnative, vs. humpback chub, 230–232; radiation effects on, 93; selective breeding of, 95–96, 99; stocked vs. native, 186
Trouvelot, Leopold, 187
Truman, Harry S., 211
turkey oak, 226
Tyson, Wayne, 190

Udall, Stewart, 142, 145, 146, 151–153, 180, 182, 286n54
ultralight aircraft, whooping crane migration guided by, 3–5
Union of Concerned Scientists, 218
UN Decade on Ecosystem Restoration, 6, 228
UN Framework Convention on Climate Change, 6, 218
UN International Conference on Peaceful Uses of Atomic Energy, 111
U.S. Army Corps of Engineers: Central & Southern Florida Project, 200–201, 208, 211–212, 213; Comprehensive Everglades Restoration Plan, 211–214; definition of waters, 215; and Disney Wilderness Preserve, 197–207; and Endangered Species Act, 156–159; and Environmental Concern, Inc., 176; Grayrocks Dam controversy, 158–159; Hoover Dike, 200; mitigation policy of, 204, 205; Section 404 permitting, 202–204; wetland regulation, 199; wetlands report, 217
U.S. Fish and Wildlife Service: agricultural focus of, 84; archives of, 8; CCC projects for, 49–54; consolidation of power in, 166; creation of, 13, 84, 252n73, 284n13; and Disney Wilderness Preserve, 206; Endangered Species Act and, 140–141, 148–156; endangered species management by, 156–167; and Everglades, 212, 213; history of, 251n54; influence of restoration thinking on, 15; mitigation policy of, 203–204; National Wildlife Refuge System, 13, 37–41, 84, 148–149, 153, 232–233, 252n74; perpetual ecological intervention by, 135; predator control policies of, 141–148, *147,* 148, 155; return of Native American land/rights, 233; review of wildlife practices (Leopold Report), 142–148; Section 404 permitting, 202–204; on selection of Bikini Atoll, 100; Shelter-Belt Program, *52,* 52–54; whooping crane program, 1–5, 148–153; work of American Bison Society and, 38–39
U.S. Forest Service: debate over role and purpose of, 77; establishment of, 13; Leopold (Aldo) and, 76–81; nature reservations sought from, 46–49; work of American Bison Society and, 34, 38–39
U.S. Geological Survey, 83
U.S. Initiative on Joint Implementation, 219
Universal Studios Florida, 209–210
University of Illinois, 47, 88
University of Washington School of Fisheries, 95–96
University of Wisconsin–Madison, 66–67, 85–90
UtiliTree, 219

Vassar College Ecological Laboratory, 73–76, *74–75,* 88, 264n41
Vitousek, Peter, 187
vivaria, 47
Volcanoes National Park, 186
von Humboldt, Alexander, 45

Walker Ranch, 206–207. *See also* Disney Wilderness Preserve
Walt Disney Company: acquisition and development for Disney World, 198–199, *200;* carbon offsetting by, 210–211; off-site mitigation by, 16, 197–211; reforestation project, 221. *See also* Disney Wilderness Preserve
"War and the Living Environment" (Wurtz), 113
Warming, Eugenius, 44
Waugh, Frank Albert, 70
Weaver, Jim, *161*
Weaver, John, 56–58, 61, 117
wetlands: change in spatial distribution, 216–217; Comprehensive Everglades Restoration Plan, 211–214, 229–230; defining, 215–216; Disney Wilderness Preserve, 16, 197–211; drainage in Florida, 198–202, *201;* Environmental Concern,

Inc. and, 176, *177*; European colonial view of, 199–200; federal spending on restoration, 6; Hambleton Island project, 175–176, *176*; mitigation banks, 205–206, 216–217; "No Net Loss," 205; nonnative species in, 187, 195; restoration of, 6; Section 404 permitting, 202–204; and whooping cranes, 149
whooping crane, 1–5, 148–153, *152*; Aransas refuge, 149; decline of, 1, 149; goal of restoration, 3; migration of captive-bred cranes, 3–5, *4*; new meaning of wildness and, 165; number of (2017), 10; Platte River compensatory mitigation for, 158–159; privatization of program, 244n31; restorationist view of, 7; as symbol of endangered species, 148
Whooping Crane Conference, 149–150
Whooping Crane Recovery Team, 3
Wichita Game Reserve, 32–35, *34*, 166, 252n74
Wichita Mountains Wildlife Refuge, 166, 252n74
Wiener, Norbert, 117
Wilcove, David, 194
wild areas, 48
Wild Botanic Garden (Minneapolis), 69–72, *72*, 88, 263n36
wilderness: half-earth, 227–228; nature reservations vs., 59, 259n57; restored ecosystem vs., 214–217, 229
Wilderness Act, 182
Wilderness Society, 182
wild fisheries, decline of, 110–111, 276n69
Wild Flower (publication), 47
wildflower excursions, 63–64, *64*
Wild Flower Preservation Society, 24–25, 47, 63–68, 88, 89; expansion and membership of, 65; immigrant programs of, 64–65; opposition to automobiles, 65; sanctuary sites of, 67–68; takeover by male botanists, 65–67; and Vassar College Ecological Laboratory, 73
"Wildlife Conservation on the Farm" (Leopold), 85
wildlife management: earliest federal, 37; funding for, 39, 82–83; Leopold (Aldo) and, 78–81; predator control, 38, 77–79, *79*, 82, 141–148, *144*, 155, 180. *See also* game management; *specific programs*
wildness: active approach to, 193–196; aesthetics of, and botanical gardens, 68–71; assisted evolution/migration vs., 225–227; captivity vs., 30, 250n45; ESA and new meanings of, 164–167; Leopold on, 84–90; narrower view of, 227
Wilson, E. O., 128–129, 134, 174, 193, 227–228
Winchester Repeating Arms Company, 83
Wisconsin Arboretum, 84–90; CCC work at, 86–87, *87*; as first major site of ecological restoration, 67–68; Leopold's vision for, 85–87, 269n91
Wisconsin Electric Power Company, 218–219
Wisconsin Prison for Women, 88
Wolcott, Robert, 46
Wolfe, John N., 116
wolves: extermination of, 77–78; reintroduction of, 165, *166*, 166–167
WONDRA flour, 108
Woodwell, George, 124–125, 128, 130, 132
wooly mammoth, 14
working landscapes, 7
Works Project Administration, 52, 88
World Resources Institute, 218
World Wildlife Fund, 162, 218
Wurtz, Robert, 113, 122, 123

Yale Forest School, 50
Yard, Robert Sterling, 38
Yellowstone National Park: bison in, 22, 26, 37; controlled burns in, 183; creation of, 7; elk culling in, 142, 180; pronghorn antelope from, 37; reintroduction of wolves in, *166*, 166–167
Yosemite National Park, 6–7

Zahniser, Howard, 182
Zon, Raphael, 51, 52–54
zoological parks, 22–23, *23*, 30
Zuni, 231–232